RECENT DEVELOPMENTS IN
RELIABILITY-BASED CIVIL ENGINEERING

RECENT DEVELOPMENTS IN
RELIABILITY-BASED CIVIL ENGINEERING

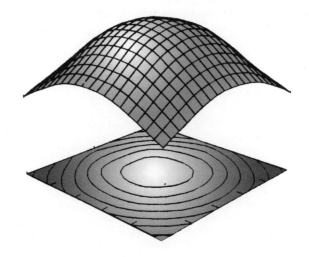

Editor

Achintya Haldar

University of Arizona, USA

 **World Scientific**

NEW JERSEY · LONDON · SINGAPORE · BEIJING · SHANGHAI · HONG KONG · TAIPEI · CHENNAI

Published by

World Scientific Publishing Co. Pte. Ltd.

5 Toh Tuck Link, Singapore 596224

USA office: 27 Warren Street, Suite 401-402, Hackensack, NJ 07601

UK office: 57 Shelton Street, Covent Garden, London WC2H 9HE

British Library Cataloguing-in-Publication Data
A catalogue record for this book is available from the British Library.

RECENT DEVELOPMENTS IN RELIABILITY-BASED CIVIL ENGINEERING

ISBN 981-256-419-5

Editor: Tjan Kwang Wei

Typeset by Stallion Press
E-mail: enquiries@stallionpress.com

Printed in Singapore by Mainland Press

PREFACE

This book covers some of the most recent developments (theoretical and with application potential) in reliability-based civil engineering analysis and design. The chapters are authored by some of the most active scholars in their respective areas. The topics represent some of the most recent research, upcoming interests, and challenges to the profession. As a result of extensive multidisciplinary efforts, the reliability-based engineering concept has matured over the last three decades. Reliability evaluation procedures with different degrees of sophistication are available. Various simulation schemes are also available to verify the theoretical developments. Reflecting these developments, most of the engineering design codes or guidelines are being modified or have already been modified to explicitly consider the presence of uncertainty. However, some recent incidents have identified some of the weaknesses that have yet to be addressed. Extensive damage during some recent earthquakes forced the profession to change the design philosophy. The design philosophy of human safety has been extended to consider damage to structures, and the concept of performance-based design is being advocated. Risk assessment and management of complicated structural systems are some of the major building blocks of this concept. The World Trade Center incident of September 11, 2001 also prompted a discussion on how to design for low-probability high-consequence events. Safety evaluation of bridges and offshore structures has generated a considerable amount of interest. The civil engineering applications domain has been extended to the reliability of fatigue and corrosion related problems. Recent trends include reliability evaluation of complicated structures using their realistic behavior using the nonlinear finite element method and the meshfree method. Life-cycle cost analysis for maintenance decision is now being advocated, although the major sources of uncertainty during the structural life have yet to be incorporated in the formulation. Health assessment of existing structures has attracted world-wide interest.

Some of these challenges are very recent and the profession has just initiated discussion on the related issues. These topics are not expected to be available in book form. However, the thoughts and recent works of some scholars who are providing leadership in developing these areas need to be readily available to benefit students (undergraduate and graduate), researchers (university and industrial), and practitioners. This edited book provides a sampling of some exciting developments in upcoming areas. It is designed for readers who are familiar with the fundamentals and wishes to study or to advance the state of the art on a particular topic. It can be used as an authoritative reference.

The book consists of thirteen chapters. They can be grouped into several theme topics including design for low-probability high-consequence events, performance-based design for building, bridges, and structure-foundation systems, risk-based design of offshore structures, reliability assessment of fatigue and corrosion, numerical methods for the reliability assessment of complicated structural systems using the nonlinear finite element method and the meshfree method, reliability assessment using information from experts, life-cycle cost analysis considering major sources of uncertainty, and the health assessment and monitoring of existing structures in the presence of uncertainty.

In Chapter 1, Corotis discusses risk in the built environment, and the role of risk and society's risk perception in making decisions for structures and infrastructure. The chapter provides a brief presentation of the inherent risks in life, the hazards that face communities, and the limitations imposed by economic and political realities in making decisions for low-probability high-consequence future events. The chapter ends with thought-provoking discussions on societal decision issues involving trade-offs, development, sustainability, inter-generational transfer, utility, discounting, probability versus uncertainty, risk versus risk perception, and political realities.

Rackwitz reports some recent developments in setting up rational public risk acceptability criteria for technical facilities exposed to natural or man-made hazards in Chapter 2. The life quality index is chosen as the basis for modern economic considerations and integrates the available information on world-wide communities. The social value of a statistical life and the corresponding willingness-to-pay are presented.

Wen introduces the performance-based design concept in Chapter 3. The consideration and proper treatment of the large uncertainty in the loadings and capacity, including complex response behavior in the nonlinear range, is essential in the evaluation and design process. A performance check of the structures can be carried out at multiple levels from immediate occupancy to incipient collapse. Lifecycle cost can be used to arrive at optimal target reliability. Performance-based evaluation and design are demonstrated for earthquake and wind hazards. The approach is expected to provide the foundation for the upcoming performance-based design guidelines.

In performance-based design, prescribed reliabilities must be satisfied at different levels. In general, a complicated system consists of many elements. Some of the elements act in series and other act in parallel, and the failure of one element may not indicate the failure of the system. Furthermore, the brittle and ductile behaviors of elements after they reach their capacities also affect the overall reliability of the system. In most cases, strength performance requirements are satisfied at the element level and the serviceability or deflection requirements are satisfied at the structural level. For the performance-based reliability evaluation procedure, the element-level probabilities of unsatisfactory performance for strength and system-level probabilities for serviceability need to be combined. Chowdhury and Haldar presented the relevant issues in detail in Chapter 4. They use complicated structure-foundation systems to demonstrate the necessity of such evaluation.

Ghosn discusses issues related to the reliability evaluation of bridges, a generally-overlooked topic of considerable recent interest, in Chapter 5. He points out that bridge design has emphasized the evaluation of the members on an individual basis while ignoring their interaction as a structural system. Ghosn presents the subject in the context of developing performance-based engineering methods that take into consideration the whole system's response range. He discusses the applicability of reliability methods for the safety assessment of bridge components as well as systems.

Offshore structures are becoming increasingly essential to maintain our way of life. The reliability evaluation of offshore structures is very specialized. Quek, Zheng, and Liaw present two frequency-domain methods to estimate the stochastic response of fixed offshore structures in Chapter 6. They address the complexity in stochastic analyses due to the presence of several sources of nonlinearity including drag and waves.

Fatigue is generally considered the most important failure mode in mechanical and structure systems. Modern reliability methods are available for managing the relatively large uncertainties in fatigue design factors. Wirsching provides a review of fatigue reliability methods in Chapter 7.

Structural deterioration and its effect on the reliability evaluation have recently become important research topics. Material loss due to corrosion and pitting corrosion are comprehensively discussed in Chapter 8 by Melchers. The variables involved in natural environments are identified. Recently-introduced probabilistic phenomenological models for marine immersion corrosion are described for both general and pitting corrosion. They show that the nature of the corrosion process changes from being controlled by oxygen diffusion to anaerobic bacterial action. The observations affect the longer-term prediction of general corrosion loss and for maximum pit depth. The effects of steel composition, water velocity, salinity, pollution and season of first immersion on corrosion are also summarized.

Reliability analysis techniques have matured in recent years, but they are not yet widely accepted by the deterministic community. One of the major objections is that the available techniques are not adequate to evaluate the risk of complex real structures. To overcome this concern, Huh and Haldar, in Chapter 9, propose a new efficient and accurate finite element-based hybrid method to evaluate the reliability of nonlinear real structures under short-duration dynamic loadings, including earthquake loading. The method explicitly considers nonlinearities due to geometry and material characteristics, boundary or support conditions, and connection conditions and the uncertainty in them. It intelligently integrates the response surface method, the finite element method, the first-order reliability method, and an iterative linear interpolation scheme. This approach can evaluate the time domain reliability of any structures that can be represented by finite elements, thus removing one of the major concerns of the deterministic community.

In Chapter 10, Rahman provides an exposition of stochastic meshfree methods that involve deterministic meshfree formulation, spectral representation of random fields, multivariate function decomposition, statistical moment analysis, and reliability analysis. Numerical results indicate that stochastic meshfree methods, employed in conjunction with dimension-reduction and response-surface methods, yield accurate and computationally efficient estimates of statistical moments and reliability. Rahman observes that although significant strides have been made, breakthrough research is essential on enhancing the speed and robustness of meshfree methods for their successful implementation into stochastic mechanics.

Reliability assessment of an engineering system depends to a great extent on performance data obtained in field or laboratory environments. In cases where such data is not readily available or cannot be obtained at a reasonable cost, data from experts can be utilized. In some cases, expert opinions may be necessary to provide direction on which methods would be most suitable for conducting a rational reliability analysis. Mohammadi and DeSantiago focus on reliability analysis using information from experts in Chapter 11. They specifically address the significance of using experts to provide information on the reliability methods and data collection process, and performance data obtained from the opinions of experts. Different statistical methods for conducting tests on such data, the identification of biases, and methods to improve the outcome of reliability analysis using information from experts are also discussed.

Life-cycle cost analysis is becoming an essential part of the maintenance decision process including inspections, repair or replacement. Since all cost must be discounted down to the decision point, a sustainable, intergenerationally acceptable discount rate must be used. A rate decaying from some market rate down to the real economic growth rate per capita in a country based on intra- and intergenerational equity is proposed by Rackwitz in Chapter 12. The renewal model established elsewhere for setting up suitable objective functions is adjusted to time-dependent interest rates. Optimal replacement strategies including inspection and repair are proposed. The theory is extended to series systems of dependent or independent deteriorating structural components.

Health assessment of existing structures in normal operating conditions or just after a natural or man-made event must be conducted as objectively as possible. The sources of uncertainty in such an evaluation are numerous, and they should be kept to a minimum. Katkhuda and Haldar propose a system identification-based nondestructive defect evaluation procedure to identify structural stiffness parameters at the element level using only limited noise-contaminated response information and completely ignoring the excitation information. The structures are represented by finite elements. The procedure detects defects by tracking the changes in the stiffness property of each element. The area is rapidly evolving and is of considerable interest to many segments of our profession.

This book is unique in terms of its content, which addresses many emerging research areas where the available information is scarce or not yet properly formulated. The recent thoughts and opinions of experts presented in this book are expected to accelerate the development of these areas. The book provides the reader with a wealth of insight and a unique global perspective. It is hoped that this book will convey to the reader the excitement, advances, promise, and challenges the reliability community faces in the near future.

Achintya Haldar
Tucson, Arizona

CONTENTS

CHAPTER 1

RISK AND RISK PERCEPTION FOR LOW PROBABILITY, HIGH CONSEQUENCE EVENTS IN THE BUILT ENVIRONMENT

ROSS B. COROTIS

Denver Business Challenge Professor University of Colorado
Boulder, Colorado 80309-0428, USA
E-mail: corotis@colorado.edu

Risk in the built environment, and the role of risk and risk perception in how society makes decisions are important aspects of safety for our structures and infrastructure. This chapter provides a brief presentation of inherent risks in life, of hazards that face communities, and of the limitations of economic and political realities in making decisions for low probability, high consequence future events. It begins with an introduction on the presence of risks, and then discusses the mathematical basis and codified approach to structural reliability. A definition of the aspects of risk and risk perception is followed by issues unique to low probability, high consequence events, including small probabilities, the convolution of probability and consequence and political accountability. There is then a section on alternative decision criteria. Finally, there is a section on societal decision issues involving trade-offs, development, sustainability, inter-generational transfer, utility, discounting, probability versus uncertainty, risk versus risk perception, and political realities.

1. Why Risk?

All activities involve a degree of risk, defined as occurrence of some outcome that is normally considered undesirable.[1] Such outcomes could be the failure of a dam to hold the water in a reservoir following a storm, the collapse of a structure during an earthquake, the inability of a bridge to support traffic loads following a collision by a boat, etc. Although not universally accepted, the most common quantification of risk is the product of the likelihood of occurrence and the effect of the event, such as change in structural state, loss in monetary value, and/or mortality and morbidity. Thus something as specific as the annual risk of death due to flooding would be the probability of occurrence of one death in one year, whereas the lifetime risk of damage due to earthquakes would be the product of some appropriate measure of the damage (perhaps monetary value) multiplied by the probability of occurrence of this damage over the design lifetime of the structure.

Every civil engineering structure entails some degree of uniqueness, either in the design or construction of the structure, the supporting soil and foundation, the environmental setting or the use. In addition, all structures are built to exist over a future intended lifetime, during which they will be exposed to the natural environment and societal effects, both intended and unintended. Therefore, the performance of every structure during its service lifetime exposes it to uncertainties that involve risk, and it is the role of building codes to ensure that structures designed in accordance with them have an acceptably low level of risk.[2] The development of these codes is based on careful evaluation of the performance of existing structures, professional judgment concerning materials, loads and analysis, and finally a consensus process that determines, at least implicitly, an acceptable level of risk.[3]

The process described above has worked remarkably well, with experience resulting in cost-effective structures for which any particular structure exhibits an annual probability of severe unintended consequence estimated to be on the order of 10^{-4}. Yet this level of performance is not the result of comprehensive probabilistic analysis or conscience effort to set the safety level, rather it has evolved from a "comfortable" level of risk determined by professionals and endorsed, at least in a taciturn way, by society. In this chapter there will be a review of the concepts of the probabilistic approach to structural safety, followed by sections on risk and risk perception, special issues of low probability — high consequence events, and finally a discussion of social and political issues surrounding the civil engineer's role in risk management in the built environment.

2. Structural Reliability

2.1. *Mathematical basis*

The basic concept of structural reliability in probabilistic terms is the determination of the likelihood that a given structure will perform as intended. This determination can be viewed as the convolution of the probability density function of the demand applied to the structure (the "loads") with the cumulative distribution function of the capacity, or fragility, of the structure (the "resistances"),

$$P = \int f_L(x) F_R(x) dx \tag{1}$$

in which both $f_L(x)$ and $F_R(x)$ are to be interpreted in general as multi-dimensional distributions, and x as a vector of parameters of both loads and resistances. In addition, in general both L and R should be interpreted as time-varying, or stochastic, random variables.

This deceptively simple equation, therefore, embodies the concept of the lifetime history, or scenario, of a structure composed of multiple structurally interdependent components exposed to demands or loads caused by the natural environment, the usage, and unintended forces. The capacity of the structure, and even the loads in some circumstances, can be altered by the particular time history of all events up to the present instant of evaluation.

Other chapters in this book, and indeed other books, are available that discuss the various modeling of load and system parameters as well as approximate solution techniques. Several computer programs have been developed to assist in this evaluation, including ones based on approximate first and second order methods, simulation and response surfaces. In all cases, the resulting evaluation of Eq. (1) leads to a stated probability of failure, with the definition of failure dependent on the framing of the problem and the definition of loads and capacities. The calculation of the probability of failure for a particular situation is the fundamental objective of traditional structural reliability.

2.2. *Mathematical reliability versus codified design*

While the evaluation of Eq. (1) has been the goal of structural reliability, there has been an understanding among researchers, sometimes explicitly stated, that the resulting probability should not be interpreted in a literal sense as the likelihood of occurrence of the particular event. In large part this dates back to the work of Alfred Freudenthal, beginning at the middle of the last century.[4,5] He claimed that the computed probabilities of structural safety were "notional," meaning that they were to be used only in a relative sense to compare alternative designs, and certainly not across different categories of risk to society. This was based on three important shortcomings associated with structural reliability theory, which all relate to the problem of dealing with extreme events. One is that since occurrences of extreme events are by definition rare, and thus there will be relatively few observations. Therefore, the confidence levels associated with the selected values of these parameters are very large, and epistemic uncertainty can overwhelm computed probabilities. A second shortcoming is that computational power limits the practical ability to utilize mechanics-based models of structural behavior that are realistic in the nonlinear response regions, which are the essence of interest for failure analysis. This was especially true during the time of Freudenthal's concern, but is becoming less of a problem due to the increasing power of computers. The third area lacking verification is the modeling of extreme events. For instance, while methods of extreme value statistics have been developed to determine the forms of asymptotic probability distributions, it is not clear that the series of events leading to unusual circumstances have been captured in models of structural reliability. While Freudenthal continued to promote the concept of notional probabilities,[6] others advocated that societal trade-offs required comparisons across alternatives.[7]

Building codes evolved not from a formal sense of structural reliability, but from the need to capture, quantify and reproduce practices that resulted in structures that performed satisfactorily in normal circumstances. Studies of live loads in the late 1800's and early 1900's[8,9] recognized that loads could be mathematically quantified, and that a probabilistic model of load intensity could be related to intended use and the extent of structurally-supported floor area. While this use of formal probability led to the proposal for a live load reduction formula, its bound was set by reference to a purely deterministic overload.[10] This lack of willingness to adopt code provisions based solely on the theories of probability continued to manifest itself in the advancements in concrete design that led to the incorporation of ultimate strength design in 1956. While the separate identification of load and strength factors can be seen as a direct transformation to account for different probabilistic variations of various loads and strengths, the method was not marketed as a probabilistically-based advancement. As with prior code changes, the new method was "calibrated" through trial designs such that sections designed with the new procedure were very similar to those resulting from the prior working stress method.

The concept of calibrating new code provisions to existing design continues as the method of assurance that code changes will not significantly affect the safety of structures. Examples are the load and resistance factor design codes for steel, masonry and timber structures. This approach is prudent, in that it protects against the reality that new models of loads and structural behavior are not "field tested" and may not be complete. Progress is still possible because the new methods can lead to increased consistency of reliability across alternative designs. Unlike all other fields of engineering, civil engineering structures are tested only through field applications of actual structures over design lifetimes. It is also consistent with Freudenthal's advocacy that structural reliability probabilities be used only to compare designs, and not as actual chances of events.

This approach has not, however, permitted an open discussion of the risks inherent in structural design codes, nor a comparison with other risks faced by society. It is in this broader construct of risks and risk trade-offs that civil engineering reliability professionals have before them an opportunity for major impacts for society.[11-13] But to engage this domain, it is incumbent upon engineers to have a fuller understanding of the way in which society perceives and acts on the risks faced by individuals.

3. Risk and Risk Perception

At the beginning of this chapter, risk was defined as the probability of occurrence of an undesirable outcome, multiplied by a quantitative measure of that outcome. This will be referred to as mathematical risk. Studies by social psychologists and others have documented that comparative levels of risk defined this way are inconsistent with the trade-off decisions that are made by society at large.[14-16] Therefore, there must be other conceptualizations of risk that are being used to form decisions. Indeed, different sets of researchers have documented separate and important issues that are used in societal decision-making. The issues related to the manner in which decisions are framed have been studied extensively by Daniel Kahneman and Amos Tversky.

Kahneman and Tversky[17,18] have investigated how outcomes depend on the manner in which questions are framed and the way in which situations are referred back to experiences. As an example of the former, they demonstrate that very different choices are made when questions are framed to emphasize the positive aspects of one choice over another. One of their classic examples (due to Maurice Allais) is described below.[17]

In Problem 1 a group of respondents is given two choices: (A) Win 2500 monetary units with a probability of 0.33, win 2400 monetary units with a probability of 0.66, or win no monetary units with a probability of 0.01, and (B) win 2400 monetary units with certainty.

In Problem 2, another group of respondents is also given two choices: (A) Win 2500 monetary units with a probability of 0.33 or no monetary units with a probability of 0.67, and (B) win 2400 monetary units with a probability of 0.34 or no monetary units with a probability of 0.66.

Studies on groups have shown that in general for Problem 1 82% of the subjects will choose alternative (B), and for Problem 2 83% of the subjects will choose alternative (A). A careful examination shows that both problems are of the same structure and *should* produce the same choice, since the only difference between Problem 1 and Problem 2 is eliminating a 0.66 probability of winning 2400 monetary units for both choices. Someone with linear utility theory who bases selections on expected value would choose (A) for both problems.

Paul Slovic[19] has investigated the characteristics by which people evaluate the risk associated with various events. He has determined that the probability of occurrence and the magnitude of outcome are only two of 20 such characteristics. In a comprehensive study he asked people to evaluate 90 common risks faced in society. He instructed them to rate the risks, and then used descriptors to determine their importance in the perception assigned by the subjects. Slovic then performed a statistical factor analysis in order to understand the primary characteristics that were important in the minds of individuals when they assessed risk. This process led to two factors that had high statistical loadings with differing characteristics (he chose to label these two factors Dread and Familiarity). A third factor was related to a single characteristic, and from subsequent studies he added two additional factors. This information is summarized below.

Table 1. Risk perception factors and the risk characteristics with which they exhibit high statistical loadings.

Factor	Characteristics w/high statistical loadings
Dread	Dreadful, catastrophic, hard to prevent, fatal, inequitable, threatening to future generations, not easily reduced, involuntary, personally threatening
Familiarity	Familiar, observable, degree of knowledge, immediacy of consequences
Number of People Exposed	
Trust	
Technological Stigma	

An important lesson from Slovic's study is that the public does not use the simple measure of risk that has been adopted by structural reliability professionals, but instead considers risk to be a multi-attribute quantity. He concludes that risk is relative, existing in the minds and cultures of society. An example of the disparity between engineering and sociological approaches is his conclusion[19] (p. 392): "... there is no such thing as 'real risk' or 'objective risk'."

Another conclusion of the work of Slovic and others[20,21] is that people translate situations about which decisions are to be made to those for which they have personal or studied experience, and then adjust for the differences. This method has three distinct steps. The first is the judgment of probability by availability. That is, the evaluation of the problem by the availability of similar instances that can be recalled. The second is anchoring, which is the process of selecting a base probability from the reference situation that has been selected by the decision-maker. This process is very dependent on how recently various situations were experienced, and their relative importance to the decision-maker. The third step is the adjustment to the probability for perceived differences between the reference situation and the current one.

4. Low Probabilities and High Consequences

Studies of public decision-making have indicated that people have a difficult time calibrating decisions to probabilities that are not around 50%.[22] Something with a very high probability will be "counted on" to occur, and something that has a very low probability will be assumed not to occur, i.e., its risk will be set to zero. Similarly, probabilities in a broad range around 50% will be classified as an "even chance," with mild qualitative adjustments to account for variations in either direction. While this characterization is clearly an over-simplification of complicated rationalizations, it illustrates the challenges of forming decisions that must contrast and compare outcomes with vastly different probabilistic levels, such as damage due to an earthquake versus increased efficiency due to construction of a new bridge.

4.1. *Issues with events of small probability*

A special challenge for trade-off decisions involving risks in the built environment is that the probabilities of occurrence of damages due to various hazards will all be "small." This means that such events will be in the realm of probabilities that are usually dismissed, making quantitative comparisons, or even recognition of possibilities, more difficult. Accurate assessment of the consequences of occurrence of rare events is likely to vary significantly with the rarity of the event.

As an example of this last difficulty, consider the case of earthquake damage to a building. In certain areas of the world (California, for instance) moderate earthquakes occur with some regularity. An earthquake that has a relatively high (say around 50%) chance of occurring over a ten-year period is likely to cause minor damage to a structure, perhaps less than 10% of its value, without long-term disruption of operation and without displacement of people or services. While a precise estimate of the damage may not be possible, the product of the probability and the estimated loss, adjusted to an annual expected risk, should form a convenient basis for decision-making. For instance, the annual risk in this case could be estimated as one-half of 1% of the value of the building. This can be contrasted due a very large earthquake, say one with a chance of 10% over a 100-year period. If one were considering only property loss, this latter earthquake might be estimated to cause damage of 100% value of the building, leading to a rather precise annual risk computed as one-tenth of 1% of the value of the building, or one-fifth as large as in the first case. Such a large earthquake, however, could cause the loss of the contents of the building and the disruption of business, valued in total at five times the value of the building itself. Then the annual risk would be one-half of 1% of the value of the building, the same as the case of the moderate earthquake. Alternatively, the collapse of the building could cause injury and/or death and disruption of life, valued at anywhere from 10 to 1000 times the value of the building, producing annual risk varying from 1% to 100% of the building value.

4.2. *Issues with the convolution of probability and consequence*

The example in the previous section indicates the difficult nature of assigning consequent costs to large-scale events. It also illustrates the complex sociological issue of comparing risks for events with vastly different probabilities and consequences. The moderate earthquake and the second example of damage due to a large earthquake had the same risk, yet members of society might not feel comfortable concluding that these two events were "comparable." In addition, if the total earthquake risk consisted of these two events, then the total risk would be equal to 1% of the building value, with 50% of the risk arising from the moderate earthquake and 50% from the large earthquake. This convolution nature of the risk combines events with differing probabilities and consequences to determine the total risk value, and in this illustration the two events would contribute equally to the total risk. This psychological distinction has been identified by Yacov Haimes and he attempts to address it by what he terms, "the partitioned multi-objective risk method".[23]

4.3. *Issues of political accountability*

Another characteristic of low probability events is related to the political reality that they are unlikely to occur during the term of office of a particular elected or appointed official. Therefore, the politically "astute" office-holder will find that he or she is much more likely to satisfy constituents by spending limited resources on activities with immediate, demonstrable returns, rather than investing in reduction of risk that has already been ignored by most people. Unfortunately, while such logic may be appropriate for a particular decision-maker, it is not often the best in the long-term. This paradox that a series of "optimal" solutions leads to an apparent contradictory long-term solution was discussed many years ago by the economist Paul Samuelson.[24] In order to overcome this paradox in the case of the built environment, it is necessary to adjust the cost and benefit reward system so that each sequential decision leads toward the optimal long-term solution.

Both the probability of occurrence aspect and the consequence aspect must be presented in such a way that the public comprehends the true risk. As stated earlier, the public has a difficult time comparing events with differing probabilities, and this is especially true for very small probabilities. Therefore, it is necessary to present the probability of occurrence in absolute or comparative forms that allow people to understand and appreciate the magnitudes. One way to do this for all probabilities to be expressed as a multiple of some common reference risk. While it would be desirable to have the reference risk be something that is time-invariant and universal, it is also important that it be something to which people can readily relate. Purely, as an example, one might consider the annual probability of a person who has attained the age of 50 years of dying from natural causes within the next year, termed a danc (death annually due to natural causes). All risks, whether annual or for longer reference terms, could then be expressed as a number of these units, such as 50 dancs, or 0.1 dancs. This has the advantage of being relatively stable and technology-independent, except for long-term trends due to health and medical advances. It can also be viewed as relatively natural, and therefore a non-controversial base.

In assessing risks for the built environment, it is important that return periods not be used in conveying information to the public at large. Their concept is often misunderstood, leading to false confidence concerning the occurrence of the next event. Thus, even if the concept of the preceding paragraph is not adopted, events such as the 100-year flood should be referenced as the annual 0.01 flood.

In addition to expressing the probabilities in a form that can be easily used for decisions, there must be an accounting of the various risks to society presented in a public manner.[25] Indeed, the detailed analysis of structural reliabilities[26,27] must be combined with societal issues in such a way that broad segments of a community can make meaningful input to risk decisions.

5. Alternative Decision Criteria

5.1. Expected value

Although not explicitly stated, the use of the term risk as the probability of occurrence multiplied by the measure of consequence is based on an expected value criterion. That is, the number is the average value of the consequence if exposure is repeated many, independent times. While there is a strong mathematical and actuarial basis for using the expected value criterion, there are a number of deficiencies in the case of the built environment, and other decision rules can be considered.

From the definition above, expected value can claim an almost indisputable logic as a decision rule when something is repeated in many, identical experiments. In this case, the expected value of the consequence for a single exposure, or for a set accounting time period such as one month or one year, can be set aside as appropriate to cover the cost of the consequence when it occurs (or used to purchase an insurance policy when supplemented by the associated operational expenses of the insurance company). In this manner it meets both emotional and fiscal needs. The decision rule would logically be to maximize the expected net benefit (benefit adjusted by cost).

The assumptions of repetition, however, are rarely met for the built environment. While the exposure to many types of hazards might be modeled as being composed of repeated, independent annual events, each structure or set of structures tends to be unique. Therefore, precise estimates of both probabilities and consequences are unlikely. In addition, when rare events are being considered the lifetime of a structure (or at least the term of duration of a single owner) might be viewed as a single experiment, and in this case there is no repetition.

Another problem with the use of expected value was demonstrated earlier with the earthquake example. People do not necessarily wish to equate two events in their decision-making just because for each the product of probability and outcome are the same. Indeed, if one accepts the multi-attribute nature of risk, then the two such events could be very different in many of their other attributes, and thus as dissimilar as two events with comparable levels of dread, familiarity and technological stigma, but vastly different probabilities and costs.

One can of course accept the concept of multi-attribute risk and still use the expected value as a decision criterion. If a measure of common utility can be applied across attributes, then the problem remains the same as before. On the other hand, most situations will not lend themselves to reduction to a common measure of attributes, in which case a multi-objective rule should be utilized. In this case, the expected value along each axis can be used as a component of the risk, but the final decision will not be clear unless one option completely dominates all others for every component.

5.2. *Minimum regret*

Another decision rule, which can be used for single or multiple objectives, is that of minimum regret. In this case, the optimal value associated with each attribute is identified, subject to the constraints. For each decision choice, the distance is measured between the optimal value of the objective(s) and the one realized for the decision. This difference is termed the regret, and a decision is selected such that the maximum regret over all objectives is minimized. This rule ensures the smallest amount of deviation between the ideal solution and the one selected. With uncertain outcomes, as being dealt with in this book, the regret most likely has to be defined in probabilistic terms, so one is likely to use minimum expected regret, combining attributes of expected value without necessarily maximizing expected net benefit.

5.3. *Mini-max and as low as reasonable*

These two terms are closely related and based on similar principles. The first is short for minimizing the probability of occurrence of the outcome with the maximum harm (cost). It was originally used for a set of decisions related to alternate designs to meet a desired goal, for which there was a fixed and definable set of undesirable outcomes. The idea is that one would select the design associated with the minimum probability for the worst of the undesirable outcomes. The expression as low as reasonably achievable (ALARA) is used for quantitative design decisions and is based on the principle of improving the design until the probability of a given undesirable outcome is as low as is reasonable given constraints of cost, practicality and the current state of knowledge. It is used in the nuclear power industry and with certain toxic chemicals, but appears to be too subjective for helpful guidance in the design of civil structures for the built environment. A similar approach is expressed as best available control technology (BACT).

These two principles appear to hold promise for a multi-attribute approach to decisions in the built environment. For instance, consider a series of undesirable outcomes (one can again use the concept of damage from a moderate and a very large earthquake). Rather than combine these in a true expected value sense, one could decide on a maximum accepted probability of damage due to each event. The optimal design would then be the one that meets this level of probability for the minimum total cost. This would require the setting of probabilistic goals for each of the outcomes, which seems entirely consistent with the current call in the civil engineering field for performance-based design. The damage level due to a moderate earthquake could also be determined based on a minimum expected cost criterion, while that for the very large earthquake could be based on an acceptable probability taking into account the direct cost, dread, indirect costs and social amplification associated with the outcome.

5.4. *Additional criteria*

There are other criteria that can be applied, especially in the case of multiple attributes. One of these is mutual elasticity, where an optimal decision point is defined where a certain percentage improvement in an objective can only be gained by a larger percentage decrease in every other objective. Another approach is to maximize the norm of the objectives, defined as the square root of the sum of the squares of the attributes. This requires that a common measure is found that can quantitatively be used across attributes. In general, it may be necessary to incorporate multiple objectives directly rather than trying to reduce the decision to a single trade-off.[28]

6. Societal Decision-Making

6.1. *Trade-offs*

All of the issues discussed previously provide the quantitative and perception information that is necessary for society to make informed and enlightened decisions among various applications of resources,[29] although it can be argued that this has not always been the case.[30] Societies will always be faced with more opportunities for improving their utility than there are resources. The issues of meeting current needs for shelter, food, and education must be balanced with investment for future security and serviceability of infrastructure. When that investment in the built environment involves a degree of recognized risk, such as due to natural and society hazards, then the balance of current needs with uncertain future risks becomes even more difficult.[31-33] Society is just beginning to take a comprehensive view of the risks from natural hazards,[34-36] and these are yet to be combined with other hazards (Paté-Cornell).[37]

6.2. *Development, sustainability and intergenerational issues*

The development goals of a region traditionally have been related to the creation of employment, and the servicing of the resultant workforce. These goals have not focused on risks and probabilities, especially long-term hazards of development, or the effects on the sustainability of the environment. As stated by Tierney *et al.*,[38] "... human settlements are based upon principles of short-term growth and profits for privileged segments of the population instead of safety and sustainability for the society as a whole." With regard to sustainability, they comment that decisions related to growth of the built environment are often based on "... policies that emphasize growth at the expense of safety and... operation of political-economic forces that depend on the exploitation of natural and environmental resources." It is upon this community framework that it is necessary to introduce a probabilistic and long-term view that includes trade-offs of risk and economic analysis over the expected service life of infrastructure, along with the aspects of sustainability.[39-41] This requires both education of the public and an appropriate supportive method for presentation of accomplishments by elected officials.[42]

For decisions that affect future risk, those affected by the decision should be involved in that decision. But building and infrastructure decisions involve economic lifetimes of 50 to 100 years, or longer in the case of certain structures such as dams and nuclear storage facilities. Similarly the ecological footprint of building projects and the sustainability of resources certainly involve generations far into the future. Therefore, it is impossible for all affected players to be at the table. This situation extends the problem of group optimization in gaming theory and extends it to the situation where each player has to represent to some degree players not present.

Today's decision-makers must be prepared to become defenders of future generations, while at the same time be willing to assume that new technologies and capabilities will permit advances not now possible. A founding principle comes from the definition adopted by the United Nations[43] that sustainable development "meets the needs of the present without compromising the ability of future generations to meet their own needs." A more specific working definition is provided in 1993 by the National Commission on the Environment[44]: Sustainability implies "a strategy for improving the quality of life while preserving the environmental potential for the future, of living off the interest rather than consuming natural capital..."

While these definitions of sustainability are meritorious, they beg the issues of how much sacrifice of resources is reasonable in light of the uncertainty of future changes in technology. Society is left with the fundamental issue of deciding in the present the risk that is reasonable for those of the future (for a good discussion, see *Risk Analysis*).[45]

6.3. *Assessing risks and benefits*

It has been established that the public uses several standards to assess future benefits and risks, and these must be taken into account in order to establish a system that will be effective in involving the public in the political decision-making process. Fischhoff *et al.*[46] have identified four basic comparisons commonly used by individuals. One of these is a cost-benefit analysis, using future benefits and expected cost. This requires a comprehensive accounting of both costs and benefits, and an ability to fully accept the concept of present discounting. Another is revealed preferences, whereby an analyst can infer the tradeoffs of current and future risks and benefits from other decisions. This is hampered by the difference of up to three orders of magnitude in willingness to accept risk when the action is viewed as voluntary. A third approach is that of expressed preferences, whereby individuals explicitly express their willingness to trade off risks and benefits, including those that involve future expenditures and rewards. This method has the political advantage of having the public state direct values, rather than trying to infer them from other actions. On the other hand, such decisions are known to be inconsistent, and there is no simple way to reconcile them. The fourth method to assess risk and reward is to use the natural environment to calibrate values that are based on some sort of natural standard. This is an especially promising approach when it comes to flood risk and other land use decisions, but may not be applicable to issues of infrastructure improvement, such as location and design of bridges.

6.4. *Economic lifetime*

All structures and infrastructure are intended to last for a significant time into the future, referred to as the economic or design lifetime. Typical values range from 50 to 100 years, depending on the nature of the project. Therefore, considerations of safety and economics must be cast in terms of the full life-cycle of the structure.[47–51]

The decision, at the design phase of a building, to announce that the facility is intended to last for a specific number of years could be highly unpopular. Currently, the economic lifespan is reflected essentially through the design loads, such as a code-specified wind or earthquake intensity with a 2% probability of exceedance in 50 years. With negligible structural deterioration, there is no specified lifetime for the building, merely a specific annual probability of a particular load being exceeded. This is important, since the public statement that a particular new building is "intended" to last, say, 50 years might be unpopular and even unacceptable, and it raises the issue of whether it really makes sense to specify a lifetime. If design is to include the consideration of the balance of initial construction cost and lifetime risk and operational costs, then there must be a specified lifetime over which these costs are incurred. This issue can be avoided by selecting an economic lifetime over which the structure is computed be economically balanced. This would not imply that the building is expected to only last a particular number of years, but that its economic balance point was set at some lifetime. It would have to be made clear that this would not necessarily imply that the building was uneconomical to continue operating past that time, just that the "break even point" of initial design had been reached.

6.5. *Utility and discounting*

A goal of individuals in society, and presumably society as a whole, is to increase utility. Therefore, the measures of consequence that have been referred to thus far need to be cast in terms of utility.[52] Both benefits and costs should be construed to mean positive and negative values of utility, introducing a degree of subjectivity.

Since utility by definition reflects the personal worth of an outcome, it is different for different people. In order to use it for decisions with respect to the built environment, it is necessary to define some societal standardization. It is important to distinguish the societal issue of risk aversion, which produces a characteristic convex shape for the utility curve, from the issue of diminishing importance of marginal increases in wealth, which also produces a convex shape. These two issues will need to be considered separately and calibrated for society, perhaps on a community level.

Since most decisions relating to the built environment involve costs far into the future, discounting to the present value is very important. Discounting, however, can also introduce subjectivity, and methods must be adopted that provide a normative basis.[53] Usually the discount rate is defined in terms of an annuity, relating the amount of money that must be invested in the present to purchase something in the future. This usual definition sets the discount rate as the investment rate of return minus inflation (all estimated for future periods). Psychologists note, however, that there is a natural human preference to receive a benefit immediately rather than at some point in the future. This preference to receive benefits in the near term produces a discounting of future rewards, completely independently of investment and inflation.[54]

One final issue considered here is the discounting of future injuries and death. Since these are statistically and not cohort-based, the characteristics of those affected presumably do not change over time. Therefore, it is not clear that it is appropriate to discount these "costs." In other words, should the loss of a life in year ten be discounted relative to a loss of life in year one (as it presumably would be for a specific individual)? While it would seem that it should not, a cogent argument can be made for such discounting based on the use of societal resources to improve conditions for the future.[55]

6.6. *Probabilistic risk versus uncertainty*

In all the discussions thus far, the term probability has been used to denote the likelihood of occurrence of an event. It has been observed, however, that the nature of the modeling of this uncertainty has an effect on the way individuals evaluate its acceptance. In terms of risk and decision for the built environment, therefore, it is necessary to distinguish among the various concepts.[56,57]

6.6.1. *Objective and subjective probability*

Probability is generally defined in terms of two different concepts, sometimes referred to as objective (or frequentist) and subjective (or Bayesian) probability. The former is formulated as the fraction of time a particular outcome occurs when an experiment is repeated independently and with identical parameters, as the number of repetitions approaches infinity. This was the sense that was used earlier in this chapter. The latter definition is a personal best estimate of the likelihood of a particular outcome occurring, and is the only way to evaluate unique situations. For instance, an estimate as to whether location A along a flood plain is at a higher elevation than location B is a Bayesian probability. It either is or is not, and in assigning a probability one is integrating knowledge to assign a probability as to whether the statement is true.

6.6.2. *Aleatoric and epistemic uncertainty*

There is an alternative formulation regarding aspects of probability, and the distinction has been observed by social psychologists as important in decision-making. The first of these relates to the inherent randomness of a process, and is referred to as the aleatoric uncertainty. For instance, with full knowledge of a completely fair coin-flipping experiment, the outcome of each experiment can still not be predicted with certainty.

In practical situations society lacks full knowledge of the conditions of the probability experiment. Thus, it is necessary to introduce additional uncertainty due to this lack of information, and this is referred to as epistemic uncertainty. Additional data and field verification of models and parameters should reduce this uncertainty, with it theoretically approaching zero with sufficient verified information.

Experiments by social psychologists have confirmed that people react differently to aleatoric and epistemic uncertainty. They generally accept the former as fact (although they have trouble conceptualizing probabilities that are far from 50%, as mentioned earlier), whereas they apply biases to the latter. In general, they apply an optimism to the epistemic uncertainty, hoping that with additional information it will increase for favourable outcomes and decrease for unfavourable outcomes. Apparently, the lack of knowledge is taken as an opportunity for subjective optimism.

6.6.3. *Psychological distinctions*

In presenting probabilities to the public for decision-making, it is important to distinguish between those probabilities that will be accepted as fact and those that will be treated as being based on judgement. As a general rule, those that are founded on objective probability and aleatoric uncertainty are taken as factual, and those that involve the subjective judgement of professionals and epistemic uncertainty are treated as estimates. It should be made clear, however, that these distinctions are in the eyes of the public, and not based on distinctions in the mathematics of probability. The fact remains that decisions made by elected officials and the public at large will be altered based on these perceptions.

6.7. *Risk or risk perception?*

It should be clear from the preceding discussion that the risk as strict probability generally presented by risk engineering professionals gives only one small aspect of the factors that are considered when society as a whole makes decisions. There are two necessary directions of change. To the extent that decisions by society utilize poorly estimated approximations to true probability, then it is incumbent upon risk professionals to help educate the public so that better estimates of probability are achieved, and perceived. It is somewhat disturbing, for instance, to note that studies have observed that only those individuals who are moderately depressed have accurate assessments of risks in society.[58] But even if society has and understands accurate estimates of risk, the fact remains that society bases its decisions on additional dimensions of risk. In order to address this there are two conflicting approaches, and it is not clear which is the preferable one.

It is easy to understand that an argument can be made that it is important to assign society's limited resources in the most effective manner possible. Following this mandate, one can only trade off alternate expenditures by comparing the actual reduction in risk as measured by probability and consequence of different actions. Therefore, to effect the most efficient utilization of resources, science-based professionals need to work to educate the political establishment and others in society to comprehend the most accurate perception of risk in terms of actual probabilities of occurrence.

On the other hand, society has shown by experience that it does not wish to allocate its resources strictly in this narrowly-defined sense of optimization. Funds invested in air travel safety (through National Transportation Safety Board in the United States, for instance) are far more than would be justified on the basis of the probability of injury or death than those invested in highway safety. Research money invested by the United States Congress through the National Earthquake Hazard Reduction Program far exceeds the money for storm and riverine flood hazards, even though the latter two cause significantly more injury, loss of life and property damage per year.[59] In these examples higher resources are dedicated to those hazards that are seen as less controllable and more spectacular (airplane crashes and earthquakes). Similarly, the fear of toxic chemicals and nuclear radioactivity have led to severe curbs in the chemical and nuclear energy fields, reflecting both dread and public distrust of experts in those industries, with much less stringent controls on common runoff toxicity and greenhouse emissions.[60]

Since truth in a democratic society can perhaps only be defined in terms of the laws and policies of that society, it is impossible to be dogmatic about a probabilistic basis. An excellent set of articles debating this subject appeared in a special issue of the journal *Reliability Engineering and System Safety*.[61]

6.8. *Political issues*

Decisions in the Western world are based on a combination of influences, chiefly among them being the courts, elected politicians, and professionals. The courts have the role of interpreting the laws and assigning responsibility and accountability. They carry a great deal of influence because fault in the case of a structural failure is most often decided in a court through the interpretation of law, and responsibility is assessed in monetary terms through that court. Since decisions are based on precedence and laws, however, it is the elected officials who have the ultimate responsibility for shaping the nature of assumed risk in society. Therefore, it is the set of measures that is used by citizens in electing its political leaders that ultimately determine the levels of risk and responsibility that will be determined by laws. Too often, however, it has been observed that groups in strong political positions have trouble seeing risk issues from the viewpoint of others.[62]

Public acceptance of someone in a position of power is directly related to the security and benefit felt by individuals. Other factors include philosophical congruence and personal appeal. The benefit felt by the public is often related to policies and laws that bring financial advantages to an individual, as well as tangible improvements in life style and local pride. It is the cost-effectiveness of these improvements from the individual's viewpoint that relates to the subject of investment in infrastructure. At the time of election in a democratic society, the populace generally evaluates a political party and its candidates based on the tangible benefits accrued since the last election. In the case of public infrastructure, such as buildings, bridges and roads, this generally will be evaluated in terms of improved level of current service due to new or significantly upgraded facilities.

When it is time for reappointment, review, or re-election of a public official, the perception of the real value of the infrastructure and immediate return-on-investment has substantially more impact than esoteric arguments of future cost savings and present discounted value. This perception of benefit is a driver in the political process. Some studies or corporations have shown that there are systematic methods to develop strategies of risk throughout a company, and that these strategies might offer a source of investigation for public risk and responsibility.[63]

It will be necessary to align accountability in the political process with the long-term goals of economics, balanced risk and sustainability. One possibility is that at the time of any election, a community or state official who is running for re-election presents a present value analysis of the public infrastructure within his or her region. This would include not only the current value of a benefit/cost analysis for any new structures, but also the change in value to existing infrastructure. If nothing had been done to improve the efficiency and lifetime safety of existing infrastructure, then during the time since the prior election there may have been a deterioration or decreased serviceability of these facilities. This loss would be combined with the benefits of new infrastructure to present a total picture to the public.

The exact nature of such a tool would need significant research and consideration, including the involvement of appropriate legislative and economic experts, but the basic concept is quite straightforward. The fundamental goal is that of an infrastructure report card, perhaps as part of a total public trust report card. Such an expectation could become part of the lexicon of the political process for all government authorities having jurisdiction of public infrastructure. The credits and debits of existing and new infrastructure would reflect the current value, including discounted future benefits and costs. Such future costs should be those associated with operation and maintenance, as well as those reflecting a probabilistic assessment of potential losses due to both natural and society-induced hazards.

Finally, it must be recognized that the professional community has not only crucial importance, but also tremendous influence on risks to society. The consensus standards process is driven and dominated by professionals with special expertise and interest. They evaluate the adequacy of current codes and procedures, and make adjustments that are then adopted, usually in their entirety, but local jurisdictions having authority (local communities within the United States, federal bodies in many other countries). Thus, changes to building codes to reflect new knowledge, new material, and new appreciation of hazards, are often implemented completely due to the decisions of the professional community. Even public review is mostly from the participation of professionals, may times reflecting special interest groups. The process leads to formal standards and codes that are deemed by the professionals to be in the best interests of society, broadly defined. The role of risks and consequences is certainly present, but often only in implicit terms.

References

1. D. Blockley, *The Nature of Structural Design and Safety*, Wiley, NY (1980).
2. R. Melchers, Rational optimization of reliability and safety policies, *Reliability and System Safety* **73** (2001) 263–268.
3. ASCE, Minimum Design Loads for Buildings and Other Structures, ASCE 7-02, American Society of Civil Engineers, Reston, VA (2002).
4. A. Freudenthal, The safety of structures, *American Society of Civil Engineers Transactions* **112** (1947) 125–180.
5. A. Freudenthal, Safety and the probability of structural failure, *American Society of Civil Engineers Transactions* **121** (1947) 1337–1397.
6. A. Freudenthal, Safety and reliability of large engineering structures, *Public Safety: A Growing Factor in Modern Design*, National Academy of Engineering, Washington, DC (1970) 82–87.
7. J. A. Blume, Civil structures and public safety: Safety in design and construciton of civil structures, *Public Safety: A Growing Factor in Modern Design*, National Academy of Engineering, Washington, DC (1970) 88–95.
8. C. H. Blackall, Live loads in buildings, *The American Architect and Building News*, August 26 (1893) 129–131.
9. C. H. Blackall, Live floor loads, *The American Architect — The Architectural Review* (1923) 6–8.
10. American Standard Building Code Requirements for Minimum Design Loads in Buildings and Other Structures, US Department of Commerce, National Bureau of Standards, Miscellaneous Publication M179, (October 1945).
11. R. B. Corotis, Risk-setting policy strategies for hazards, *Life-Cycle Cost Analysis and Design of Civil Infrastructures, Proceedings of March 2003 Conference*, Lausanne, Switzerland, ASCE (2003) 57–64.
12. B. R. Ellingwood, Probability-based structural design: Prospects for acceptable risk bases, *Applications of Statistics and Probability*, R. E. Melchers and M. G. Stewart, Editors, A. A. Balkema, Rotterdam (2000) 11–18.
13. D. Rosowsky and S. Schiff, What art our expectations, objectives, and performance requirements for wood structures in high wind regions? *Natural Hazards Review* **4**, 3 (2003) 1–5.
14. P. Stern and H. Fineberg, Editors, *Understanding Risk*, National Academy Press, Washington, DC (1996).
15. M. Rabin, Diminishing marginal utility of wealth cannot explain risk aversion, in *Choices, Values, and Frames*, D. Kahneman and A. Tversky, Editors (2000) 202–208.
16. A. Tversky and M. Bar-Hillel, Risk: The long and the short, *Journal of Experimental Psychology: Learning, Memory, and Cognition*, Vol. 9, No. 4, October, American Psychological Association (1983) 713–717.
17. D. Kahneman and A. Tversky, Editors, *Choices, Values, and Frames*, Cambridge University Press, Cambridge (2000) 1–16.
18. A. Tversky, and D. Kahneman, Rational choice and the framing of decisions, in *Choices, Values, and Frames*, D. Kahneman and A. Tversky, Editors (2000) 209–223.
19. P. Slovic, *The Perception of Risk*, Earthscan Publications, Sterling, VA (2000).
20. P. Slovic, B. Fischhoff and S. Lichtenstein, Rating the risks, *Environment* **21**, 3 (2000) 14–20 and 36–39.
21. P. Slovic, B. Fischhoff and S. Lichtenstein, Facts versus fears: Understanding perceived risk, in *Judgment Under Uncertainty: Heuristics and Biases*, D. Kahneman, P. Slovic and A. Tversky, Editors, Cambridge University Press, Cambridge, United Kingdom (1982) 463–489.

22. D. Kahneman and A. Tversky, Prospect theory: An analysis of decision under risk, *Choices, Values, and Frames*, D. Kahneman and A. Tversky, Editors, Cambridge University Press, Cambridge (2000) 17–43.

23. Y. Haimes, *Risk Modeling, Assessment, and Management*, Wiley Interscience, NY (1988).

24. P. A. Samuelson, Risk and uncertainty: A fallacy of large numbers, *Scientia*, Vol. XCVIII, Annus LVII, MCMLXIII, International Review of Scientific Synthesis, Amministrazione della Rivista: ASSO (Como), Bologna, Italy (1963) 108–113.

25. H. Kunreuther and R. J. Roth, Sr., Editors, *Paying the Price*, Joseph Henry Press, Washington, DC (1988).

26. M. Jiang, R. B. Corotis and J. H. Ellis, Optimal life-cycle costing with partial observability, *Journal of Infrastructure Systems*, ASCE **6**, 2 (2000) 56–66.

27. Y. Li and B. R. Ellingwood, Risk analysis of light-frame wood construction subjected to extreme winds, *Applications of Statistics and Probability in Civil Engineering*, A. Der Kiureghian, S. Madanat and J. Pestana, Editors, Millpress, Rotterdam, The Netherlands (2003).

28. R. L. Keeney and H. Raiffa, *Decision with Multiple Objectives: Preferences and Value Tradeoffs*, John Wiley & Sons, NY (1976).

29. IBHS, *Are We Planning Safer Communities?* Institute for Business and Home Safety, Washington, DC (2000).

30. P. Szanton, *Not Well Advised*, Authors Choice Press, Lincoln, NE (2001).

31. R. J. Burby, *Cooperating with Nature*, Joseph Henry Press, Washington, DC (1988).

32. T. Cohn, T. K. Gohn and W. Hooke, *Living with Earth's Extremes*, Institute for Business and Home Safety, Tampa, FL (2001).

33. D. R. Godschalk, T. Beatley, P. Berke, D. J. Brower and E. J. Kaiser, *Natural Hazard Mitigation: Recasting Disaster Policy and Planning*, Island Press, Washington, DC (1999).

34. D. Mileti, *Disasters by Design*, Joseph Henry Press, Washington, DC (1999).

35. HAZUS, HAZUS, FEMA's Tool for Estimating Potential Losses from Natural Disasters, Federal Emergency Management Agency, Washington, DC (1999).

36. HAZUS 99, Estimated Annualized Earthquake Losses for the United States, Federal Emergency Management Agency, Washington, DC (2001).

37. E. Paté-Cornell, Numerical safety goals for engineering risk management, in *Risk Analysis, Proceedings of the Symposium*, August 11–12, 1994, University of Michigan, Ann Arbor, MI (1995) 175–191.

38. K. Tierney, M. Lindell and R. Perry, *Facing the Unexpected: Disaster Preparedness and Response in the United States*, Joseph Henry Press, Washington, DC (2001) 252.

39. A. A. Guggemos and A. Horvath, Strategies of extended produced responsibility for buildings, ASCE, *Journal of Infrastructure Systems* **9**, 2 (2003) 65–74.

40. *Living with Risk: A Global Review of Disaster Reduction Initiatives*, United Nations, preliminary version, July, http://www.unisdr.org/unisdr/Globalreport.htm (2002).

41. F. J. Moavenzadeh, J. Janaki and P. Baccini, Editors, *Future Cities: Dynamics and Sustainability*, Kluwer Academic Publishers (2002).

42. R. B. Corotis, The political realities of life cycle costing, *First International Conference on Bridge Maintenance, Safety and Management*, J. Casas, D. Frangopol and A. Nowak, Editors, International Center for Numerical Methods in Engineering, Barcelona, Spain (2002).

43. World Commission on Environment and Development, *Our Common Future*, Oxford University Press, Oxford (1987).

44. National Commission on the Environment, *Choosing a Sustainable Future*, Island Press, Washington, DC (1993).

45. *Risk Analysis*, Vol. 20, No. 6, December, Blackwell, Malden, MA.

46. B. Fischhoff, P. Slovic and S. Lichtenstein, Weighing the risks, *Environment* **21**, 4 (2000) 17–20 and 32–38.

47. D. M. Frangopol, J. S. Kong and E. S. Gharaibeh, Reliability-based life-cycle management of highway bridges, *Journal of Computing in Civil Engineering*, ASCE **15**, 1 (2001) 27–34.

48. D. M. Frangopol, D. Maute and K. Maute, Life-cycle reliability-based optimization of civil and aerospace structures, *Computers and Structures*, Pergamon, article in press (2003).

49. Z. Tao, R. B. Corotis and J. H. Ellis, Reliability-based bridge design and life cycle management with Markov decision processes, *Structural Safety* **16** (1994) 111–132.

50. Z. Tao, J. H. Ellis and R. B. Corotis, Reliability-based life cycle costing in structural design, *Structural Safety and Reliability*, G. I. Schueller, M. Shinozuka and J. T. P. Yao, Editors, Proceedings of ICOSSAR'93. Balkema, Rotterdam (1994) 685–692.

51. Y.-K. Wen and Y. Kang, Minimum building life-cycle cost design criteria. I: Methodology and II: Applications, *Journal of Structural Engineering* **127**, 3 (2001) 330–346.

52. J. von Neumann and O. Morgenstern, *Theory of Games and Economic Behavior* (second edition 1947), Princeton University Press, Princeton, NJ (1944).

53. R. Rackwitz, Discounting for optimal and acceptable technical facilities involving risks, extended version, *Ninth International Conference on Statistics and Probability in Civil Engineering*, July 6–9, San Francisco, CA (2003).

54. C. Heath, Escalation and *de*-escalation of commitment in response to sunk costs: The role of budgeting in mental accounting, *Organizational Behavior and Human Decision Processes* **62**, 1 (1995) 38–53.

55. E. Paté-Cornell, Discounting in risk analysis: Capital vs human safety, *Risk, Structural Engineering and Human Error*, M. Grigoriu, Editor, University of Waterloo Press, Waterloo, Ontario, Canada (1984).

56. R. B. Corotis, Socially relevant structural safety, *Applications of Statistics and Probability in Civil Engineering*, A. Der Kiureghian, S. Madanat and J. Pestana, Editors, Milpress, Rotterdam, The Netherlands (2003) 15–24.

57. B. R. Ellingwood, Acceptable risk bases for design of structures, *Progress in Structural Engineering and Materials,* 3, John Wiley and Sons, Ltd. (2001) 170–179.

58. S. Taylor and J. Brown, *Psychological Bulletin*, American Psychological Association **103**, 2 (1988) 193–210.

59. J. H. Wiggins, *Building Losses from Natural Hazards: Yesterday, Today and Tomorrow*, J. H. Wiggins Company, Redondo Beach, CA (1976).

60. S. Pacca and A. Horvath, Greenhouse gas emissions from building and operating electric power plants in the Upper Colorado River Basin, *Environmental Science and Technology*, American Chemical Society, in press (2002).

61. *Reliability Engineering and System Safety*, Issue 59(1), January, Elsevier, Oxford, England (1998).

62. E. Enarson and B. H. Morrow, *The Gendered Terrain of Disaster: Through Women's Eyes*, Laboratory for Social and Behavioral Research, Florida International University, Miami, FL (1998).

63. D. Murphy and E. Paté-Cornell, The SAM framework: Modeling the effects of management factors on human behavior in risk analysis, *Risk Analysis* **16**, 4 (1996) 501–515.

CHAPTER 2

SOCIO-ECONOMIC RISK ACCEPTABILITY CRITERIA

RÜDIGER RACKWITZ

Institut fuer Baustoffe und Konstruktion
Technische Universitaet Muenchen, Arcisstr. 21
D - 80290 Muenchen, Germany
E-mail: rackwitz@mb.bv.tum.de

This contribution reports about some recent developments in setting up rational public risk acceptability criteria for technical facilities exposed to natural or man-made hazards. The life quality index is chosen as basis for modern economic considerations. The societal value of a statistical life and the corresponding willingness-to-pay are presented.

1. Introduction

The public requires safety from technical installations and natural environment if human life is at risk. Traditionally, risk acceptance criteria are implicit in codes of practice, standards or regulations since well-defined fields of application are calibrated against past and present practice. This is all but satisfying. Not only is it unclear whether present rules are already optimal, extrapolations into new fields of application are also extremely difficult due to a missing rational basis. From many empirical studies it is known that the efforts spent in saving a life in different areas by different interventions vary from a few hundred US$ to more than a hundred million US$ just by illustrating the unpleasant situation. The public must require that criteria for setting limits should be uniform, affordable and efficient.

The following considerations are valid only for public reduction of involuntary risks of an anonymous member of society. Attributes like life expectancy, age, work and leisure time, income and consumption will be incorporated. Risk reduction is a primary concern of society, but it is not the only one as it generally involves cost. Thus, the cost expended for risk reduction must be balanced against other competing needs in view of limited resources. If we wish to recommend rational choices for risk reduction, then monetary valuations of human life cannot be avoided. However they must be compatible with ethical standards such as the right for life, freedom of personal development and intra- as well as intergenerational equity as laid down in our constitutions and elsewhere.[1] Cantril[2] and similar more recent studies conclude from empirical sociological investigations that long life and wealth are among the primary concerns of humans in a modern society — among others such as good family relationship, personal well being, a good cultural and ecological environment, etc., all parameters which define the "quality of life". Long life (in good health), wealth and time for leisure will, in fact, be the main ingredients of our developments.

2. Theoretical Developments

According to Yaari[3] and Usher[4] the quality of life in an economical sense can be measured by the life time utility derived from the level of consumption. Denote by $c(\tau) > 0$ the consumption rate at age τ and $u(c(\tau))$ the utility derived from consumption between age a and t:

$$U(a,t) = \int_a^t u(c(\tau))d\tau. \tag{1}$$

Individuals tend to undervalue a prospect of future consumption as compared to that of present consumption. This is taken into account by (continuous) discounting. The life time utility for a person at age a until she/he attains age t is

$$U(a,t,\gamma) = \int_a^t u(c(\tau)) \exp\left[-\int_a^\tau \gamma(\vartheta)d\vartheta\right]d\tau, \tag{2}$$

where the time preference rate (discount rate), $\gamma(t)$, should be decomposed into a subjective time preference rate ρ and the rate δ is related to real economical growth per capita. In particular, we will decompose the time preference rate as

$$\gamma(t) = \rho(t) + \varepsilon\delta > 0, \tag{3}$$

as proposed in economic growth theory,[5,6] assuming $\varepsilon\delta$ as the characteristic constant for each country and taking the subjective rate $\rho(t)$ as time-dependent in sufficient generality. The parameter ε is close to one and will be further discussed below. There is a general agreement to use special rates associated with health and life extension (one cannot save good health at age 30 for consumption at age 75). Exponential population growth with rate n should be considered taking into account that families are by a factor $\exp[nt]$ larger at a later time $t > 0$. Constant exponential population growth can be verified for the last 100 years in good approximation from the data collected in Ref. 7. The economics literature also states that if no such "discounting" is applied more emphasis is placed on the well being of future generations rather than improving the welfare of those alive at the present, assuming economic growth. Future generations are simply wealthier, so therefore one should add the real, exponential growth rate ζ of an economy. Exponential economic growth at a nearly constant rate can again be verified from the data in Ref. 7 for at least the last 100 years if short term fluctuations are averaged out. The growth rate per capita then is $\delta = \zeta - n$. Following Bayer[8] we assume constant preference rates ρ for living generations in an overlapping generation model but omit ρ for all future generations as being ethically indefensible.[5] The demand for intergenerational equity lets us then define an equivalent time-variant rate for easy understanding and analytical convenience.[9] This can be approximated by

$$\gamma(t) = \rho_0 \exp[-0.013t] + \varepsilon\delta. \tag{4}$$

The expected remaining present value life time utility at age a (conditional on having survived until a) is then[3,10,11]

$$L(a) = E\left[U(a, a_u)\right] = \int_a^{a_u} \frac{f(t)}{S(a)} U(a, t)dt$$

$$= \int_a^{a_u} \frac{f(t)}{S(a)} \int_a^t u(c(\tau)) \exp\left[-\int_a^\tau \gamma(\vartheta)d\vartheta\right] d\tau \, dt$$

$$= \frac{1}{S(a)} \int_a^{a_u} u(c(t)) \exp\left[-\int_a^t \rho(\vartheta)d\vartheta + \varepsilon\delta(t-a)\right] S(t)dt \quad (5)$$

where $f(t)dt = (\mu(t)\exp[-\int_o^t \mu(\tau)d\tau])dt$ is the probability of dying between age t and $t + dt$ which can be computed from life tables containing the age-dependent mortalities $\mu(t)$. $S(t) = \exp[-\int_o^t \mu(\tau)d\tau]$ is the survival probability from which the (unconditional) life expectancy at age a, $e(a) = \int_a^{a_u} S(t)dt$ can be obtained. Life expectancy conditional on having survived until a is $e(a)/S(a)$. a_u is some upper bound age. In economics this function is used to find the optimal consumption path $c^*(t)$ by dynamic optimization given a realistic function for the earnings, rational behavior of the consumer and suitable budget constraints. The introduction of a constant consumption rate $c^*(t) = c$ independent of t can be shown to be optimal under perfect market conditions because insurance annuities can compensate for outliving one's assets or temporary overconsumption and because a surplus of earnings over consumption is best invested into an insurance,[10] simplifies [Eq. (5)].

$$L(a) = u(c)e_d(a, \rho, \delta), \quad (6)$$

where $e_d(a, \rho, \delta)$ is the "discounted" remaining life expectancy at age

$$e_d(a, \rho, \delta) = \frac{1}{S(a)} \int_a^{a_u} \exp\left[-\int_a^t \rho(\vartheta)d\vartheta + \varepsilon\delta(t-a)\right] S(t)dt. \quad (7)$$

Shepard/Zeckhauser[10] assume that there are no legacies or bequests. At birth there is no wealth and there is no wealth left at death at some old age. During years of none or low earnings the individual can borrow (life-insured) money against future earnings and can invest it into (fair) life insurances during periods of larger earnings in order to cover the expenses after retirement. They, in fact, find that the maximum and optimal consumption level is constant over lifetime. In the real world perfect markets do not exist. Shepard/Zeckhauser therefore considered the other extreme, the so-called "Robinson Crusoe" case, where the individual is absolutely self-consistent leading to an inverted U-shaped consumption pattern. Shepard/Zeckhauser take those cases as extremes. Since we are considering the whole life of an individual we may say slightly more realistically that the family and possibly the public provides through redistribution support for consumption during childhood and education and life insurances, public or other pension systems provide sufficient support during retirement. This acts implicitly as the supposed perfect market although the assumption of a perfectly constant consumption throughout one's lifetime will be valid only in approximation.

The expected remaining present value residual life time utility $L(a)$ varies with age as $e_d(a, \rho, \delta)$. For constant discounting it is a monotonically decreasing function. This is confirmed in many theoretical and, in part, also empirical studies. Time variant discounting, however, enables it to grow until an age of about 25 years and then decrease.

Next, we have to choose an appropriate utility function $u(c)$. This is essentially the question of how much consumption and other aspects of life quality to be sacrificed in favor of more life years which can be achieved by some payment for risk reduction. Income is produced by work, all considered at a national level. More income will be produced by more work but this will leave less time for leisure. It will be proportional to productivity of work defined as earnings per time unit of work. Some recent ideas to apply these aspects are repeated here in all brevity. Nathwani *et al.*[12] hypothesized that "... people on the average work just enough so that the marginal value of wealth produced, or income earned, is equal to the marginal value of the time they lose when at work" (also denoted as the work-leisure optimization principle) and defined a measure for life quality as $L = f(k)h(t)$ where k is consumption, $t = (1 - w)e(0)$ is leisure time with $e(0)$ life expectancy at birth, $w(0 < w < 1)$ the fraction of life expectancy lost due to (paid) work and $f(k)$ and $h(t)$ two unknown function of these quantities. Switching to small relative changes dL/L and imposing the requirement of indifference of the relative impact of k and t on life quality with respect to the actual values of k and t they find two simple differential equations and derive the following life quality index:

$$L = k^r t^s. \tag{8}$$

In order to determine the unknown exponents they applied the work-leisure optimization principle. Writing $L = (pw)^r((1 - w)e(0))^s$ with p the labor productivity and solving $\frac{dL}{dw} = 0$ together with $r + s = 1$ gives

$$L = (pw)^w(1 - w)^{1-w}e(0)^{1-w}, \tag{9}$$

where $c = pw$ is the yearly consumption rate. Clearly, the work-leisure optimization will be performed by those who work, i.e. produce wealth, possibly together with their families. Division by w removes a minor inconsistency of the original form for all practical purposes because persons with the same c and $e(0)$ but larger w would have higher life quality so that finally

$$L = \frac{1}{w}c^w(1 - w)^{1-w}e(0)^{1-w}. \tag{10}$$

For later convenience we take the $(1 - w)$-root and divide by $q = w/(i - w)$ so that

$$L = \frac{1}{q}c^q e(0)(1 - w). \tag{11}$$

Therefore $u(c) = \frac{c^q}{q}(1 - w)$, where the last factor may be condensed into a constant, $e(0)$ may be replaced by the discounted remaining life expectancy $e_d(a, \rho, \delta)$ at age a and Eq. (11) is fully determined. If this is interpreted as a utility function it will belong to the class of so-called constant relative risk aversion (CRRA) functions according to Arrow-Pratt for constant q. The parameter $\varepsilon = \frac{cd^2u(c)/dc^2}{du(c)/dc}$ as introduced before is the elasticity of marginal consumption and is taken as a constant over time. The risk aversion parameter is q or $\varepsilon = 1 - q$. A large q indicates low risk aversion and preference for large consumption c and vice versa. For an average of $q = 0.12$ implying about 1600 yearly working hours in a life working time of 45 years one finds $\varepsilon \approx 0.87$, a value consistent with or a little larger than empirical findings based on other concepts. The validity of the work-leisure optimization principle and other assumptions in Ref. 12 are extensively discussed in Ref. 13.

Finally, we have to choose a value for the consumption rate c. Despite ongoing discussions, especially with respect to sustainability, the presently best measure for the wealth-related aspects of the quality of life (standard of living) is the gross domestic product (GDP) of a country or region which is available everywhere and can be updated frequently. The real (i.e. net of inflation) gross domestic product per capita and year is a common indicator of the economic status and behavior in a country. The GDP is roughly the sum of all incomes created by labor and capital (stored labor) in a country during a year.[14] In most developed countries about 60% of the GDP is used privately, 15% by the state (e.g. for military, police, jurisdiction, education, etc.) without transfer payments (social security and other welfare payments) and without health-related or risk-related expenses. The rest is used for (re-)investments into deteriorating production means. The GDP also creates the possibilities to "purchase" additional life years through better medical care, improved safety in road and railway traffic, more safety in or around building facilities, more safety from hazardous technical activities, more safety from natural hazards, etc. It does not matter whether those investments into "life saving" are carried out individually, voluntarily or enforced by regulation or by the state via taxes. Neither the share for the state nor the investments into depreciating production means can be reduced appreciably because they form the (external) conditions for the individual to enjoy life in high quality, now and in the future. Therefore, only the part for private use is available for risk reduction. The part available for risk reduction is then $g = c \approx 0.6\,\text{GDP}$ as a lower bound and a first approximation. The exact share for risk reduction must be determined for each country or group in a country separately and requires great care. The public must decide how much it is willing to spend on risk reduction, how much it is willing to give up on public services and how much it wants to use for private consumption. A realistic estimate is determined from the GDP by leaving the investments untouched but taking the government share without any transfer payments, tax-paid social benefits and, possibly, any payments for public risk reduction programs. We then have $g = c \approx 0.7$ GDP as a crude estimate. The full GDP is the upper bound, of course.

We intend to apply the above concepts to event-type hazards where a representative cross-section of the population is affected by the event. This requires averaging over the proportions in which different ages are present. The density of the age-distribution of a (demographically stable or stationary) group is

$$h(a, n) = \frac{\exp[an]\, S(a)}{\int_0^{a_u} \exp[an]\, S(a)da} \approx \frac{S(a)}{e(0)}. \tag{12}$$

Age-averaging finally gives the *societal life quality index*[15] which is to be interpreted as an age-averaged expected remaining present value life time utility:

$$SLQI = \frac{g^q}{q} \int_0^{a_u} h(a, n) e_d(a, \rho, \delta) da = \frac{g^q}{q} \bar{e}(\rho, \delta). \tag{13}$$

All quantities like g, w, the life tables and at least δ are observable. It should be clear by its derivation that the special risk aversion parameter q must only be used in the context of risk reduction. It has nothing to do with the same parameter used in economics in connection with intertemporal consumption choices.

Willingness-to-pay (WTP) measures a person's willingness to sacrifice one desired attribute, wealth or consumption in order to obtain another desired attribute, or improved survival. Let de denote a marginal change in life expectancy and dg the marginal change in consumption. Shepard/Zeckhauser[10] and others introduced the {willingness-to-pay} as defined by the indifference of L with respect to loss (gain) of consumption or gain (loss) of life expectancy

$$L = L(a, g) = \text{constant}, \tag{14}$$

or, equivalently, that the total differential of the present value remaining life time utility $L(a, g)$ [Eq. (14)] is zero:

$$dL(a, g) = \frac{\partial L(a, g)}{\partial g} dg + \frac{\partial L(a, g)}{\partial e(a)} de(a) = 0 \tag{15}$$

so that after rearrangment:

$$\text{WTP}(a) = dg = -\frac{\frac{\partial L(a)}{\partial e(a)}}{\frac{\partial L(a)}{\partial g}} de(a) = -\frac{g}{q} \frac{de(a)}{e(a)}. \tag{16}$$

Using discounted life expectancies and performing age-averaging gives the societal willingness-to-pay

$$\text{SWTP} = dg = -E_A \left[\frac{g}{q} \frac{de_d(A, \rho, \delta)}{e_d(A, \rho, \delta)} \right]. \tag{17}$$

The criterion is independent of the constants in $L(a)$ and also of monotone transformations. It is possible and many times useful to replace the quantities like $\frac{de(a)}{e(a)}$ under the criteria of Eq. (16) by Taylor expansions to first order (at $x = 0$):

$$E \left[\frac{de(x)}{e(x)} \right] = E \left[\frac{\frac{d}{dx} e(x)|_{x=0} x}{e(x)} \right] = -C_x x. \tag{18}$$

For example, for a mortality reduction scheme reducing mortality by a constant small quantity Δ at all ages, i.e. $\mu_\Delta(a) = \mu(a) + \Delta$ one finds

$$C_{\Delta\bar{e}}(\rho, \delta) = E \left[\frac{\frac{de_d(A, \rho, \delta, \Delta)}{d\Delta}}{e_d(A, \rho, \delta)} \right]. \tag{19}$$

This allows us to define a new constant which we will denote by "societal value of a statistical life (SVSL)}". To assign a monetary value to human life, on whatever basis, to a known or anonymous, young or old, rich or poor, healthy or sick individual is a very controversial issue. In fact, a monetary value of life does not exist. "...the value of human life is infinite and beyond measure,...".[16] It must rather be understood as a formal constant in a relation expressing the "willingness-to-pay" for risk reduction given the financial capabilities of a society. The "societal willingness-to-pay" per unit risk reduction finally is:

$$\text{SWTP} = G_{\Delta \bar{e}} \Delta \tag{20}$$

with $G_{\Delta \bar{e}} = \frac{g}{w} C_{\Delta \bar{e}}$ which is denoted as "societal value of a statistical life". The constant C_x depends on the specific mortality reduction scheme, i.e. the age dependency of a mortality reduction. The influence of such schemes is significant. It depends on the mortality reduction scheme of a particular intervention, for example whether the intervention reduces mortality proportional to age-dependent mortality, only in certain age ranges or simply as a constant at all ages. In the following only constant mortality changes denoted by scheme Δ will be considered. For the applications we have in mind this is intragenerationally acceptable. Other mortality reduction schemes are discussed in Ref. 17. The criterion Eq. (20) is necessary, affordable and efficient from a societal point of view.[12]

3. SWTP's for Some Selected Countries

Setting risk acceptability criteria requires intergenerational equity because the next or even the far future is concerned. Period life tables have been used in Refs. 13 and 18. So-called cohort life tables certainly are more realistic as they reflect the common trend towards larger life expectancies and more compact age distributions. Since we are interested in future risks we have to extrapolate into the future. Time- and age-dependent mortalities can be obtained by extrapolating from a sequence of historical period life tables so that

$$\mu_{\vartheta y}(a) = \mu_y(a) b(a)^{\vartheta + a - y}, \tag{21}$$

where y is the reference year, i.e. the last year for which a period table is available and θ is the year of birth. Unfortunately, cohort life tables do exist only for a few countries. Predictive cohort table must be constructed. Results are collected in Table 1 for 13 developed countries. An uninterrupted sequence of period life tables must be available for at least the last 40 years so that extrapolations for the age dependent mortalities can be performed. Some tables reach back into the 18th and 19th century. The data used are all from Ref. 19. Clearly, such extrapolations are based on the assumption that the observed demographic trends continue throughout the next 100 years. Trends in all other demographic parameters are not taken into account but must be expected. Their effect, however, is small. For example, if population growth n is set to zero resulting in a small change in the age structure of a population no noticeable change in Eq. (20) is observed.

Table 1. Data for several countries.

Country	GDP[(1)], g[(2)]	δ[(3)]	n[(4)]	e[(5)], e[(6)]	q[(7)]	C_Δ[(8)], $C_{\Delta e}$[(9)]	G_Δ[(10)], $G_{\Delta e}$[(11)]
CAN	27330, 16040	2.0	0.99	78, 84	0.13	43, 17	4.6, 1.8
USA	34260, 22030	1.8	0.90	77, 86	0.15	44, 18	5.6, 2.3
D	25010, 14460	1.9	0.27	78, 87	0.12	44, 17	4.9, 1.8
F	24470, 14660	1.9	0.37	78, 85	0.12	43, 16	4.7, 1.8
S	23770, 12620	1.9	0.02	79, 82	0.12	42, 15	3.8, 1.4
CH	29000, 17700	1.9	0.27	79, 85	0.12	43, 16	5.8, 2.2
UK	23500, 15140	1.3	0.23	78, 87	0.13	40, 16	4.1, 1.7
JAP	26460, 15960	2.7	0.17	89, 92	0.13	46, 15	4.8, 1.6

[(1)] in PPPUS\$ in 1999,[22] [(2)] private consumption in PPPUS\$,[23] [(3)] yearly economic growth per capita in % for 1870–1992,[7] [(4)] population growth (2000) in %,[24] [(5)] life expectancy in 2000,[(6)] life expectancy for those born in 2000,[(7)] estimates based on Ref. 20 including 1 hour travel time per working day and a life working time of 45 years,[(8)] without any discounting and age-averaging,[(9)] with discounting according to Eq. (4) and age-averaging,[(10)] SVSL*10^6 without any discounting and age-averaging, [(11)] SVSL*10^6 with discounting according to Eq. (4) with $\rho_0 = 0.03$ and $\varepsilon = 1 - w$.

Numerical values for various economic and demographic quantities are given in Table 1. Only those countries which are economically and demographically compatible with complete data are chosen. The data for GDP, g (as proportion of GDP for private consumption), δ , n and w collected in Table 1 are in part from different sources. Some minor inconsistencies in the data could not been removed, however. Monetary quantities are given in PPPUS\$ (corrected for purchasing power parity). The trends in time in g are taken into account by discounting. Trends in w are neglected. If the life working time is made time-dependent, i.e. by a suitable function fitting the trends in Refs. 7 and 20 sufficiently well one finds that $G_{\Delta \bar{e}}$ goes up by at most 0.2×10^6. But the future development of $w(t)$ depends on many factors which are very difficult to predict.[13] Therefore, the values in Table 1 are those for the year 2001 in official statistics based on a life working time of 45 years. They also include 1 hour travel time per working day because this time is not available for leisure. For all considered countries $\rho_0 = 0.03$ is assumed in Eq. (4) but reasonable variations of ρ change the results only insignificantly. The results for $G_{\Delta \bar{e}}$ show remarkable consistency among the countries considered. The values of $G_{\Delta \bar{e}}$ without any discounting and age-averaging can be easily computed. The demographic constant $C_{\Delta \bar{e}}$ varies only around $16 \pm 10\%$ as compared to $C_\Delta \approx 43$ suggesting that one can also ignore the demographic differences for practical purposes. Those last conclusions do not hold for countries with significantly differing demographic and economic conditions.

Even if one considers the variations in g and w for the different countries it is clear that Table 1 presents a much smaller spread of the constants than empirical estimates as collected in Viscusi/Aldy.[21] Comparing our results with the empirical estimates shows excellent agreement in the mean in the value of VSL (or SVSL) although some uncertainty exists about the precise meaning of the reported values of VSL, for example whether they take into account of the age-dependency, consider age-averaging and the particular mortality reduction scheme. It is worth nothing that a factor of say, 2, makes hardly a difference in applications.

4. Application to Event-Type Hazards from Technical Facilities and the Natural Environment

For technical facilities the mortality change must be replaced by a change in the asymptotic failure rate, i.e. Ref. 17:

$$dm = kdr(p), \tag{22}$$

where k is the probability of being killed in or by the facility in an adverse event and $r(p)$ is the (asymptotic) failure rate dependent on a parameter set p. k must be estimated taking into account of the number of persons endangered by the event N_{PE}, the cause of failure, the severity and suddenness of failure, possibly availability and functionality of rescue systems, etc. The probability k can vary between less than $1/10000$ and 1. For example, estimates show that $k = 0.01$ to 0.1 for earthquakes,[25] $k = 10^{-6}$ to 0.1 for floods,[26] $k = 0.1$ for building fires[27] and $k = 0.02$ or less to 0.7 for large fires in road tunnels.[28] A summary of the estimation methodology is given in Ref. 29.

Making use of Eq. (20) this results in:

$$dC_Y(p) = -dg = -\frac{g}{q}C_{\Delta\bar{e}}dm = -G_{\Delta\bar{e}}dm = -G_{\Delta\bar{e}}kN_{PE}r(p), \tag{23}$$

where $dCY(p)$ are the incremental yearly cost for a risk reduction by $dr(p)$. Age-averaged willingness-to-pay is the correct quantity to use as it must be assumed that a representative cross-section of the population is endangered by the event. $C_Y(p)$ may also be interpreted as cost of financing of single investments into risk reduction $dC(p)$ at time $t = 0$ such that $dC_Y(p) = dC(p)\frac{\gamma\exp[\gamma t_s]}{\exp[\gamma t_s]-1}$ where t_s is the credit period. g in $G_{\Delta\bar{e}}$ also grows in the long run approximately exponentially with δ, which is the rate of effective economic growth in a country. If it is taken into account by the same type of discounting the effects of the two discounting schemes cancel asymptotically. Consideration of the time preference rate ρ is, as mentioned, ethically indefensible if future generations are concerned, i.e. for a long planning horizon. The acceptability criterion for individual technical projects is then (discount factor for discounted erection cost moved to the right hand side):

$$\frac{dC(p)}{dr(p)} = -\frac{e^{\gamma t_s} - 1}{\gamma e^{\gamma t_s}}G_{\Delta\bar{e}}\frac{\delta e^{\delta t_s}}{e^{\delta t_s} - 1}kN_{PE}, \tag{24}$$

but

$$\frac{dC(p)}{dr(p)} = \lim_{t_s \to \infty} -\frac{e^{\gamma t_s} - 1}{\gamma e^{\gamma t_s}}G_{\Delta\bar{e}}\frac{\delta e^{\delta t_s}}{e^{\delta t_s} - 1}kN_{PE} = -\frac{\delta}{\gamma}G_{\Delta\bar{e}}kN_{PE}. \tag{25}$$

The equivalent time-dependent $\gamma(t)$ also approaches δ (or $\varepsilon\delta$) for $t_s \to \infty$ so that there is no discounting asymptotically. It is important to note that the additional risk reducing investment $dC(p)$ depends on the change $dr(p)$ of the yearly failure rate and not on its absolute value.

5. Summary and Conclusions

A new concept for determining rational risk acceptability criteria based on earlier work by Nathwani *et al.*[12] and Pandey/Nathwani[15] is presented. Life time utilities from consumption discounted down to the decision point are used. This is weighted by the age-distribution in a population. By requiring indifference of small changes in lifetime consumption and small life expectancy extensions by payments for risk reduction a risk acceptability criterion is derived. This is applied to technical facilities under man-made or natural risks. The societal willingness-to-pay is computed for a number of countries.

References

1. UN World Commission on Environment and Development (known as the Brundland Commission), Our Common Future (1987).
2. H. Cantril, *The Pattern of Human Concerns*, New Brunswick, NJ, Rutgers University Press (1965).
3. M. E. Yaari, *Uncertain Lifetime, Life Insurancs and the Theory of the Consumer*, Review of Economical Studies **32**, 2 (1965) 137–150.
4. D. Usher, An imputation to the measure of economic growth for changes in life expectancy, published in *National Accounting and Economic Theory: The Collected Papers of Dan Usher Vol. 1*, 105–144. Edward Elgar Publishing Ltd., UK (1994).
5. F. P. Ramsey, A mathematical theory of saving, *Economic Journal* **38**, 152 (1928) 543–559.
6. R. M. Solow, Growth Theory, Clarendon Press, Oxford (1970).
7. A. Maddison, Monitoring the World Economy 1820–1992, OECD, Paris (1995).
8. S. Bayer, Generation-adjusted discounting in long-term decision-making, *Int. J. Sustainable Development* **6**, 1 (2002) 133–149.
9. R. Rackwitz, A. Lentz and M. Faber, Sustainable civil engineering infrastructures by optimization, *Structural Safety* **27**, (2005) 187–229.
10. D. S. Shepard and R. J. Zeckhauser, *Survival versus Consumption, Management Science* **30**, 4 (1984) 423–439.
11. W. B. Arthur, The economics of risks to life, *American Economic Review* **71** (1981) 54–64.
12. J. S. Nathwani, N. C. Lind and M. D. Pandey, *Affordable Safety by Choice: The Life Quality Method*, Institute for Risk Research, University of Waterloo, Waterloo, Canada (1981).
13. R. Rackwitz, Optimal and acceptable technical facilities involving risks, *J. Risk Analysis* **24**, 3 (2004) 675–695.
14. R. J. Barro, Macroeconomics, Wiley, New York (1987).
15. M. D. Pandey and J. S. Nathwani, Life quality index for the estimation of societal willingness-to-pay, *J. Structural Safety* **26**, 2 (2004) 181–200.
16. I. Jakobovits, Jewish Medical Ethics, New York (1975).
17. R. Rackwitz, Acceptable risks and affordable risk control for technical facilities and optimization, submitted for publication in *J. Reliability Engineering and Systems Safety* (2003).
18. R. Rackwitz, Optimization and risk acceptability based on the life quality index, *J. Structural Safety* **24**, 1 (2002) 297–331.

19. Human Mortality Database. University of California, Berkeley (USA), and Max Planck Institute for Demographic Research (Germany), www.mortality.org.
20. OECD, Employment Outlook and Analysis, Paris (2001).
21. W. K. Viscusi and J. E. Aldy, *The Value of a Statistical Life: A Critical Review of Market Estimates Throughout the World, Risk & Uncertainty* **27**, 1 (2003) 5–76.
22. World Development Indicators database, www.worldbank.org/data/
23. United Nations, Human Development Report 2001, www.undp.org/hdr2001.
24. CIA-factbook 2001, www.cia.gov/cia/publications/factbook/
25. A. Coburn and R. Spence, *Earthquake Protection*, J. Wiley & Sons, Chichester (1992).
26. S. N. Jonkman, Global perspectives of loss of human life caused by floods, *Natural Hazards* **34** (2005) 151–175.
27. M. Fontana, C. Lienert, J.-P. Favre and T. Maag, Statistische Untersuchungen zu Gebäudebränden, VdS Schadensverhütung, S+S Report **1** (2002).
28. OECD, *Safety in Tunnels: Transport of Dangerous Goods through Road Tunnels*, Paris (2001).
29. A. Lentz and R. Rackwitz, *Loss-of-Life Modelling in Risk Acceptance Criteria, Proc. PSAM7* (2004).

CHAPTER 3

RELIABILITY IN STRUCTURAL PERFORMANCE EVALUATION AND DESIGN

Y. K. WEN

Department of Civil and Environmental Engineering
University of Illinois at Urbana-Champaign, Urbana, IL 61801, USA
E-mail: y-wen@uiuc.edu

Structural performance in recent earthquakes and hurricanes have exposed the weakness of current design procedures and shown the need for new concepts and methodologies for performance evaluation and design. The consideration and proper treatment of the large uncertainty in the loadings and the building capacity including complex response behavior in the nonlinear range is essential in the evaluation and design process. A reliability-based framework for analysis and design is most suitable for this purpose. Performance check of the structures can be carried out at multiple levels from immediate occupancy to incipient collapse. To arrive at optimal target reliability, lifecycle cost criteria can be used. The methodology is demonstrated by application to performance evaluation and design of structures under earthquake and wind hazards. The inconsistency of design against these hazards in current codes is pointed out.

1. Introduction

Building performance has been less than satisfactory in recent natural hazards such as Northridge (US), Kobe (Japan), and Chi-Chi (Taiwan) earthquakes and hurricanes Hugo, Andrews, and Ivan in United States. These events bring to focus how the structural engineering profession treats the large uncertainty in the natural hazards and structural capacity and what reliability existing buildings have against future hazards. Although the uncertainty of loads has been well recognized by the profession, until recently, the incorporation of uncertainty in most building code procedures has been limited to the selection of design loads based on return period. This design load is then used in conjunction with a series of factors reflecting the influence of structural period, loading characteristics, site soil condition, structural inelastic behavior, importance of the structure etc. These factors are largely determined based on judgment and experience and often calibrated in such a way that the resultant designs do not deviate significantly from the acceptable practice at the time. Recent studies of the reliability of buildings designed for seismic loads according to current code procedures in US, Japan, and Taiwan[1] have shown that there are large differences implied in the reliability, which can be attributed to the differences in hazard characteristics (e.g. seismcity, wind climate), design philosophy, construction practice, and code provisions (e.g. importance factor, ductility reduction factors, gust response factor). It clearly points out the need for a statistical and reliability framework for development of performance-based design considering properly the uncertainty in the demand and capacity, and balance of cost and benefit. In the following, a framework for reliability and performance based design is described.

2. Reliability-Based Performance-Oriented Design

In developing a reliability framework for performance evaluation and design, some critical issues have been identified from recent research.

2.1. *Performance goal*

The performance of a structure can be described in terms of specific limit states being reached. For example, in design against earthquake, a commonly accepted specification[2] has immediate occupancy (IO), life safety (LS), and collapse prevention (CP) each associated with a design earthquake of given return period. The actual reliability corresponding to these limit states vary due the system capacity uncertainty and conservatism built into the various factors in the codes. To strictly enforce reliability performance goals, the target probabilities need to be set directly for the limit states rather than for the design earthquake.[1] In evaluation, the performance of the structure is satisfactory if the limit state probabilities are below the target values. In developing reliability-based design formats, one starts from this target reliability goals corresponding to physical limit states such as incipient damage and incipient collapse and develop the required deterministic design format, which will yield a design that satisfies these goals. An example of such bi-level performance acceptance criteria based on this procedure in terms of annual interstory drift probabilities is shown in Fig. 1. Such a procedure, for example, has been proposed in the US SAC Steel Project.[3-5]

2.2. *Reliability evaluation*

To determine the reliability for performance evaluation and design of structures under severe loads such as seismic excitations, factors like yielding, fracture, and deterioration of members and connections in stiffness and/or strength need to be considered. The well-known First Order Reliability Method (FORM) in combination with static or equivalent static response analysis of the structures has been successfully used as basis for reliability-based design formats in most recent code procedures.

When the structural responses become dynamic, nonlinear, and hysteretic caused by inelastic deformation and brittle fractures; however, the problem is generally not amenable to FORM since it is difficult to describe the limit state function. For problems of such complexity, the direct time history/simulation method may be more practical. To evaluate the reliability, however, one needs to generate a large number of ground motions that represent all future events in the surrounding region

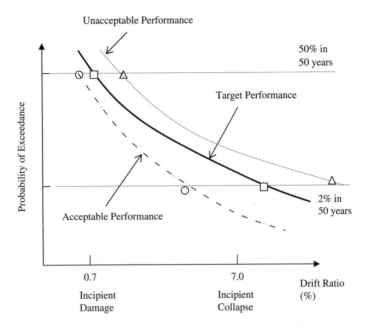

Fig. 1. Bi-level performance check in terms of 50-year limit state probability.

that have an impact on the building. It can become computationally unmanageable, especially for nonlinear analyses of structures of large sizes.

To reduce the computation, a small number of earthquake ground motions can be generated using the values of seismic parameters that contribute most to the limit state probability. For example, in the SAC Steel Project, recorded and broadband simulated ground motions were scaled such that the median (50 percentile) values of the resulting elastic response spectra of these motions matches that of the uniform hazard response spectrum at the site, i.e. spectrum corresponding to a given probability of exceedance.[6]

In locations where records are scarce such as Mid-America, one can generate suites of ten ground motions according to regional seismicity and the latest ground motion models.[7] These ground motions represent what we expect to occur at these sites, corresponding to a specified level of probability of exceedance are therefore referred to as uniform hazard ground motions (UHGM). The structural response corresponding to the probability level can be obtained as the median response of structures under such suites of ground motions using a nonlinear time history analysis. This process is then repeated for several probability levels to obtain the performance curve. Figure 2 shows an example of diaphragm slip demand on a 2-story, un-reinforced masonry building. For example, for a diaphragm with an anchor of 2 inches, the 50-year probability of diaphragm falling from the wall (slip exceeding 2 in) is 0.015.

The accuracy of this method for nonlinear inelastic systems has been proved to be very good in comparison with results of full simulations for several Mid-America

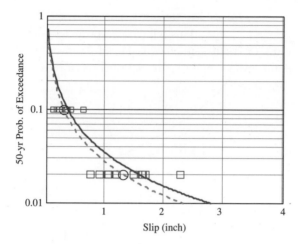

Fig. 2.   URM building diaphragm slip demand curve using 10% and 2% in 50 years uniform hazard ground motions at Memphis: □ denotes data point corresponding to each UHGM, ○ denotes median value, effect of uncertainty due to record-to-record variability across UHGMs considered (solid line), and not considered (dashed line).

cities.[7] The uniform hazard ground motion approach is therefore an efficient method of accounting for excitation uncertainty. The effect of capacity uncertainty and other modeling errors can be included in the above analysis by randomizing the building capacity. Alternatively, one can account for this type of uncertainty by using a uncertainty correction factor as will be discussed in the following section.

The performance can also be evaluated using a scalar excitation intensity measure and relating the structural response and limit state to this measure via repeated time history response and regression analyses. The uncertainty in the excitation is then described by the intensity hazard curve and the limit state or reliability analysis of the structure can be performed based on the hazard curve and the results of the response regression analysis. Simple power laws can be used in modeling both the intensity hazard and the intensity/response relationship such that the limit state probability can be evaluated in closed form with all uncertainties explicitly considered. This approach has been used in the recent FEMA/SAC Steel Project[8] in which the seismic hazard in terms of annual probability of exceedance of spectral acceleration is described by

$$H_t(s) = k_0 s^{-k} \tag{1}$$

in which s is spectral acceleration, k_0 and k are respectively site-dependent frequency and scatter parameters determined from seismic hazard analysis. For example, k is around 3 for Western United States (WUS) and around 1 for Central and Eastern United States (CEUS) indicating much larger scatter in the latter region. The nonlinear displacement demand (D) and spectral acceleration relationship can be described by

$$D = aS^b \tag{2}$$

in which a and b are determined by regression analysis of dynamic structural responses under ground accelerations which also provides the uncertainty in this relationship in terms of coefficient of variation of D given S, $\delta^2_{D|S=s}$. A distinctive advantage of this approach is that it yields closed form reliability estimates of demand D exceeding d as

$$P_t(D > d) = H_t(s^d) \exp\left[\frac{1}{2}\frac{k^2}{b^2}\delta^2_{D|S=s}\right], \tag{3}$$

where S^d is the intensity level corresponding to the demand d according to Eq. (2) when the demand uncertainty can be modeled by a lognormal random variable. Also simple, explicit tracking of uncertainty propagation can be done as shown in the following section.

2.3. *Effect of capacity and epistemic uncertainty*

Reliability evaluation of structures against natural hazard in the past has been mostly concerned with the inherent variability (aleatory uncertainty or *randomness*) of the excitation and structural capacity against a given limit state. The uniform hazard ground motion approach is an efficient method of capturing this aleatory uncertainty in the excitation. In addition, there are also modeling errors (epistemic uncertainty or simply *uncertainty*) which need to be taken into consideration. Capacity variability can be attributed to material variability, non-structural component contribution to resistance such as claddings and partitions, and structural modeling errors such as the highly unpredictable brittle failure of steel connections. The total uncertainty (coefficient of variation) on the capacity side has been estimated to be 40% or higher,[9] which is significant in itself but still small compared with that of the seismic load which is generally in excess of 80%. An interesting implication is that because of this dominance of the uncertainty on the loading side, the effect of the uncertainties can be de-coupled as an approximation. As a result, one can first ignore the uncertainty by using the mean value in the reliability analysis as is done in the above time history/simulation and recover it by a multiplier called uncertainty correction factor. This correction factor can be used in both reliability analysis and reliability-based design.

The general effect of uncertainty is a decrease in reliability. It can be incorporated by a correction factor defined as the ratio of the limit state probabilities with and without consideration of the uncertainty. It is a very convenient tool in both reliability evaluation and development of reliability-based design when uncertainty is present but cannot be accurately quantified at the time. For example, one can treat the uncertainty separately by first neglecting its effects in the reliability evaluation and design and recover it using the correction factor which can be continually updated as more information on model uncertainty becomes available. The method has been demonstrated[3] that the correction factor for limit state probability evaluation due to capacity and epistemic uncertainties is given approximately by

$$C_F \approx 1 + \frac{1}{2} S^2 \delta_R^2, \tag{4}$$

in which S is the sensitivity coefficient depending on the seismic hazard and reliability level and δ_R is the measure of the total uncertainty in terms of coefficient of variation. S generally increases with the uncertainty in the hazard and reliability level. This factor has been used to account for effect of record-to-record variability of the UHGM on the structural median response estimate in Fig. 2. In a reliability-based design, the correction factor for design capacity to incorporate capacity and epistemic uncertainty for a given target reliability is

$$C_D \approx 1 + \frac{1}{2} S \delta_R^2. \tag{5}$$

The net effect of capacity and epistemic uncertainty is an increase in the required design capacity.

In the FEMA/SAC approach the uncertainty has been incorporated in a more comprehensive manner.[8] For example, if the system capacity can also be modeled by a lognormal random variable with median m_C and dispersion δ_C, a point estimate of the limit state probability, $D > C$, one obtains

$$P_t = H(s^C) \exp[(k/b)^2 (\delta_{DR}^2 + \delta_C^2)/2], \tag{6}$$

in which the substitution $\delta_{DR} = \delta_{D|S=s}$ has been made for brevity and $s^C = (m_C/a)^{1/b}$ = spectral acceleration that corresponds to the median capacity. Equation (6) considers the effect of aleatory uncertainty (*randomness*). A point estimate of risk that includes the epistemic uncertainty (*uncertainty*) can be described by the *mean* limit state probability, $E[P_t]$, determined as follows:

$$E[P_t] = E[H_t(s^{\tilde{C}})] \exp\left[\frac{1}{2}\frac{k^2}{b^2}(\delta_{DR}^2 + \delta_{DU}^2 + \delta_{CR}^2 + \delta_{CU}^2)\right], \tag{7}$$

in which

$$E[H_t(s^{\tilde{C}})] = H_t(s^{\tilde{C}}) \exp\left[\frac{1}{2}\delta^2 HU\right],$$

where $s^{\tilde{C}}$ is the spectral acceleration corresponding to the median displacement capacity, $\tilde{C}$, determined from Eq. (2). Subscripts DR, DU, CR, CU, and HU on coefficient of variation δ indicate demand randomness, demand uncertainty, capacity randomness, capacity uncertainty, and hazard uncertainty, respectively.

2.4. *IDA analysis of capacity against incipient collapse*

In structural performance evaluation the most important limit state is collapse and yet dynamic behavior of structure near collapse is nonlinear and complicated that the capacity against collapse is very complex and highly dependent on the excitation. A method of determining the capacity against incipient collapse under seismic excitation based on an incremental dynamic analysis (IDA) has been proposed[10] for determining and application to a variety of structural systems.[11] A structure response analysis is performed under a selected ground motion representing likely excitation at the site from a future earthquake. The ground motion intensity is then increased incrementally and the structural response behavior (e.g. inter-story or global drift) is monitored. The transition point at which there is a large increase in the response at a small increase of the excitation intensity signals the onset of instability is therefore recorded as the capacity (either in terms of drift or excitation intensity) against incipient collapse. The procedure is then repeated for different ground motions and statistics of the incipient collapse capacity are then obtained. For example, for post-Northridge steel moment frames the mean capacity in terms of inter-story drift is approximately 8% with a coefficient of variation of about 30%.[12] It is pointed out that although capacity uncertainty of the order of 30% or more seems large, its contribution to the limit state probability may be small when demand randomness and uncertainty dominate as can be seen from Eq. (7). This is generally the case for seismic excitation, i.e. δ_{DR} and δ_{DU} are much larger than δ_{CR} and δ_{CU}.

2.5. *Reliability-based design*

To meet a prescribed probabilistic performance goal while taking the uncertainty into consideration, a reliability-based design is to determine the required structural median capacity, $\tilde{C}$, to satisfy a prescribed target mean limit state probability, $E[P_t] = P_0$. This inverse problem can be solved as follows[8]:

$$\left\{\exp\left[-\frac{1}{2}\frac{k}{b}(\beta_{CR}^2 + \beta_{CU}^2)\right]\right\}\tilde{C} \geq \left\{\exp\left[\frac{1}{2}\frac{k}{b}(\beta_{DR}^2 + \beta_{DU}^2)\right]\right\}\tilde{D}^{P_0}. \tag{8}$$

Note that this is not a rearrangement of Eq. (7); therefore it takes a slightly different form. It can be rewritten into a more familiar load and resistance design (LRFD) format,

$$\phi\tilde{C} \geq \gamma\tilde{D}^{P_0}, \tag{9}$$

where ϕ is the capacity (resistance) factor and γ is the demand (load) factor. $\tilde{D}^{P_0}$ = median demand corresponding to $S_a^{P_0}$, a spectral acceleration of exceedance probability of P_0. From Eq. (2), one obtains

$$\tilde{D}^{P_0} = a(S_a^{P_0})^b, \tag{10}$$

in which $S_a^{P_0}$ is solved from Eq. (1). Note that from Eq. (1), smaller P_0 (higher reliability) gives larger $\tilde{D}^{P_0}$ and from Eq. (8) larger randomness and uncertainty in the demand and capacity give larger γ and smaller ϕ leading to larger design capacity $\tilde{C}$. The various sources of randomness and uncertainty are therefore considered explicitly.

2.6. *Target reliability level*

In a reliability-based performance-oriented design the starting point is the target reliability for various limit states. Determination of the target reliability levels requires broader social-economical considerations. They can be determined by comparison of risks of limit states with other societal risks. Alternatively, one can compare the notional (calculated) probability of limit states with those implied in current designs and adjust accordingly. This approach has been used in the past. For example, Ellingwood *et al.*[13] calibrated the target reliability of structural members against practice acceptable at the time in developing the AISC LRFD design recommendations, which have been adopted in the ASCE-7.[14] The need for a more rational approach to determine target reliability and acceptable risk has received serious attention by researchers and engineers recently.[15] One of such approaches is based on consideration of cost and benefit by minimization of expected lifecycle cost as given in the following section.

3. Minimum Lifecycle Cost Design Criteria

Design procedure based on optimization considering cost and benefit is generally referred to as level IV reliability-based design. Rosenblueth[16] had made convincing arguments for the profession to move from a semi-probabilistic, second moment, or full distribution design format to one based on optimization since it is the only rational procedure to ensure long term benefit to the society. Review of optimization-based design procedures can be found, e.g. in Frangopol and Corotis.[17] The method based on minimum lifecycle cost by Wen and Kang[18,19] is described in the following section.

3.1. *Design based on optimization*

The major considerations in a life cycle cost analysis of a constructed facility are loading and resistance uncertainties, limit states, and costs. The random occurrence in time and the intensity variability of the hazards are modeled by random process models. Costs include those of construction, maintenance and operation, repair, damage and failure consequence (loss of revenue, deaths and injuries, etc.), and discounting of future loss/cost over time. It is reasonable to assume there are only a small number of limit states to be considered and the loadings that can cause the facility to reach these limit states are due to severe natural and man-made hazards which occur infrequently. Over a time period (t), which may be the design life of a new facility or the remaining life of a retrofitted facility, the expected total cost can be expressed as a function of t and the design variable vector X as follows:

$$E\left[C(t,X)\right] = C_0(X) + E\left[\sum_{i=1}^{N(t)}\sum_{j=1}^{k}C_j e^{-\lambda t_i}P_{ij}(X,t_i)\right] + \int_0^t C_m(X)e^{-\lambda\tau}d\tau, \quad (11)$$

in which C_0 = the construction cost for new or retrofitted facility; X = design variable vector, e.g. design loads and resistance, or load and resistance factors associated with nominal design loads and resistance; i = number occurrences and joint occurrences of different hazards such as live, wind, and seismic loads; t_i = loading occurrence time; a random variable; $N(t)$ = total number of severe loading occurrences in t, a random variable; C_j = cost in present dollar value of jth limit state being reached at time of the loading occurrence including costs of damage, repair, loss of service, and deaths and injuries; $e - \lambda t$ = discounted factor of over time t, λ = constant discount rate per year; P_{ij} = probability of jth limit state being exceeded given the ith occurrence of a single hazard or joint occurrence of different hazards; k = total number of limit states under consideration; and C_m = operation and maintenance cost per year. The design criteria are determined by the minimization of the total expected lifecycle cost with respect to the design variable vector X. Additional constraints in the form of reliability and/or resistance may be also introduced in the minimization problem. The above formulation allows tractable solution of the minimization problem and facilitates the sensitivity study of the optimal design decision parameters. Details can be found in Wen and Kang.[18,19]

3.2. *Design against earthquakes*

The method is applied to design of a 3×5 bay, 9-story special moment resisting frame steel office building in downtown Los Angeles. The building is designed for a wide range of base shear and meeting the drift and other requirements of NHERP1997.[20] The system strength is measured by a system yield force coefficient (system yield force determined from a static pushover analysis using DRAIN2D-X divided by the system weight). Five limit states in terms of story-drift are used according to the performance levels of FEMA 273.[21] The empirical seismic hazard procedure of FEMA 273 is used to calculate the ground excitation demand for a given probability level. To obtain the drift ratio from the spectral acceleration, the method based on uniform hazard response spectra and an equivalent nonlinear single degree of freedom system (SDOF) is used. The drift ratio is then multiplied by correction factors to incorporate building capacity uncertainty and converted to damage factor according to FEMA-227.[22] The maintenance cost is not considered in this study. Initial costs are estimated according to Building Construction Cost Data.[23] The nonstructural items were not considered since they are not functions of the design intensity. The damage cost, loss of contents, relocation cost, economic loss (dollar/sqft), cost of injury ($1000/person for minor and $10 000/person for serious injury) and cost of the human fatality ($1 740 000/person) are estimated based on FEMA reports.[22] All costs are given in 1992 US dollars. A constant annual discount rate λ of 0.05 is assumed.

Following the procedure outlined above and using Eq. (11), the optimal system yield force coefficients (system yield force divided by system weight, a measure of system strength) are found to be 0.194 and 0.189 with and without considering

Table 1. Life-cycle cost (LCC) based design system yield coefficient against earthquakes, winds, and both hazards and comparison with current design (NEHRP 1997).

Location Hazard (Design Basis)	Los Angeles	Seattle	Charleston
Earthquake (NEHRP 1997)	0.140	0.100	0.075
Earthquake (LLC)	0.198	0.109	0.097
Wind (LLC)	0.073	0.073	0.121
Earthquake and Wind (LLC)	0.198	0.115	0.146

human injury and death costs respectively, both higher than the system yield coefficient of 0.14 according to 1997 NHERP. The design is then extended to Seattle, Washington and Charleston, South Carolina with proper adjustment of the cost due to regional variation. The results are shown in Table 1 in which the designs according to NEHRP1997 are shown for comparison in the first two rows. It is seen that the lifecycle cost (LCC) based designs are generally higher. The difference is large at Los Angeles, moderate at Charleston, and small at Seattle. Since some of the important decision parameters such as structural life span, discount rate, injury and death costs, and system capacity uncertainty are difficult to estimate accurately, a sensitivity of the optimal design was also carried out. The results (see Fig. 3, EQ) show that the optimal design intensity depends moderately on discount rate, and increases fast with structural life for $t < 20$ years but slowly for $t > 50$ years. It is insensitive to cost estimate of injury and death at Los Angeles but quite sensitive at Charleston due to the proportionally much larger contribution of these costs to the overall cost at Charleston. The seismic hazard curve at Charleston has a rather flat tail due to larger uncertainty in the high intensity range. As a result, the ratio of expected cost of death and injury to that of damage and economic losses is much higher at Charleston than at Los Angeles. Although not shown in Fig. 3, the results are found insensitive to the structural capacity uncertainty due to the dominance of the hazard uncertainty.

3.3. Design against earthquake and wind hazards

The application of the method is then extended to design in the three cities under both wind and earthquake hazards. It is of interest of compare the contribution of different hazards in the lifecycle cost design since at Angeles and Seattle seismic loads dominate whereas at Charleston wind loads play a very important role. The wind hazard and structural response analyses are based on provisions in ASCE-7.[14]

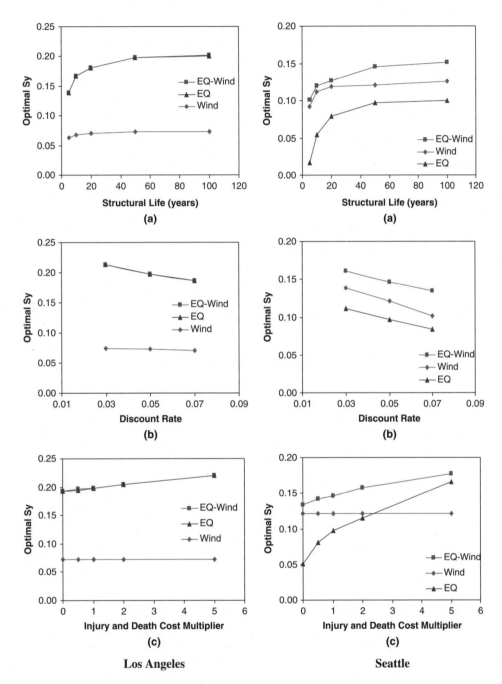

Fig. 3.   Sensitivity of optimal design to structural life, discount rate, and injury/death cost multiplier.

The optimal design of structural strength under both hazards is again expressed in terms of the system yield force coefficient S_y. Figure 3 shows the sensitivity of the optimal S_y under a single hazard and under both hazards to the structural design life, discount rate, and the multiplier applied to the cost of death and injury. At Los Angeles, the design is dominated by seismic load. The wind load contribution is so small that there is practically no difference between the design for both winds and earthquakes and that for earthquakes only. The design is sensitive to structural life for $t < 20$ years and becomes almost constant for $t > 50$ years. It is moderately dependent on the discount rate. The insensitivity the optimal S_y increases from 0.121 under winds only to 0.146 when seismic load is also considered. The optimal design is more sensitive to the multiplier due to the larger uncertainty associated with the large seismic events causing larger contribution to the total expected lifecycle cost. Considering again all these factors, the optimal design S_y may reasonably estimated to be 0.15. It is found that in both locations, the reliability against winds is much higher than that against earthquakes. It is therefore concluded that uniform reliability against different hazards is not required in an optimal design. The optimal target reliabilities are higher at Los Angeles than at Charleston primarily due to the much higher seismic hazard at Los Angeles.

4. Application to Vulnerability and Retrofit Analysis

The application of the above reliability evaluation procedure to vulnerability analysis and retrofit decision with proper consideration of consequences is demonstrated by retrofit of un-reinforced masonry (URM) structures in Mid-America. A large portion of the building stock of the design to the cost multiplier is due to the dominance of expected lifecycle cost of damage and economic losses since the probability of death and injury is extremely small by comparison. Considering the difficulty of assessing these factors, the optimal design coefficient is reasonably estimated to be 0.20. At Charleston, the wind hazard becomes more important. Seismic loads, however, still contribute. In other words, wind load does not "control," as would be the case in traditional design procedure. For example, for a design life of 50 years, in Mid-America, there are URM structures and many essential facilities such as fire stations and police stations that utilize URM construction.[24] Such buildings are vulnerable to seismic forces and their failures may cause serious consequences. The response of such buildings to seismic actions is nonlinear and complex and the uncertainty in their capacity is large.

To demonstrate the risk-benefit approach to decisions regarding rehabilitation, a study of a URM building is carried out. The building selected is a two-story residential/commercial building of 1930 vintage located in Memphis, TN. A finite element model based on ABAQUS was developed to model the nonlinear dynamic response behavior. The diaphragm/wall connections were modeled either as flexible with gravity friction, bolts, and nails for a typical connection details for such as structure, or assumed to be rigid after retrofit. Details can be found in Wen *et al.*[11]

Table 2. 50-year limit state probability of URM building.

Limit state	Immediate Occupancy (IO)	Life Safety (LS)	Collapse Prevention (CP)	Incipient Collapse (IC)
Reference	FEMA 273	FEMA 273	FEMA 273	IDA
Drift Ratio Capacity (%)	0.3	0.6	1	1.3(F) 1.74 (R)
St. Dev. of Capacity (%)	0.3	0.3	0.3	0.37 (F) 0.33 (R)
L. S. P. (F)	1.15×10^{-1}	7.51×10^{-2}	5.43×10^{-2}	4.65×10^{-2}
L. S. P. (R)	3.75×10^{-2}	2.11×10^{-2}	1.35×10^{-2}	8.36×10^{-3}
St. Dev. of Capacity	0.6	0.6	0.6	
L. S. P. (R)	4.06×10^{-2}	2.30×10^{-2}	1.495×10^{-2}	

IO = Immediate Occupancy; LS = Life Safety; CP = Collapse Prevention; IC = Incipient Collapse; F = Flexible wall/diaphragm connections; R = Rigid wall/diaphragm connections; L. S. P. = Limit State Probability.

and Ellingwood and Wen.[25] The system performance levels selected were immediate occupancy (IO), life safety (LS), and collapse prevention (CP), corresponding to maximum drifts of 0.3%, 0.6% and 1.0%, respectively, as described in FEMA 273 (1997).

IDA analyses are performed to determine the capacity against incipient collapse (IC) and the results are fitted by a lognormal distribution. The 50-year limit state probabilities of IO, LS, CP and IC were also calculated and shown in Table 2 where capacity uncertainties due to material properties and other sources are also shown. These values were doubled to check the sensitivity (the last two rows of Table 2 for the case of rigid connections). As expected, the rigid connections between the walls and diaphragms reduce the limit state probabilities by approximately a factor of four indicating that such an intervention measure would be quite effective. Notice also that the vulnerabilities are not particularly sensitive to the capacity uncertainties because of the dominance of the uncertainty in the seismic excitation. The relations between limit states and damage and between damage and loss can be established by regression analyses of cost on damage using data from past earthquake damage surveys.[26] Since such data are generally unavailable in Mid-America, one must use similar statistics elsewhere such as in Western US where URM buildings damages in past earthquakes are more readily available. In FEMA 273,[21] there are qualitative descriptions of damage and likely damages in consequences associated with each limit state. These are summarized in Table 3. More factual information is required to establish a functional relationship and estimate of uncertainty between limit states and moreover, the damage-to-loss estimate requires knowledge of repair cost, economic loss, cost of injury and death, etc. Large uncertainties exist in these relationships.

To demonstrate the procedure, an approximate analysis is performed using a liberal interpretation of the FEMA 273 performance state descriptions and consequences in terms of mean damage index, mean injury and death rate given the

Table 3.   Limit state/damage/expected life-cycle cost relationship.

Limit state (Performance level)	(IO)	(LS)	(CP)	(IC)
Damage level	Very limited structural damage, very low risk of life-threatening injury (FEMA 273)	Significant damage, low risk of life-threatening injury (FEMA 273)	Substantial damage, significant risk of injury due to falling hazards (FEMA 273)	Collapse is imminent (IDA analysis)
Damage index	0.05	0.2	0.4	0.9
Injury rate	0	0.1	0.2	0.6
Death rate	0	0.01	0.04	0.12
Cost given limit state	$22 000	$244 000	$794 000	$2 251 000
Contribution to 50-year expected life-cycle cost	$355 (before) $138 (after)	$2015 $671	$2302 $1419	$39323 $7212

occurrence of each performance state shown in Table 3. The costs considered include damage repair cost ($85/ft^2), content cost ($30/ft^2), injury (10^3–10^4/person) and death (2×10^6/person). These costs are to be multiplied by the floor area and then a damage index and expected injury and death rates respectively depending on limit state as shown in Table 3. The 2-story URM building has a foot print dimension of 83 ft by 23 ft and total floor area of 3818 ft^2. An occupancy rate of 2 persons/1000 ft^2 is used. To calculate the life-cycle cost, a discount rate $\lambda = 5\%$ per year and a life of 50 years are assumed. The expected total life-cycle cost can be calculated according to Eq. (11).

Using the cost assumptions given above, the 50-year expected total losses to the existing building are $43 995 before retrofit and $9440 after retrofit. Such information can be used in design/retrofit decision making. For example, if the cost of retrofitting the connections is less than the difference of the expected total lifecyle cost before and after retrofit, i.e. $34 555 (or approximately 10% of replacement cost) the retrofit decision is justified. The last row of Table 3 shows that all limit states contributed to the expected total lifecycle cost, although the IC limit state dominates due to the large fatality cost and flat seimic hazard curve at Memphis. In other words, the retrofit is justified primarily because it saves lives. The results are sensitive to the cost estimates as well as the relationship between costs and damage and limit states, i.e. the mean estimates given in the third row of Table 3.

Additional research is needed to obtain accurate estimates of these numbers and relationships. In addition to the mean life-cycle cost, the probability as function of loss can be also calculated such information would be useful for decision making. It requires uncertainty in the relationships between limit state to damage and damage to cost, i.e. in terms of coefficient of variations of damage index, injury and death rate and type of distributions for given each limit state and the uncertainty in the various estimates of costs per square footage.

5. Reliability and Redundancy

Redundancy of structural systems has attracted much attention of engineers after the large number of failures of structural systems in recent earthquakes. Most define redundancy according to the structural configuration. For example, in ASCE-7,[14] a reliability/redundancy factor ρ is define as function of the floor area and maximum element-story shear ratio. The allowable range of ρ for the seismic lateral force can vary by as much as 50%. There has been much criticism by engineers on the rationale behind this provision and as a result, a new ρ factor with more consideration of the structural response behavior has been proposed and currently under review by ASCE-7 Committee on Minimum Design Loads for Buildings and Other Structures.

In view of the large uncertainty of the seismic excitation and structural resistance, the redundancy of a structural system under seismic load cannot be treated satisfactorily without a careful consideration of the uncertainty. For example, a simple deterministic system of a given strength and identical parallel members under tension force will fail when the strength is exceeded regardless of the number of the parallel members since all members will reach the collapse threshold at the same time. Therefore, there is no advantage of having more members. The situation is drastically different if there is uncertainty in both loading and member strength. It has been shown[27] that under random static loads, the parallel systems have significant redundancy (much higher reliability) if there is adequate number of members, moderate degree of ductility, low strength correlation among members, and small load variability compared with that of the member resistance. It is clear that all these factors have not been considered in redundancy study of structures thus far, especial under random dynamic loads. The same is true in the case of the largely empirical redundancy factors ρ proposed in code procedures. The redundancy considering the structural nonlinear response behavior and uncertainty in excitation and structural capacity has been investigated in Wang and Wen,[28,29] Wen and Song,[30] and Liao and Wen.[31]

The results of redundancy of steel moment frames are shown here. The nonlinear behavior of the members and connections including inelastic and brittle fracture failures are accurately modeled according to test results. The ground motions corresponding to three different probability levels (50%, 10%, and 2% in 50 years) developed in the SAC Steel Project[6] are used as excitation. To realistically model the load redistribution and effect of possible unsymmetric failure of members, 3D response analysis methods were developed which allow evaluation of torsional response. Member resistance is modeled by random variables. The findings are summarized and implications on the ρ factors examined in the following.

5.1. *Steel moment frames of different connection ductility capacities*

Two low-rise steel moment frame buildings in the LA area, of 2 stories and 2 by 3 bays, and 3 stories and 3 by 5 bays, were designed according to current code procedures. The diaphragms are assumed to be rigid in its own plane but flexible

Table 4.   50-year incipient collapse probability and uniform risk redundancy
factor for 3-story steel frame building.

Assumption of excitation, connection behavior, and accidental torsion	Drift ration capacity (%)	P_f	R_R	R
Uni-directional, ductile, no torsion	0.08	0.004	1	8
Uni-directional, brittle, no torsion	0.05	0.021	0.986	7.88
Bi-directional, ductile, no torsion	0.08	0.037	0.828	6.63
Bi-directional, brittle, no torsion	0.05	0.124	0.546	4.37
Bi-directional, ductile, torsion	0.08	0.034	0.850	6.80
Bi-directional, brittle, torsion	0.05	0.129	0.538	4.31
Bi-directional, brittle, column damage and torsion	0.05	0.182	0.471	3.77

out of plane. Plastic hinges can form at the ends of beams and columns. Fracture
failures of connections occur when the capacity as a function of ductility and cumulative energy dissipation is exceeded. The capacity is modeled as random variable
based on test results. The smooth hysteresis model[32] was extended and used to
describe the post-yielding ductile and brittle behavior of the members and connections and calibrated against test results. It reproduces the nonlinear behavior well.
A 3D response analysis method is then developed based on these element models. Response statistics under the SAC ground motions were obtained. The 50-year
probabilities of maximum column drift ratio of the 3-story steel building exceeding
incipient collapse capacity under various assumptions of the ground excitation and
structural response behavior are shown in the third column of Table 4. The incipient
collapse thresholds in terms of column drift ratio are assumed to be 8% for ductile
systems and 5% for systems with possible brittle connection failures, based on SAC
research results by Yun *et al.*[12] It can be seen that the brittle fracture failure of
the connections have only moderate effects compared with ductile systems under
uniaxial excitation. The coupling of such failures with biaxial interaction and torsional response due to possible unsymmetric member failures however, significantly
increases the displacement demand on the structures (inter-story and global drift)
and the probability of incipient collapse.

5.2. *Uniform-risk redundancy factor R_R*

It is obvious from the above that redundancy can be accurately defined only in terms
of system reliability; therefore, it can be incorporated into design only through a

reliability-based design procedure. To achieve uniform reliability in the current design procedure, for systems with different degrees of redundancy, a uniform-risk redundancy factor, R_R, may be used in conjunction with the widely used response modification factor, R, to determine the required design force. The R value for special moment frames is 8 according to ASCE-7-02.[26] R_R is defined as the ratio of the system spectral acceleration capacity corresponding to the actual that required to achieve the allowable (target) P_f. For example, one can set the target P_f to be 2% in 50 years and calculate these two spectral acceleration values based on the foregoing reliability analysis. Details can be found in Wang and Wen.[28,29] For a system with inadequate reliability/redundancy, R_R will be smaller than unity. When the system has adequate reliability/redundancy; in other words, P_f is lower than the allowable value, R_R is equal to unity. R_R, therefore, functions as an adjustment factor to assure that the target (allowable) reliability will be achieved. The design seismic force is reduced by a factor of R multiplied by R_R. The R and R_R for the 3-story building under various assumptions of component ductility capacity and excitation characteristics are shown in the third and fourth columns of Table 4. For example, if an allowable value of P_f is 2% in 50 years, R_R varies from 1 under the assumption of uniaxial excitation and ductile connections, to 0.471 ($R = 3.77$) under biaxial excitation with possible brittle connection failures. The results indicate that for a structure of a given configuration, ductility capacity and structural 3D response (bi-axial and torsional motions) can greatly change the structural reliability/redundancy and lead to a large increase in the required design force which has not been considered in current code procedures.

5.3. *Moment frames of different configurations*

The recently proposed provision according to NEHRP 2003[33] is such that the redundancy factor ρ is 1.0 when loss of moment resistance at the beam-to-column connections at both ends of a single beam would not result in more than 33% reduction in story strength, nor create an extreme torsional irregularity, and otherwise it is 1.3. The torsional effects as mentioned in the previous section have been incorporated. It is an improvement over the existing ρ factor. The recommended ρ factors for some typical configurations are shown in Fig. 4 for 5 × 5 bays moment frame buildings with different numbers and locations of moment frames investigated in NEHRP 2003. The recommended values of either 1.0 or 1.3, perhaps necessary in code provisions for simplicity, need to be examined within the framework of the above reliability-based procedure.

In Liao and Wen[31] the investigation has been extended to low-rise (3-story) and medium-rise (12-story) prototype moment frame buildings shown in Fig. 4. The buildings are designed according to NEHRP 2003. The FEMA/SAC ground motions are used again. In the response analysis, additional factors considered include the number and layout configuration of the moment frames, effect of panel zones, and contribution of the gravity frames. The IDA analysis is also extended to three

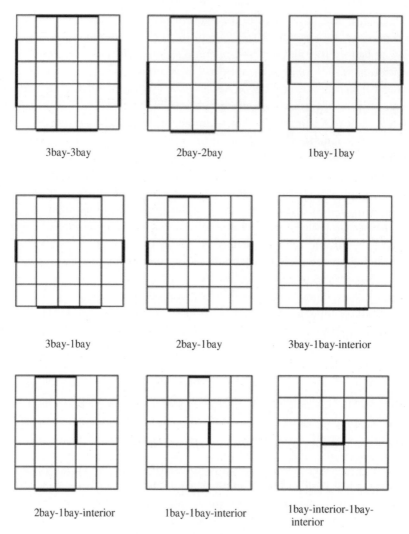

Fig. 4. Nine plan configurations of 3-Story and 12-Story buildings (bold lines represent the moment frames).

dimensional. A biaxial spectral acceleration, defined as the maximum vector sum of the spectral accelerations in the two principal directions is used as an intensity measure for better correlation with the 3D structural response in the uniform-risk redundancy factor analysis.

Typical IDA analysis results of a 12-story moment frame building are shown in Fig. 5 in which each curve represents the IDA curve of the building under a FEMA/SAC ground motion and the building properties are randomized across the IDA curves under different ground motion to including the effect of excitation and building material and component capacity variability. The transition points

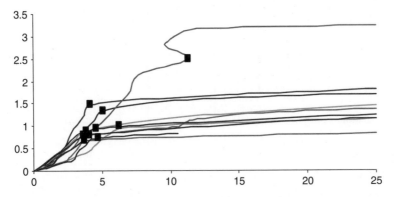

Fig. 5. IDA curves of a 12-story building with 3 bays of moment frames in both directions, transition points to incipient collapse are shown by square points. Building properties are randomized across IDA curves each under a different ground motion.

to instability are shown by black squares. The capacity against incipient collapse can be measured in terms of either structural response (interstory drift ratio) or excitation intensity (biaxial spectral acceleration). Base on the statistics obtained from the IDA curves, the capacity against incipient collapse is modeled by a lognormal distribution for reliability analysis and determination of the uniform-risk redundancy factor.

The uniform risk redundancy factor these prototype buildings are obtained. Table 5 shows the R_R factor, the equivalent $\rho(1/R_R)$, and NEHRP 2003 ρ values for 12-story moment frame buildings of different configurations given in Fig. 4. The results show that as expected, the HEHRP ρ factor tends to oversimplify the complicated redundancy problem and is not risk-consistent. It is, however, a significant improvement over the current ρ factor. Further refinement of the ρ factor is possible using the reliability-based procedure to make it more risk-consistent.

Table 5. Uniform-risk redundancy factors (R_R) and corresponding ρ factors $(1/R_R)$ for a 2% in 50 years target incipient collapse probability of 12-story buildings of various configurations and comparison with NEHRP (2003) ρ values.

Building configuration	R_R	$(1/R_R)$	NEHRP ρ
3bay-3bay	1.0	1.0	1.0
2bay-2bay	0.91	1.1	1.0
3bay-1bay	0.95	1.05	1.0
2bay-1bay	0.91	1.1	1.0
3bay-1bay-interior	0.83	1.2	1.0
1bay-1bay	0.81	1.23	1.3
2bay-1bay-interior	0.77	1.29	1.3
1bay-1bay-interior	0.75	1.33	1.3
1bay-interior-1bay-interior	0.73	1.37	1.3

6. Concluding Remarks

Recent developments are reviewed and a framework is discussed on reliability-based performance evaluation and design under natural hazards. The major factors in design are considered and properly treated in this framework including uncertainty in hazard demand and structural capacity, nonlinear structural response behavior, redundancy, balance of costs and benefits, and target reliability in design for a single or multiple hazards. Sensitivity studies are carried out to identity most important system and cost parameters. Examples are given on design of multistory buildings against earthquakes and winds.

It has been shown that currently available structural response and reliability analysis methods have the capability of treating these major factors in design and allowing development of risk-based, comprehensive, and yet practical design procedures familiar to engineers. It is also shown that structural design based on such a procedure is highly dependent on consequence of structural limit states and the minimum expected lifecycle cost is a viable approach to setting reliability and performance goals. For multiple hazards, uniform reliability against different hazards is not required and hazards of large uncertainty and high consequence generally dominate. The proposed method is a viable tool for vulnerability analysis and retrofit decision making.

Finally, the largely empirical reliability/redundancy factor in current codes has been proved to be inadequate and may yield inconsistent results. It needs to be considered in the framework of reliability-based design. The uniform-risk redundancy factor proposed is one possible such approach to ensure that the structures of different redundancies will meet desirable target reliability. It provides a basis that can be used to develop more rational codes and standards.

Acknowledgments

The support of National Science Foundation through Grant CMS-95-10243, CMS 02-18703, and NSF Earthquake Engineering Research Centers Program through MAE Center CEE-9701785 is appreciated.

References

1. Y. K. Wen, Building reliability and code calibration, *Earthquake Spectra* **11**, 2 (1995) 269.
2. Structural Engineers Association of California (SEAOC), *Performance Based Seismic Engineering of Buildings*, Vol. I, Vision 2000 Report, California (1995).
3. Y. K. Wen and D. A. Foutch, *Proposed Statistical and Reliability Framework for Comparing and Evaluating Predictive Models for Evaluation and Design and Critical Issues in Developing Such Framework*, SAC Joint Venture Report No. SAC/BD-97/03 (1997).
4. D. A. Foutch, *State of Art Report-Performance Prediction and Evaluation of Moment-Resisting Steel Frame Structures*, FEMA 355F (2000).

5. Federal Emergency Management Agency, *Recommended Seismic Design Criteria for New Steel Moment-Frame Buildings*, FEMA350 (2000).
6. P. Somerville, N. Smith, S. Punyamurthula, and J. Sun, *Development of Ground Motion Time Histories for Phrase 2 of the FEMA/SAC Project*, SAC/BD-97/04 (1997).
7. Y. K. Wen and C. L. Wu, Uniform hazard ground motions for mid-America cities, *Earthquake Spectra* **17**, 2 (2001) 359.
8. C. A. Cornell, F. Jalayer, R. O. Hamburger and D. A. Foutch, Probabilistic Basis for 2000 SAC Federal Emergency Management Agency Steel Moment Frame Guidelines, *J. Struct. Eng.*, ASCE **129**, 4 (2002) 526.
9. R. H. Sues, Y. K. Wen and A. H. S. Ang, Stochastic evaluation of seismic structural performance, *Journal of Structural Engineering*, ASCE **111**, 6 (2002) 1204.
10. D. Vamvatsikos and C. A. Cornell, Incremental dynamic analysis, *Journal of Earthquake Engineering and Structural Dynamics* **31**, 3 (2002) 491.
11. Y. K. Wen, B. R. Ellingwood and J. Bracci, *Vulnerability Function Framework for Consequence-Based Engineering*, Report DS-4, Mid-America Earthquake Center, University of Illinois at Urbana-Champaign, http://mae.ce.uiuc.edu (2004).
12. S. Y. Yun, R. O. Hamburger, C. A. Cornell and D. A. Foutch, Seismic performance evaluation for steel moment frames, *Journal of Structural Engineering*, ASCE **128**, 4 (2002) 534.
13. B. R. Ellingwood, T. V. Galambos, J. G. MacGregor and C. A. Cornell, Probability-based load criteria: Load factors and load combinations, *Journal of Structural Division*, ASCE **108**, 5 (1982) 978.
14. American Society of Civil Engineers, Minimum Design Loads for Buildings and Other Structures, ASCE-7-02 (2003).
15. B. R. Ellingwood, Probability-based structural design: Prospects for acceptable risk bases, Keynote lecture, *Proc. 8th International Conference on Applications of Statistics and Probability*, Sydney, Australia (1999).
16. E. Rosenblueth, Towards optimal design through building codes, *Journal of the structural Division*, ASCE **102** (1976) 591.
17. D. Frangopol and R. B. Corotis, Editors, Reliability-based structural system assessment, design and optimization, Special Issue, *Journal of Structural Safety* (1994).
18. Y. K. Wen and Y. J. Kang, Minimum building life-cycle cost design criteria. I: Methodology, *Journal of Structural Engineering*, ASCE **127**, 3 (2001) 330.
19. Y. K. Wen and Y. J. Kang, Minimum building life-cycle cost design criteria. II: Applications, *Journal of Structural Engineering*, ASCE **127**, 3 (2001) 338.
20. NEHRP, Recommended Provisions for Seismic Regulations for New Buildings and Other Structures, Federal Emergency Management Agency, Washington D. C. (1997).
21. Federal Emergency Management Agency (FEMA), NEHRP Guidelines for the Seismic Rehabilitation of Buildings, FEMA-273 (1997).
22. Federal Emergency Management Agency (FEMA), *A Benefit-Cost Model for the Seismic Rehabilitation of Buildings*, Vol. 1 and 2, FEMA-227 & FEMA-228 (1992).
23. Building Construction Cost Data (*BCCD*), 54th Annual Edition (1996).
24. S. French, Inventory of Memphis Buildings, MAE Center Report (2004).
25. B. R. Ellingwood and Y. K. Wen, *Risk-Benefit Based Design Decision for Low Probability/High Consequence Earthquake Events in Mid-America* (in press) Progress in Structural Engineering and Material (2004).
26. K. A. Porter and A. S. Kiremidjian and J. S. LeGrue, Assembly-based vulnerability of buildings and its use in performance evaluation, *Earthquake Spectra*, EERI **17**, 2 (2001) 291.

27. S. Gollwitzer and R. Rackwitz, On the reliability of daniels systems, *Structural Safety* **7** (1990) 229.

28. C.-H. Wang and Y. K. Wen, Evaluation of pre-Northridge low-rise steel buildings — Part I: Modeling, *Journal of Structural Engineering*, ASCE (2000).

29. C.-H. Wang and Y. K. Wen, Evaluation of pre-Northridge low-rise steel buildings — Part II: Reliability, *Journal of Structural Engineering*, ASCE (2000).

30. Y. K. Wen and S.-H. Song, Structural reliability/redundancy under earthquakes, *Journal of Structural Engineering*, ASCE **129** (2003) 56.

31. K.-W. Liao and Y. K. Wen, *Redundancy of Steel Moment Frame Systems under Seismic Excitation,* Structural Research Series No. 636, University of Illinois at Urbana-Champaign (2004).

32. Y. K. Wen, Methods of random vibration for inelastic structures, *Applied Mechanics Review* **42**, 2 (1989).

33. NEHRP (FEMA-450), NEHRP Recommended Provisions for Seismic Regulations for New Buildings and Other Structures, Federal Emergency Management Agency, Washington DC (2004).

CHAPTER 4

PERFORMANCE-BASED RELIABILITY EVALUATION OF STRUCTURE-FOUNDATION SYSTEMS

MOSTAFIZ CHOWDHURY

Weapons and Materials Division, Ordnance Materials Branch
US Army Research Laboratory, 2800 Powder Mills Rd
Adelphi, MD 20783-1145, USA
E-mail: Mchowdhury@arl.army.mil

ACHINTYA HALDAR

Department of Civil Engineering and Engineering Mechanics
University of Arizona, Tucson, AZ 85721, USA
E-mail: Haldar@u.arizona.edu

The most recent professional expectation is that the reliability of complicated mechanical or structural systems should be evaluated using their performances. In general, a complicated system consists of many elements, some of which act in series and others act in parallel. The failure of one element may not indicate the failure of the system. Furthermore, the brittle and ductile behaviors of elements after they reach their capacities also affect the overall reliability of the system. In most cases, strength performance requirements are satisfied at the element level and the serviceability or deflection requirements are satisfied at the structural level. For the performance-based reliability evaluation procedure, the element-level probabilities of unsatisfactory performance (UP) for strength and system-level probabilities of UP for serviceability need to be combined. A procedure is presented here to calculate the lower and upper bounds of the probability of failure of the system. A complicated structure-foundation system consisting of concrete superstructure and pile-supported foundation is considered to illustrate the procedure. The discussions indicate that the estimation of the probability of failure at the element level, particularly for the strength performance function often practised in the profession, may not be sufficient to develop performance-based design criteria.

1. Introduction

The performance evaluation of mechanical or structural systems is a challenge to the profession. The task is more challenging if the system is in operation. A typical redundant structural or mechanical system consisting of many subsystems or components and they can affect the performance of the system in a many different ways. Functional disorder of a system can be defined as unsatisfactory performance (UP), i.e. the system can no longer serve its intended purpose. Thus, a UP may or may not cause a catastrophic failure in a physical sense. Identifications of some of the significant UP modes are thus essential elements of the performance evaluation of existing operational systems.

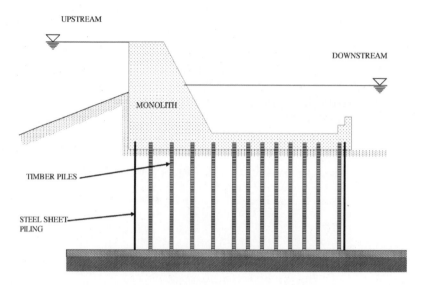

Fig. 1.   A typical pile-supported dam.

For the ease of discussion, consider a cross-section of a structure-foundation system representing a dam shown in Fig. 1. In this case, the concrete super-structure is
supported by aged timber piles. It represents a typical redundant system consisting
of many parallel components. This complicated system may not be able to function
as intended in many different ways. Several structural elements may behave poorly
and the system may not be able to resist the applied loads. Some structural elements
may perform unsatisfactorily first, followed by excessive deformation, causing functional disorder of the system due to insufficient stiffness. In some cases, the intact
structure may produce inadmissible deformation rendering the system inoperable.
To maintain the operational status of such a system without compromising the
reliability, it is essential to evaluate the reliability at the system level considering
all major potential loads and load combinations, and different performance modes
in terms of strength and serviceability.

A relatively simple representation of the pile-supported system is shown in
Fig. 2. In this representation, the rigid pile cap transfers the loads from the super-
structure to the pile group. The loads, concrete superstructure (rigid cap), and
group of piles supporting the superstructure are shown in Fig. 2, in an idealized way.

In a typical pile-supported structure, the redundant resistance provided by the
multiple piles provides a parallel resistance mechanism so that UP of a single pile
will not usually result in total distress of the overall system. Several piles need to
perform unsatisfactorily to cause an overall strength failure or an excessive lateral
movement of the rigid pile cap beyond an acceptable limit, producing global distress
to the structure supported on the pile cap.[1] The overall concept is shown in Fig. 3.
In massive navigation structures, "global distress" can be regarded as the movement of the rigid cap that can substantially limit their operation or serviceability

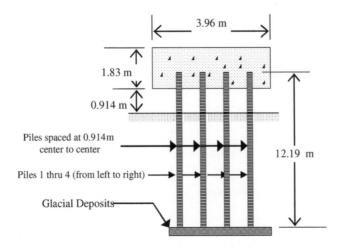

Fig. 2.   A cross-sectional view of a pile-group.

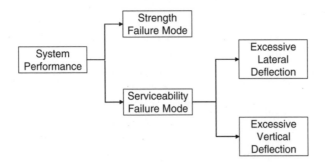

Fig. 3.   Performance modes of the system illustrated in Fig. 1.

requirements. In defining performance of these structures, thus, UP is appropriately used to represent the global distress that may not cause physical damage or failure in the structure.

The complicated system considered here consists of many sub-systems arranged in parallel and series, as shown in Fig. 4. The behavior of a sub-system, in this case piles, after reaching the load carrying capacity will also dictate the reliability on the system. A pile may not carry any load after reaching its capacity, indicating brittle behavior. In some cases, a pile can carry some load after reaching its capacity, indicating ductile behavior. Thus, brittle or ductile behavior of structural elements is also expected to influence the behavior of the system. Strength requirements may be local[2] or global,[3] and may be related to reserve and residual strength.[4] Some of the piles may perform unsatisfactorily at first, followed by excessive deformation and causing functional disorder of the system due to insufficient stiffness. Thus, the strength failure may lead to serviceability failure.

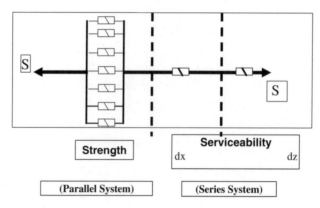

Fig. 4. A system model (S) showing the failure modes of the structure-foundation system shown in Fig. 1. Redundant supports provide a parallel system in strength mode. Serviceability modes are denoted as dx and dz. Lateral (dx) and vertical (dz) deflections may well depend on the available strength capacity of redundant members in subsequent loading stages in a parallel system.

In evaluating the reliability of complicated structures, it is essential to identify potential performance modes (PMs) and their predetermined acceptance criteria. Acceptance criteria are generally established considering nature, importance, and consequences of failure of the structure under consideration. Engineering models are routinely used for identifying performance or failure modes of systems. Effectiveness of such identification depends on the validity of the inherent assumptions and its ability to accommodate uncertainties in the predictive model. A probabilistic approach is a natural choice for bridging the gap between the reality and the predictability of engineering solution. Probabilistic approaches requiring identification of some of the most significant performance modes are described in this chapter.

2. Reliability-Based Approaches

Reliability-based assessment methodologies, explicitly addressing major uncertainties in engineering analysis, are desirable, and it is important to standardize such procedures and implement them in practice.[5] In spite of the progress in the implementation of reliability-based procedures, the rate of its acceptance by practitioners has been slow. This is primarily due to lack of familiarity in the general areas of risk and reliability and the analytical intricacy of the system reliability evaluation. The inclusion of reliability assessment procedures into the available deterministic analytical approaches increases the chances of acceptance by the users, in easing the transition from the deterministic to the probabilistic domain. Discussions made in this chapter are expected to help in this transition process.

The state-of-the-art in the reliability evaluation of civil engineering systems is very advanced. The uncertainties associated with most major variables in any civil engineering problem have already been identified or quantified. If the acceptance or allowable criteria are known, the strength and serviceability-related limit state or performance functions can be defined, as will be discussed later. For a known performance function, the reliability can be estimated using several well-established reliability evaluation techniques,[6,7] including the first-order reliability method (FORM), second-order reliability method (SORM), and several schemes of Monte Carlo simulation techniques. FORM is very frequently used. In this approach, the reliability is generally estimated in term of the reliability index β and the corresponding probability of failure P_f is estimated as $P_f = 1.0 - \Phi(-\beta)$, where Φ is the standard normal cumulative distribution function. The probability of failure is higher when β is smaller. Several procedures with various degrees of complexities are available to evaluate the probability of failure. They will not be discussed further. FORM is used to estimate the reliability index in all the subsequent discussions.

3. Performance Functions

As mentioned earlier, the reliability of a civil engineering system is generally estimated considering the strength and serviceability performance functions. They are separately discussed next.

3.1. *Strength performance functions*

Strength requirements may be local or global, and may be related to instability, reserve and residual strength, and fatigue damage. Since the axial load effect on a pile could be tensile or compressive, and a pile can be distressed either in pure axial load or combined axial load and bending moment, the following four strength performance functions for a two-dimensional structure are considered in this study:

Compressive load:

$$g(x) = 1.0 - \left(\frac{F}{AC} \right) \left(\frac{1}{OSF} \right), \tag{1}$$

$$g(x) = 1.0 - \left[\frac{F}{ACC} + \frac{|M|}{AM} \right] \left(\frac{1}{OSF} \right). \tag{2}$$

Tensile load:

$$g(x) = 1.0 - \left(\frac{-F}{AT} \right) \left(\frac{1}{OSF} \right), \tag{3}$$

$$g(x) = 1.0 - \left[\frac{-F}{ATT} + \frac{|M|}{AM} \right] \left(\frac{1}{OSF} \right), \tag{4}$$

where F is the applied axial load, M is the applied bending moment, AC and AT are the nominal axial compressive and tensile load capacity of a pile, respectively, AT is assumed to be $AC/2$ in this study, ACC is the nominal axial compressive load capacity of a pile when the combined effect of axial compressive load and bending moment is considered, ATT is the nominal axial tensile load capacity of a pile when the combined effect of axial tension and bending moment is considered, AM is the nominal bending moment capacity of a pile, and *OSF* is the allowable overstress factor. For unusual or short duration load cases, $OSF = 1.33$ is generally used.[8]

Using a deterministic linear computer pile group analysis (CPGA) program[8] and the Taylor Series Finite Difference (TSFD) method, the means and the standard deviations of axial load and bending moment (F and M) acting on each pile can be calculated. The statistical characteristics of the resistance-related random variables, AC and AT, can be estimated from pile load-tests.[9] For round timber piles, ACC, ATT, and AM are estimated as:

$$ACC = F_{ac}A = F_{ac}\pi\left(\frac{d^2}{4}\right), \quad ATT = F_{at}A = F_{at}\pi\left(\frac{d^2}{4}\right), \tag{5}$$

and

$$AM = F_b S = F_b\pi\left(\frac{d^3}{32}\right), \tag{6}$$

where F_{ac}, F_{at} and F_b are the allowable axial stresses in compression, tension and bending, respectively, and d, A, and S are the diameter, cross-sectional area, and the section modulus of a circular pile, respectively. They are assumed to be random variables in this study. The statistical characteristics of F_{ac}, F_b, and d were evaluated by Mlaker and Stough,[10] and are summarized in Table 1. Using the statistical information of these independent variables, the means and standard deviations of random variables ACC, ATT, and AM are estimated. In Table 1, N_h is introduced to represent the coefficient of horizontal sub-grade reaction. The parameter is required to calculate the load distribution in piles. All the variables in Table 1 are assumed to be lognormal.

3.2. *Serviceability performance functions*

Serviceability requirements depend on the intended functional requirements of the structure. They may include excessive vibration,[11] and excessive deflection.[12] Only issues related to the horizontal and vertical deflections are considered in this study. Excessive global movement of the rigid cap is used to define the serviceability limit states of the pile system. The limit state equation for the serviceability requirement can be written as:

$$g(x) = 1.0 - \frac{\delta}{\delta_a}, \tag{7}$$

Table 1. Statistical characteristics of timber piles.

Variable	Mean	Standard deviation
N_h	10.630×10^3 kN/m^3	2.519×10^3 kN/m^3
d (diameter)	0.363 m	0.006 m
F_b	39.65 MPa	6.83 MPa
F_{ac}	19.58 MPa	3.79 MPa
AT	311.39 kN	40.48 kN
AC	622.78 kN	80.96 kN

where δ is a random variable representing the lateral or vertical deflection of the pile cap produced by the applied loads and δ_a is the corresponding allowable deflection. The mean and standard deviation of δ are evaluated using the TSFD estimation procedure. The reliability index for the serviceability performance function, represented by Eq. (7), can be evaluated for any allowable deflection limit using the FORM method.

4. System Reliability

An essential step in the system reliability evaluation is the estimation of the element-level reliabilities considering all applicable strength and serviceability-related performance limit states. In general, system reliability evaluation is very complicated and depends on many factors. Some of the important factors are (i) the contribution of the component failures to the system's failure, (ii) the redundancy in the system, (iii) the post-failure behavior of a component and the rest of the system, (iv) the statistical correlation between failure events, and (v) progressive failure of components.

For the structure-foundation system under consideration, the major task is to evaluate the system reliability considering all possible failure modes, including the individual strength of the piles and the lateral and vertical deflections of the pile cap. The element-level reliability for each pile needs to be calculated considering several limit states. Then the system reliability for the strength PM needs to be calculated considering a parallel structural system with brittle and ductile pile behavior. Since lateral and vertical deflections of the pile cap are system-level parameters representing the global behavior, the corresponding reliabilities will give directly the system-level reliabilities for the serviceability limit states. Thus, the element-level reliability for strength and system-level reliability for serviceability need to be combined to evaluate the system reliability. Since it is difficult to determine the joint probabilities of more than two failure events except by using Monte Carlo simulation or numerical integration, several approximate bounds have been proposed for the system reliability evaluation. Since the statistical dependencies between the PMs are unknown, the system reliability, P_{SS}, is defined by the lower and upper bounds[1] as:

$$\prod_{i=1}^{N} P(\bar{E}_i) \leq P_{SS} \leq \min_{i} P(\bar{E}_i). \tag{8}$$

The corresponding bounds for the probability of failure of the system, P_{fs}, can be expressed as:

$$\max_{i} P(E_i) \leq P_{fs} \leq 1 - \prod_{i=1}^{N} [1 - P(E_i)], \tag{9}$$

where $P(E_i)$ denotes the probability of UP of the ith component, $P(\bar{E}_i)$ is the probability of survival of the ith component, and N is the total number of PMs.

If the $P(E_i)$'s are relatively small, the upper bound in Eq. (9) can be evaluated as the summation of the individual probabilities of failure, i.e. $\sum_{i=1}^{N} P(E_i)$.

The lower bound in Eq. (9) represents the system failure probability if all the events are perfectly dependent. The upper bound indicates if the events are statistically independent. The foregoing first-order bounds could be very wide. In that situation, the second-order bounds considering the joint occurrences of two events can be considered.[13] However, it will not be discussed further here.

Thus, the evaluation of probability of system UP is essentially the calculation of the two bounds considering all possible PMs for a given load case. The lower bound of the system probability of UP is related to the most significant failure mode. The upper bound estimation requires consideration of the probability of UP of all significant failure modes. With the help of an example, a conceptual procedure to identify some of the most significant PMs that contribute to the estimation of bounds of system probability of UP will be presented in Sec. 5.

A robust system reliability method proposed by the authors is discussed next. The system reliability evaluation can be carried out by following the basic five steps:

Step 1: Identify the magnitude and the uncertainty associated with the external loads and the load combinations the structure will be subjected to during its lifetime.

Step 2: Evaluate the element-level reliabilities for the strength limit states for all the piles considering brittle and ductile behavior.

Step 3: Evaluate the system-level reliabilities for the serviceability limit states considering brittle and ductile behavior of piles.

Step 4: Evaluate the lower bound of the system reliability by identifying the most significant PM for each load case, and

Step 5: Evaluate the upper bound of the system reliability by considering the contributions of all related significant PMs.

For the system reliability evaluation of the structure-foundation system considered here, the brittle and ductile behavior of a pile needs to be considered properly. In a brittle system, if a pile reaches its capacity, it is assumed to have distressed and is removed from the pile group so that the applied load is redistributed to the remaining piles. In reality, however, even after a pile has reached its capacity, it will continue to carry its share of the applied load, causing a ductile effect. Consideration of the realistic ductile behavior of piles in the system reliability evaluation is thus essential and will reduce the compounding conservatism and biases in the computation scheme. A conservative estimate could inadvertently mislead by concluding that a structure's performance is severely inadequate.[3] System reliability evaluations considering brittle and ductile behavior of piles are discussed separately next.

4.1. *Performance modes (PM) — Brittle behavior of piles*

Performance modes which are interchangeably used as failure modes in this chapter are primarily a system behavior. A system consisting of components and subsystems can have numerous performance modes as illustrated below. However, it is not possible to postulate a particular PM in advance. For the structure-foundation system under consideration, the strength behavior of piles is expected to be similar. This will necessitate consideration of a large number of significant PMs considering both the strength and serviceability requirements. The identifications of some of the most significant performance modes based on strength and serviceability requirements are the essence of the reliability based methodology described in this chapter.

Suppose, there are m piles in a system shown in Fig. 1. UP due to the strength of such a well-designed structure is expected to require more than one deficient pile. The sequence of UP of piles and the redistribution of the applied loads need to be considered to estimate the system reliability. Conceptually, the UP of L piles out of a total of m piles can be considered to constitute the strength UP of the pile group. The value of L will depend on the reliability of the individual pile, brittle or ductile behavior, and the total number of piles in a group. The system can become undependable in strength by having L piles developing UP in strength. The system can perform unsatisfactorily by having excessive lateral or vertical deflection without distressing a pile, or one or some of the piles can get distressed in strength first followed by excessive lateral or vertical deflection, as shown in Fig. 4.

The performance mode approach (PMA), also known as the failure mode approach,[1] can be used to assess the system reliability. In the PMA, a PM is defined as a path that would lead to the UP of the system considering strength of the piles or the lateral and vertical deflection of the pile cap. Obviously, for a complicated problem, the numbers of possible PMs are numerous and the task of evaluation the system reliability becomes very cumbersome. A procedure to identify the significant PMs considering brittle behavior of piles is discussed next.

For a given load case, the structure-foundation system is subjected to a horizontal load P_x, a vertical load P_z, and a moment M_y. Using the CPGA program, the load distribution in each pile can be calculated. Statistical characteristics of all the load and resistance-related random variables in the strength and serviceability limit states can be established using the available information. Considering several limit states identified earlier, the strength reliability indices β_i's for piles are evaluated using FORM. Similarly, considering the allowable lateral and vertical deflections of the pile cap, the corresponding reliability indices, β_{dx} and β_{dz} are evaluated. The following four sequential tasks are to be completed.

Task 1: Analyze the intact structure using the CPGA program and obtain the strength reliability index β_{ij}, for the ith pile corresponding to the jth strength limit state using FORM. Since more than one limit states need to be considered for the ith pile, the controlling reliability index for the ith pile would be the one with the lowest β_{ij} value. Here, β_i represents the lowest β_{ij} value for the ith pile. The corresponding probability of UP, P_{fi}, can be estimated as $P_f = 1.0 - \Phi(-\beta)$. β_{dx} and β_{dz} and the corresponding P_{fx} and P_{fz} can be calculated similarly using the allowable values for the pile cap lateral and vertical deflections.

Task 2: Remove the pile in compression with the minimum β_i. This will give the maximum load distribution. Reanalyze the system using the CPGA program. Obtain new values for β_i', P_{fi}', β_{dx}', β_{dz}', P_{fdx}' and P_{fdz}', where $i = 1, 2, \ldots, (m - 1)$.

Task 3: Remove the next pile in compression with the minimum β_i' and reanalyze the system. Obtain another set of values for β_i'', P_{fi}'', β_{dx}'', β_{dz}'', P_{fdx}'' and P_{fdz}'', where $i = 1, 2, \ldots, (m - 2)$.

Task 4: Repeat Step 3 until L piles are removed. As piles are removed one by one, β_i's and β_d's values will start going down. When all $\beta_i^{(L+1)}$, $\beta_{dx}^{(L)}$, and $\beta_{dz}^{(L)}$ fall below a pre-selected reliability index, the system can be assumed to have no functional use. The reliability index evaluation can be stopped at this stage.

The information thus generated can be used to identify the significant PMs for the system reliability evaluation as discussed next.

4.1.1. *Evaluation of the most significant performance mode*

Following the four tasks discussed in the previous section, the total number of PMs can be shown to be:

$$
\begin{array}{ll}
(1) & D_x \\
(2) & 1 - D_x \\
(3) & 1 - 2 - D_x \\
& \cdots \\
& \cdots \\
(L+1) & 1 - 2 - \cdots - L - D_x, \\
(L+2) & D_z \\
(L+3) & 1 - D_z \\
(L+4) & 1 - 2 - D_z \\
& \cdots \\
& \cdots \\
(2L+2) & 1 - 2 - \cdots - L - D_z \\
(2L+3) & 1 - 2 - \cdots - L - (L+1)
\end{array}
\tag{10}
$$

where D_x and D_z denote the UP for lateral and vertical deflections of the pile cap, 1 represents the strength UP of the first pile, 2 represents the strength UP of the second pile given that one pile has performed unsatisfactorily, and so on. For an example, the third PM in Eq. (10) indicates that after the UP in strength of two piles, the pile cap produces excessive lateral deflection causing the UP of the structure. It can also be noted that the PMs (1) to $(L+1)$ cause the UP in the lateral deflection, the PMs $(L+2)$ to $(2L+2)$ cause the UP in the vertical deflection, and the last PM $(2L+3)$ is the most dominant strength PM. To identify the most significant PM among $(2L+3)$, the corresponding system probability of UP P_{fsi} for each PM can be estimated as:

$$
\begin{aligned}
&(1) &&P_{fs1} = P_{fdx} \\
&(2) &&P_{fs2} = \max(P_{fi}) \cdot P'_{fdx} \\
&(3) &&P_{fs3} = \max(P_{fi}) \cdot \max(P'_{fi}) \cdot P''_{fdx}
\end{aligned}
$$

$\dots$

$\dots$

$$
\begin{aligned}
&(L+1) &&P_{fs(L+1)} = \max(P_{fi}) \cdot \max(P'_{fi}) \cdots \max\left(P_{fi}^{(L-1)}\right) \cdot P_{fdx}^{(L)} \\
&(L+2) &&P_{fs(L+2)} = P_{fdz} \\
&(L+3) &&P_{fs(L+3)} = \max(P_{fi}) \cdot P'_{fdz} \\
&(L+4) &&P_{fs(L+4)} = \max(P_{fi}) \cdot \max(P'_{fi}) \cdot P''_{fdz}
\end{aligned}
\tag{11}
$$

$\dots$

$\dots$

$$
\begin{aligned}
&(2L+2) &&P_{fs(L+2)} = \max(P_{fi}) \cdot \max(P'_{fi}) \cdots \max\left(P_{fi}^{(L-1)}\right) \cdot P_{fdz}^{(L)} \\
&(2L+3) &&P_{fs(L+3)} = \max(P_{fi}) \cdot \max(P'_{fi}) \cdots \max\left(P_{fi}^{(L)}\right) \cdot \max\left(P_{fi}^{(L+1)}\right)
\end{aligned}
$$

The most significant PM will be the one with the maximum probability of UP, i.e. $\max(P_{fsi})$. This $\max(P_{fsi})$ will give the lower bound of system probability of UP.

4.1.2. *Upper bound evaluation of the system probability of UP*

The estimation of the upper bound system reliability using the PMA requires the identification of all significant PMs. The information available from the previous two sections can be used for this purpose. However, all possible sequences of UPs need to be considered. Thus, Eq. (10) needs to be modified as:

$$
\begin{aligned}
&(1) &&D_x \\
&(2) &&(\text{any 1st}) - D_x \\
&(3) &&(\text{any 1st}) - (\text{any 2nd}) - D_x
\end{aligned}
$$

$\dots$

$\dots$

$$
\begin{aligned}
&(L+1) &&(\text{any 1st}) - (\text{any 2nd}) - (\text{any Lth}) - D_x \\
&(L+2) &&D_z \\
&(L+3) &&(\text{any 1st}) - D_z \\
&(L+4) &&(\text{any 1st}) - (\text{any 2nd}) - D_z
\end{aligned}
\tag{12}
$$

$\dots$

$\dots$

$$
\begin{aligned}
&(2L+2) &&(\text{any 1st}) - (\text{any 2nd}) - (\text{any Lth}) - D_z \\
&(2L+3) &&(\text{any 1st}) - (\text{any 2nd}) - (\text{any Lth}) - (\text{any } (L+1)\text{th})
\end{aligned}
$$

For each of the $(2L+3)$ PMs, using the available values of P_{fi}, P_{fdx}, and P_{fdz}, the approximate system probability of UP can be shown to be:

$$
\begin{array}{ll}
(1) & P_{u1} = P_{fdx} \\
(2) & P_{u2} = \Sigma(P_{fi}) \cdot P'_{fdx} \\
(3) & P_{u3} = \Sigma(P_{fi}) \cdot \Sigma(P'_{fi}) \cdot P''_{fdx} \\
\cdots & \\
\cdots & \\
(L+1) & P_{u(L+1)} = \Sigma(P_{fi}) \cdot \Sigma(P'_{fi}) \cdots \Sigma(P_{fi}^{(L-1)}) \cdot P_{fdx}^{(L)} \\
(L+2) & P_{u(L+2)} = P_{fdz} \\
(L+3) & P_{u(L+3)} = \Sigma(P_{fi}) \cdot P'_{fdz} \\
(L+4) & P_{u(L+4)} = \Sigma(P_{fi}) \cdot \Sigma(P'_{fi}) \cdot P''_{fdz} \\
\cdots & \\
\cdots & \\
(2L+2) & P_{u(2L+2)} = \Sigma(P_{fi}) \cdot \Sigma(P'_{fi}) \cdots \Sigma(P_{fi}^{(L-1)}) \cdot P_{fdz}^{(L)} \\
(2L+3) & P_{u(2L+3)} = \Sigma(P_{fi}) \cdot \Sigma(P'_{fi}) \cdots \Sigma(P_{fi}^{(L)}) \cdot \Sigma(P_{fi}^{(L+1)})
\end{array}
\tag{13}
$$

The upper bound of the system probability of UP will be the union of the $(2L+3)$ PMs given in Eq. (13). Since the probability of UP is very small for all these cases, the upper bound can also be calculated approximately by taking the summation of these numbers.

The bounds for the system probability of UP can be shown to be:

$$
\max(P_{fsi}) < P_f < \sum_{i=1}^{2L+3} P_{ui}.
\tag{14}
$$

4.2. Performance modes (PM) — Ductile behavior of piles

As discussed earlier, the UP of one pile out of a total of m is not expected to cause overall UP of the pile group; the UP of L piles needs to be considered. The sequential distressing of piles and the subsequent load redistribution among the remaining piles using ductile behavior need to be considered at this stage. To incorporate the ductile element behavior, an incremental loading approach[14,15] is used in this study. In this approach, an incremental loading path of the system is constructed using idealized elasto-plastic element behavior for the pile elements, as shown in Fig. 5(a). The pile group is loaded incrementally such that when the most distressed pile has just attained its capacity, it continues to bear its share of loads with ductile behavior until the system has attained its limiting capacity. In general, system performance can well be explained by constructing a load-displacement history for the system as in Fig. 5(b). As seen in the figure, the first load increment $\Delta P_1 (\equiv P_1)$ distresses the first element, $\Delta P_2 [\equiv (P_2 - P_1)]$ the second, and

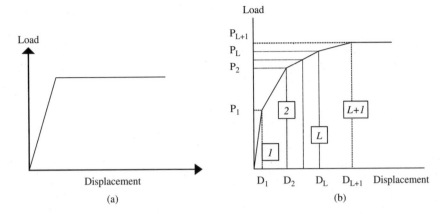

Fig. 5. (a) An idealized ductile element behavior; (b) Incremental system loading history.

$\Delta P_L \left[\equiv (P_{L+1} - P_L) \right]$ the Lth, and so on. In matrix form, the incremental loading equations can be written as:

$$
\begin{bmatrix} R_1 \\ R_2 \\ \vdots \\ R_L \end{bmatrix} = \begin{bmatrix} a_{11} & 0 & \cdots & 0 \\ a_{21} & a_{22} & \cdots & 0 \\ \cdots & \cdots & \cdots & \cdots \\ a_{L1} & a_{L2} & \cdots & a_{LL} \end{bmatrix} \begin{bmatrix} \Delta P_1 \\ \Delta P_2 \\ \cdots \\ \Delta P_L \end{bmatrix},
\tag{15}
$$

where R_i is the resistance-related factor for the ith element, and a_{ij} is the force-related factor of the ith element corresponding to a unit incremental load ($\Delta P_j = 1.0$). The computational scheme for element-level probability of UP indices of individual piles for strength PM using Eq. (15) is further discussed later with the help of an example.

As discussed in Sec. 4.1 for piles with brittle behavior, Eq. (7) can be used to evaluate the system level probability of UP for the serviceability limit states considering ductile behavior of the piles. If the pile cap produces excessive deflection without pile strength failure, then the probability of UP of the pile cap in serviceability can be estimated using FORM. However, if several piles must produce UP in strength first, followed by excessive deflection of the pile cap, then the ductile behavior of the piles needs to be considered. As shown in Fig. 5(b) the deflections of the pile cap are expected to increase during an incremental loading of the structure. Thus, using the relationship in Eq. (7) and Fig. 5(b) the probability of UP for the system based on the ductile element-level behavior can also be evaluated for the deflection limit states.

4.2.1. *Evaluation of the most significant performance mode*

As discussed in the previous section, the total number of the performance modes could be numerous. The first PM may represent a system in which the UP develops due to critical deflection (horizontal, vertical, or both). The second PM may denote the strength UP for one pile followed by excessive deflection. The third PM may indicate that, after the UP in strength of two piles, the pile cap produces excessive deflection, and so on. The probability of UP of the most significant PM will give the lower bound of the unsatisfactory system probability. The tasks are first to identify n number of significant PMs out of a total of N, followed by the evaluation of the corresponding probability of UP considering the ductile behavior of piles. A procedure to identify significant PMs is discussed below.

Significant modes are those that have relatively large probabilities of UP. Here, an arbitrary truncation limit is set to define the significant PMs by using the following relationship:

$$\frac{(P_f)_{j+1}}{(P_f)_j} = \lambda, \tag{16}$$

where $(P_f)_{j+1}$ and $(P_f)_j$ are the probabilities of UP for the $(j+1)$th and jth PM. When λ, the truncation ratio, is less than 10^{-4}, the PM corresponding to the $(j+1)$th is neglected in this study, giving n number of significant PMs. Out of a total of n significant PMs thus obtained, the PM that will produce the highest probability of UP will give the lower bound of the system probability of UP.

4.2.2. *Upper bound evaluation of the system probability of UP*

The significant PMs identified in the previous section can be used to compute the upper bound of the system reliability or system UP probability. The upper bound of UP can be easily calculated using Eq. (9) or approximately, by taking summation of individual probability of UP.

5. Illustrative Examples

The structure-foundation system shown in Fig. 2 is considered. The pile group consists of four identical vertical piles of 12.19 m length evenly spaced at 0.914 m. The piles are made of Coastal Douglas Fir and are located in alluvial sand deposits with coefficients of horizontal subgrade reactions[10] given in Table 1. A series of tests were conducted on this structure. The group was first preloaded by a vertical load of $P_z = 1067.62$ kN or 266.9 kN per pile. Along with this vertical load, a horizontal load $P_x = 133.45$ kN was slowly applied to the cap while the response of the system was measured. The lateral loads, with an eccentricity e, applied at the pile cap produced moments, M_y, of magnitude $(e \times P_x)$. From Fig. 2, e is found to be 0.914 m producing a moment $M_y = 122$ kN-m. The allowable horizontal and vertical deflections are assumed to be 25.4 mm and 6.35 mm, respectively.

The task is to evaluate the system probability of UP considering all possible ways the structure may behave unsatisfactorily considering brittle and ductile behavior of piles.

Table 2. Probability of failure — brittle behavior of piles.

Pile	Intact		Remove 1		Remove 2	
	β	P_f	β	P_f	β	P_f
1	4.19	1.427×10^{-5}				
2	5.68	6.729×10^{-9}	-1.27	0.8983		
3	7.52	2.85×10^{-14}	4.32	7.704×10^{-6}	-3.07	0.99894
4	7.63	1.152×10^{-14}	8.44	4.382×10^{-18}	-2.69	0.99644
D_x	9.43	2.053×10^{-21}	2.29	1.097×10^{-2}	< -10	≈ 1.0
D_z	9.46	1.495×10^{-21}	< -10	≈ 1.0	< -10	≈ 1.0

5.1. *System reliability bounds considering brittle behavior of piles*

Using the CPGA computer program and FORM, the reliability indices and the corresponding probabilities of UP due to strength and deflections are evaluated. The uncertainties associated with all the random variables were summarized in Table 1. The results are summarized in Table 2.

L is 2 in this case. A total of seven PMs are identified for the evaluation of the most significant PM. Among the seven PMs, three correspond to the horizontal deflection UP mode, three correspond to the vertical deflection UP mode, and the last one is due to the strength UP mode. These seven modes and the corresponding probabilities of failure are given below.

1. D_x $P_{fs1} = 2.053 \times 10^{-21}$ (can be ignored)
2. $(1) - D_x$ $P_{fs2} = 1.427 \times 10^{-5} \times 1.097 \times 10^{-2} = 1.5654 \times 10^{-7}$
3. $(1) - (2) - D_x$ $P_{fs3} = 1.427 \times 10^{-5} \times 0.89831 \times 1.0 = 1.2819 \times 10^{-5}$
4. D_z $P_{fs4} = 1.495 \times 10^{-21}$ (can be ignored)
5. $(1) - D_z$ $P_{fs5} = 1.427 \times 10^{-5} \times 1.0 = 1.427 \times 10^{-5}$
6. $(1) - (2) - D_z$ $P_{fs6} = 1.427 \times 10^{-5} \times 0.89831 \times 1.0 = 1.2819 \times 10^{-5}$
7. $(1) - (2) - (3)$ $P_{fs7} = 1.427 \times 10^{-5} \times 0.89831 \times 0.99894 = 1.2805 \times 10^{-5}$

From the above calculations, the $\max(P_{fsi}) = P_{fs5} = 1.427 \times 10^{-5}$ gives the lower bound of the probability of UP for this problem. To evaluate the upper bound of probability of UP, the contributions of the first and fourth PMs are ignored. The contributions to the upper bound from the remaining five PMs can be evaluated as:

1. $(\text{any } 1) - D_x$ $P_{u1} = \Sigma(P_{fi}) \cdot P'_{fdx} = 1.4277 \times 10^{-5}$
$\times 1.097 \times 10^{-2} = 1.5662 \times 10^{-7}$

2. $(\text{any } 1) - (\text{any } 2) - D_x$ $P_{u2} = \Sigma(P_{fi}) \cdot \Sigma(P'_{fi}) \cdot P''_{fdx} = 1.4277 \times 10^{-5}$
$\times 0.89832 \times 1.0 = 1.2825 \times 10^{-5}$

3. $(\text{any } 1) - D_z$ $P_{u3} = \Sigma(P_{fi}) \cdot P'_{fdz} = 1.4277 \times 10^{-5}$
$\times 1.0 = 1.4277 \times 10^{-5}$

4. $(\text{any } 1) - (\text{any } 2) - D_z$ $P_{u4} = \Sigma(P_{fi}) \cdot \Sigma(P'_{fi}) \cdot P''_{fdz} = 1.4277 \times 10^{-5}$
$\times 0.89832 \times 1.0 = 1.2825 \times 10^{-5}$

5. $(\text{any } 1) - (\text{any } 2) - (\text{any } 3)$ $P_{u5} = \Sigma(P_{fi}) \cdot \Sigma(P'_{fi}) \cdot \Sigma(P''_{fi}) = 1.4277 \times 10^{-5}$
$\times 0.89832 \times 1.99538 = 2.5591 \times 10^{-5}$.

Then, the upper bound of probability of UP becomes:

Upper bound $\approx P_{u1} + P_{u2} + P_{u3} + P_{u4} + P_{u5} = 6.5676 \times 10^{-5}$.

Considering brittle behavior of the piles, the bounds of the system probability of UP are:

$$1.427 \times 10^{-5} \leq P_f \leq 6.5676 \times 10^{-5}.$$

For this example, one of the vertical deflection PMs gave the lower bound of the probability of UP. If the allowable deflection is different, the most significant PM is expected to be different.[1]

5.2. *System reliability bounds considering ductile behavior of piles*

The reliability indices and the corresponding probabilities of failure for the four piles and the pile caps given in Table 2 for the intact system will remain the same. Element-level probability of UP for the strength limit states requires the construction of incremental load history for the system as in Fig. 5(b). Considering the ductile behavior of the piles in the intact structure, in which none of the pile has distressed yet, ΔP_1 or P_1, as shown in Fig. 5(b), is calculated using Eq. (15). The first row in Eq. (15), i.e. $R_1 = a_{11}\Delta P_1$, can be used for this purpose. In this case, ΔP_1 represents the first incremental horizontal load that will cause the most critical pile, already pre-loaded with vertical load, to attain its allowable capacity. Thus, the computation of ΔP_1 requires that a_{11} and R_1 are evaluated. This can be achieved by a backward calculation of the limit state equations as discussed below.

5.2.1. *Calculation of incremental load,* ΔP_1

The pile group is already pre-loaded axially with $P_z = 1067.62\,\text{kN}$ (240 kips). Now, a horizontal load P_x of $4.4484\,\text{kN}$ (1.0 kip) is applied to the pile cap. It will produce a bending moment, M_y, of $4.4484 \times 0.914 = 4.07\,\text{kN-m}$ (3.0 kip-ft). With this load condition, and using the mean values of all the random variables involved in the problem, the mean values for the axial loads and bending moment acting on each pile cap are determined using the CPGA program. The results are shown in Table 3. In this case, Pile 1 is the most critical pile. The application of P_x of $4.4484\,\text{kN}$ has resulted in an increase in the axial load by $3.115\,\text{kN}$ [$\{(1067.62/4) - 270.02\} = 3.115$]. Considering P_x to be $1\,\text{kN}$, the value of a_{11} can be estimated as $3.115/4.4484 = 0.7$.

Considering the axial limit state equation represented by Eq. (1), and the mean value of AC shown in Table 1 as $622.78\,\text{kN}$, ΔP_1 can be estimated as [$622.78 - (1067.62/4)]/0.7 = 508.39\,\text{kN}$. This completes the determination of the lateral load that would cause the first pile to attain its axial capacity.

Following the similar procedures and considering the combined axial and bending moment limit state given by Eq. (2) for the critical pile, it can be shown that

$$R_1 = AM(ACC - F) \quad \text{and} \quad a_{11} = [AM(AC - F) + ACC \times M]. \tag{17}$$

Table 3.   The CPGA results for a unit lateral load on the pile cap ($P_x = 4.4484$ kN (1 kip), $P_x = 1067.62$ kN, and $M_y = 4.07$ kN-m).

Variable	Axial load				Moment
	Pile 1 kN	Pile 2 kN	Pile 3 kN	Pile 4 kN	kN-m
Mean value	270.02	267.79	266.01	263.79	−1.40

Table 4.   Results of CPGA calculation when $P_x = \Delta P_1 = 426.52$ kN.

Variable	Axial load				Moment
	Pile 1 kN	Pile 2 kN	Pile 3 kN	Pile 4 kN	kN-m
Mean value	570.73	368.33	165.48	−36.92	−134.07

Considering the mean values of all the parameters in Eq. (17), a_{11} and R_1 can be shown to be $(3.4 \times 10^9)/4.4484 = 0.7643 \times 10^9$ N^2-m/kN and 3.26×10^{11} N^2-m, respectively. Thus, $\Delta P_1 = [(3.26 \times 10^{11})/(0.7643 \times 10^9)] = 426.52$ kN. Considering both the axial and the combined axial and bending moment limit states, $\Delta P_1 = 426.52$ kN.

Using $P_x = 426.52$ kN, $P_z = 1067.62$ kN, and $M_y = 122.03$ kN-m, another CPGA analysis is conducted, and the results are shown in Table 4. Denoting the horizontal load in each pile as F_1, it can be estimated as $426.52/4 = 106.63$ kN. It is interesting to note that if the axial load F and moment M values for Pile 1 given in Table 4 are used to evaluate Eq. (2), it will become 1, indicating that the first pile has just reached its distressed state.

Upon reaching the yield limit, the first distressed pile continues to carry its limiting constant load with no remaining elastic resistance against the next increment of global load. A new state of equilibrium will be attained by the remaining piles based on the elastic distribution of the global loads on the pile cap. Thus, the total deflection d_x at the new state of equilibrium, after attaining the first incremental load, consists of two components, d_{x1} and d_{x2}. The first one is for the ΔP_1 and the other is due to the redistributed elastic deformation for the new global loads that can be obtained from the static equilibrium equations. It can be shown that

$$P_x = 133.45 - 106.63 = 26.82 \text{ kN}, \quad P_z = 1067.62 - 570.73 = 496.89 \text{ kN},$$
$$\text{and} \quad M_y = 122.03 - 570.73 \times 1.37 + 134.07 = -526.37 \text{ kN-m}. \tag{18}$$

These loads are now considered to act on the pile cap. Again, using the CPGA program, the axial load and bending moment acting on the remaining three piles, and the horizontal and vertical deflections in terms of d_{x1}, d_{x2}, d_{z1} and d_{z2} are determined. d_{x1} and d_{x2} represent the deflection due to ΔP_1, and d_{z1} and d_{z2} represent deflection due to the redistributed load given in Eq. (18). Thus, the total horizontal and vertical deflections can be calculated as $dx = d_{x1} + d_{x2}$ and

Table 5.   Reliability indices of the pile group after the first incremental load.

| | Results of CPGA for redistributed global pile loads, (Eq. 18) | | | | | Total deflection | |
	Pile 1 (kN)	Pile 2 (kN)	Pile 3 (kN)	Pile 4 (kN)	M (kN-m)	dx (mm)	dz (mm)
Mean		35.59	165.48	295.37	−246.62	−32.412	6.529
Stand. Dev.		4.97	0.00	0.70	3.76	2.921	0.203
Reliability Evaluation							
β (Eq. 1)		21.98	10.24	5.76			
β (Eq. 2)		13.04	11.74	9.12			
β (control)		13.04	10.24	5.76		−2.65	−0.1075
P_f		3.868^{-38}	6.717^{-25}	4.176^{-9}		0.99596	0.543

$dz = d_{z1} + d_{z2}$, respectively. Corresponding values of dx and dz are also summarized in Table 5. The results indicate that the probability of UP for the critical pile is so small that it is unnecessary to calculate the second incremental load for computing the PM for the system probability of UP evaluation. In case of a significant strength mode, the next incremental load could be obtained from the second set of equation in Eq. (15). The probability of system UP evaluation for the system is discussed next.

The information on the probability of the intact structure is summarized in Table 2. The same information for the damaged structure (one pile distressed in ductile way) is summarized in Table 5. The information now can be used to estimate the lower and upper bounds of the system probability of failure.

In Table 6, seven PMs are identified. Among these, the first three are due to horizontal deflection, the second three are due to vertical deflection, and the last one is controlled by the strength mode. In Table 6, D_x and D_z represent the PMs for the horizontal and vertical deflections of the pile cap, (1) represents strength UP for the first pile, (2) indicates strength UP of the second pile, and so on, and P_{fsi} represents the system probability of UP for the ith mode. Only 2 modes are found to be significant using Eq. (16). From the above calculations, the max(P_{fsi}) = $P_{fs2} = 1.4212 \times 10^{-5}$ gives the lower bound of probability of UP for the system.

Table 6.   Performance modes for ductile behavior.

Performance mode	Associated probability of UP
(D_x)	$P_{fs1} = 2.053 \times 10^{-21}$ (can be ignored)
$(1) - (D_x)$	$P_{fs2} = 1.427 \times 10^{-5} \times 0.99596 = 1.4212 \times 10^{-5}$
$(1) - (4) - (D_x)$	$P_{fs3} = 1.427 \times 10^{-5} \times 4.176 \times 10^{-9} \times 1.0 = 5.9592 \times 10^{-14}$
	(can be ignored)
(D_z)	$P_{fs4} = 1.495 \times 10^{-21}$ (can be ignored)
$(1) - (D_z)$	$P_{fs5} = 1.427 \times 10^{-5} \times 0.543 = 7.7486 \times 10^{-6}$
$(1) - (4) - (D_z)$	$P_{fs6} = 1.427 \times 10^{-5} \times 4.176 \times 10^{-9} \times 1.0 = 5.9592 \times 10^{-14}$
	(can be ignored)
$(1) - (4) - (2)$	$P_{fs7} = 1.427 \times 10^{-5} \times 4.176 \times 10^{-9} \times \cdots < 10^{-14}$
	(can be ignored)

To evaluate the upper bound of system probability of UP, all possible sequences of PM are to be considered. The authors[1,15] suggested an approximate procedure for this purpose. The probability of failures of all PMs except 2 and 5 in Table 6 are negligible. Thus, these two most significant PMs are expected to contribute to the estimation of the upper bound of the system reliability. Their contributions can be shown to be:

$$\text{(any 1)} - D_x; \quad P_{u2} = \sum (P_{fi}) \cdot P'_{fdx}$$
$$= 1.4277 \times 10^{-5} \times 0.99596 = 1.4219 \times 10^{-5} \tag{19}$$

$$\text{(any 1)} - D_z; \quad P_{u5} = \sum (P_{fi}) \cdot P'_{fdz}$$
$$= 1.4277 \times 10^{-5} \times 0.543 = 7.7524 \times 10^{-6}.$$

In Eq. (19), P_{ui} represents the upper bound of system probability of UP for the ith mode and P'_{fdx} represents the new system probability of UP for deflection along ith direction. Thus, using Eq. (9), the bounds of system probability of UP using ductile pile behavior are:

$$1.4212 \times 10^{-5} \le P_f \le 2.1971 \times 10^{-5}.$$

When the brittle behavior of the piles was considered, the corresponding bounds were found to be:

$$1.427 \times 10^{-5} \le P_f \le 6.5676 \times 10^{-5}.$$

These results indicate that the lower bounds for the brittle and ductile systems are almost identical; however, the range between lower and upper bounds is narrower for ductile behavior of piles. This is expected since the consideration of the ductile element behavior in the system probability of UP reduces the compounding conservatism and biases in the computation scheme.

6. Conclusions

Performance mode-based approaches are presented to evaluate the system reliability of complicated structure-foundation systems. In general, they consist of many elements. Some of the elements act in series and other act in parallel, indicating that the failure of one element may not indicate the failure of the system. Furthermore, the brittle and ductile behaviors of elements after they reach their capacities also affect the overall reliability of the system. In most cases, strength performance requirements are satisfied at the element level and the serviceability or deflection requirements are satisfied at the structural level. For the performance-based reliability evaluation procedure, the element-level probabilities of unsatisfactory performance (UP) for strength and system-level probabilities of UP for serviceability need to be combined. Procedures are presented here to calculate the lower and upper bounds of the probability of failure of the system. The lower bound for the system reliability is estimated by identifying the most significant performance mode and the upper bound is estimated by considering all of the significant performance modes. A complicated structure-foundation system consisting of concrete superstructure and pile supported foundation is considered to illustrate the procedure. For the example considered in this study, the bounds of probability of UP are narrower for ductile behavior of piles than that of brittle behavior. The discussions indicate that the estimation of the probability of failure at the element level, particularly for the strength performance function, often practice in the profession, may not be sufficient to develop performance-based design criteria. The deflection performance modes could be more critical.

Acknowledgments

The methodologies presented here are available in the public literature and had been a part of the investigation sponsored by the Research and Development Directorate, Headquarters, US Army Corps of Engineers. The procedures described in this chapter have no relation to those used by the US Army Research Laboratory for reliability evaluation of systems or materials.

References

1. D. Wang, M. R. Chowdhury and A. Haldar, System reliability evaluation considering strength and serviceability requirements, *Computers & Structures* **62**, 5 (1997) 883–896.
2. US Army Corps of Engineers (USACE), *Engineering and Design: Reliability Assessment of Navigation Structures*, ETL 1110-2-532, Department of the Army, Washington, DC (1992).
3. H. Banon, Assessing fitness for purpose of offshore platforms. II: Risk management, maintenance, and repair, *Journal of Structural Engineering*, ASCE **120**, 12 (1994) 3613–3633.
4. H. Banon, R. G. Bea, F. J. Bruen, C. A. Cornell, W. F. Krieger and D. A. Stewart, Assessing fitness for purpose of offshore platforms. I: Analytical methods and inspections, *Journal of Structural Engineering*, ASCE **120**, 12 (1994) 3595–3612.
5. Probabilistic Connection: An SAE NEWSLETTER for Probabilistic Method, G-11 committee, 3, Society of Automotive Engineers (SAE), Warrendale, PA (1996).
6. P. Wirsching and K. Ortiz, Reliability methods in mechanical and structural design, *14th Annual Seminar and Workshop on Modern Reliability Technology for Design Engineers*, Tucson, AZ (1993).
7. A. Haldar and S. Mahadevan, Chapter 3 — First-order/second-order reliability methods (FORM/SORM), *Probabilistic Structural Mechanics Handbook*, C. Sundararajan, Editor, Van Nostrand, New York (1993).
8. USAEWES, *User's Guide: Pile Group Analysis (CPGA) Computer Program*, Technical Report No. ITL-89-3, Vicksburg, MS (1989).
9. J. Y. Perez and D. M. Holloway, *Results and Interpretation of Pile Driving Effects Test Program*, US Army Engineer District St. Louis, St. Louis, MO (1979).
10. P. F. Mlakar and W. R. Stough, Jr., *Condition Analysis and Evaluation of Navigation Structures: Pile Foundations*, Report No. J650-92-014/1767, US Army Engineer Waterways Experiment Station, Vicksburg, MS (1992).

11. W. M. Bulleit, D. V. Rosowsky, K. J. Fridley and M. E. Criswell, Reliability of wood structural systems, *Journal of Structural Engineering*, ASCE **119**, 9 (1993) 2629–2641.
12. C. Q. Li, Time-dependent structural serviceability analysis of concrete structures, *Structural Journal*, ACI **90**, 2 (1993) 204–209.
13. A. Haldar and S. Mahadevan, *Probability, Reliability and Statistical Methods In Engineering Design*, John Wiley & Sons, New York, NY (2000).
14. Y. C. Zhang, Rigidity reliability analysis of structural systems, *Computers & Structures* **45**, 3 (1993) 505–510.
15. M. R. Chowdhury, D. Wang and A. Haldar, Reliability assessment of pile supported structural system, *Journal of Structural Engineering*, ASCE **124**, 1 (1998) 80–88.

CHAPTER 5

APPLICATION OF PROBABILISTIC METHODS
IN BRIDGE ENGINEERING

MICHEL GHOSN

Department of Civil Engineering
The City College of New York/CUNY
New York, NY 10031, USA
E-mail: ghosn@ccny.cuny.edu

Structural reliability theory has been used for several years to account for the uncertainties associated with estimating bridge load capacity and the applied loads during the calibration of safety factors for bridge design and evaluation specifications. The emphasis has been on evaluating the members on an individual basis and ignoring their interaction as a structural system. Although the existing approach has worked well in providing society with safe bridges, recent trends in structural design have focused on developing performance-based engineering methods that take into consideration the whole system's response range. Such a new approach, which would more accurately represent the behavior of bridge structures, requires the application of system reliability techniques. The purpose of this chapter is to illustrate the applicability of reliability methods for the safety assessment of bridge components as well as systems.

1. Introduction

The aim of structural reliability theory is to account for the uncertainties encountered while evaluating the safety of structural systems or during the calibration of load and resistance factors for structural design codes. The concepts are applicable for any type of structure and in particular to highway bridge structures.

The uncertainties associated with predicting the load carrying capacity of a structure, the intensities of the loads expected to be applied, and the effects of these loads may be represented by random variables. The value that a random variable can take is described by a probability density function which for certain typical distributions can be completely defined in terms of the mean and standard deviation, or the coefficient of variation (COV). For a random variable R, these statistics will be denoted hereafter as $\bar{R}$, σ_R, and V_R, respectively. Structural design codes often specify nominal values for the variables used in the design equations. A nominal value is related to the mean value through a bias. If R is the member resistance, then $\bar{R}$ is related to the nominal value R_n using a bias, b_r, such that:

$$\bar{R} = b_r R_n, \tag{1}$$

where, b_r is the resistance bias and R_n is generally specified by the design code.

In structural reliability, safety may be described as the situation where capacity (strength, resistance, fatigue life, etc.) exceeds demand (load, moment, stress ranges, etc.). Probability of failure, P_f, which is the probability that capacity is less than applied load effects, is generally represented in term of the reliability index, β, through:

$$P_f = \Phi(-\beta), \tag{2}$$

where Φ is the cumulative distribution function of the standard normal distribution.

Several methods with different degrees of complexities are available to evaluate the reliability index. Some of the commonly used methods are the first order reliability method (FORM), the second order reliability method (SORM), and several schemes of the Monte Carlo simulation technique. Detailed presentations of the basic principles of structural reliability are available in the literature.[1-4]

The reliability index has been used by bridge code writing groups throughout the world to express structural risk. β in the range of 2 to 4 is usually specified for different structural applications. For example, $\beta = 3.5$ was used for the calibration of the Strength I limit state in AASHTO LRFD Specifications.[5] These values usually correspond to the failure of a single component. If there is adequate redundancy, overall system reliability indices will be higher. However, generally speaking previous code calibration efforts focused on the reliability of individual members because of the general lack of easy to apply techniques for considering structural system reliability. This chapter reviews some practical methods that have been recently applied for analyzing the reliability of bridge systems and the calibration of system factors for implementation in bridge design codes.

2. Code Calibration

Generally speaking, the reliability index has not yet been used in bridge engineering practice for making decisions regarding the safety of a particular design or existing structure. Rather, β, is used by code writing groups for calibrating load and resistance safety factors for bridge design or evaluation specifications. The most commonly used calibration approach is based on the principle that each type of structure should have uniform or consistent member reliability levels over the full range of applications. For example, load and resistance factors should be chosen to produce similar member β values for bridges of different span lengths, number of parallel members, simple or continuous spans, roadway categories, etc.

Some engineers and researchers are suggesting that higher values of β should be used for more important structures such as bridges with longer spans, bridges that carry more traffic, or bridges that, according to their owners, are classified as critical for "social/survival or security/defense requirements." Since higher β levels would require higher construction costs, the justification should be based on a cost-benefit analysis whereby target β values are chosen to provide a balance between cost and risk. This latter approach is still under development in order to establish proper safety criteria, outline suitable calibration methods, and determine appropriate cost functions.[6,7]

Recent bridge design and load rating codes have recommended the use of system factors to account for the behavior of the whole structural system while designing new highway bridges or verifying the safety of existing bridges.[8,9] This would require the application of system reliability techniques and the adoption of the system reliability index as the safety criterion during the code calibration process.[10]

Bridge code calibration efforts have proposed load and resistance as well as system factors which were calibrated to match appropriate member or system reliability index target values. These target values are deduced based on the performance of existing designs. That is, if the safety performance of bridges designed according to current standards has generally been found satisfactory, then the reliability index obtained from current designs is used as the target that any new design should satisfy. The aim of the calibration procedure is to minimize designs that deviate from the target indices.[5,10] Such calibration with past performance also helps minimize any inadequacies in the database.[10,11]

3. Application to Bridge Engineering

To execute the calculations for the reliability index, one needs to obtain the statistical data for all the random variables that affect the safety margin that compares the load capacity of the structure to the applied loads. This requires the statistical modeling of member and system resistances and the modeling of the input loading and load effects.

3.1. *Resistance modeling*

Experimental and simulation studies have developed statistical estimates of member resistances for different types of traditional bridge structural members such as those made of steel or reinforced concrete. These models have accounted for the variability and uncertainties in estimating the material properties; modeling errors; differences between predicted member capacities and measured capacities; human error and construction control.[1,12] For example, a bridge member resistance capacity can be represented by a variable R that is the product of several variables, such that:

$$R = MFPR_n, \tag{3}$$

where M = material factor representing properties such as strength, modulus of elasticity, capacity to resist cracking, and chemical composition; F = fabrication error including geometry, dimensions, and section modulus; P = analysis factor such as approximate models for estimating member capacity; idealized stress and strain distribution models; R_n = predicted member capacity using code specified methods. Equation (3) can be used to find the mean of R using Eq. (1) if the total resistance bias, b_r, is set to be equal to the product of the mean values of M, F and P.

3.2. *Load modeling*

For a bridge member (or structural system) to be safe, the resistance should be large enough to withstand the maximum load effect that could occur within the structure's service life. Estimating the effects of the maximum loads involves a number of random variables, which may often be associated with large levels of modeling uncertainties. In particular, the intensities of the maximum loads are time-dependent random variables in the sense that longer service lives imply higher chances that the structure will be subjected to a large extreme load level.

On the other hand, the projection of limited load intensity data, collected from previous measurements over short periods of time, to future return periods is associated with various levels of statistical and modeling uncertainties. Similarly, modeling the structure's response to the applied loads and estimating the variables that control the effects of the loads on the structure are associated with high levels of uncertainty that are independent of the return period. These modeling uncertainties are often represented by time-independent random variables. Thus, the effect of the applied load of intensity Q on a structural member may be represented by an equation of the form:

$$S = \lambda f(\lambda_Q C_j Q) \tag{4}$$

where S is the load effect (say moment effect or shearing force at a particular point of the structure), λ is the analysis modeling factor that accounts for differences between measured load effects and predicted load effects; $f(\ )$ represents the analysis prediction model that converts load intensities into load effects. Q is the projected intensity of the applied load for the return period of interest. λ_Q is the statistical modeling variable that accounts for the limitations in predicting the value of Q. C_j represents the analysis variables such as bridge material and geometrical properties required for executing the structural analysis. Several such variables, each represented by the index, j, may be required to execute the analysis. All the variables in Eq. (4) may be considered as random where Q is a time-dependent random variable and the remaining variables are time-invariant.

The probability density of the load intensity, Q, for a given return period, t, can be calculated by studying the probability that Q will exceed a given value within t. Assuming that the load events follow a Poisson model, the probability that the load intensity will exceed a value x, within a period, t is represented by $(1 - F_{Q,t}(x))$, which may be approximated as:

$$\Pr(Q > x;\ T < t) = 1 - F_{Q,t}(x) = 1 - e^{(-tp)}, \tag{5}$$

where p is the rate of exceedance per unit time.

For extreme values of x when the values of $F_Q(x)$ are close to 1.0, and when p is calculated for one unit of time while the return period, t, consists of m units of time, Eq. (5) would lead to:

$$\Pr(Q < x;\ T < t) = F_{Q,t}(x) \approx F_Q(x)^m. \tag{6}$$

Equations (4) through (6) have been used to develop live load models for vehicular traffic on short to medium span bridges for the AASHTO LRFD code calibration and to model the effects of seismic motions, wind pressures, scour, and ship collisions forces on bridge systems.[5,11,13]

3.3. *Example*

This section illustrates the application of basic reliability concepts to a simplified example in bridge engineering. The example analyzes a two-girder bridge system formed by two continuous spans as shown in Fig. 1. The applied loads are equally distributed to the two main members. The two spans are 45.75 m (150 ft) and 61 m (200 ft) long. The maximum lifetime load effect is modeled by the crossing of a point load, which has a mean weight equal to 2.07 the weight of the AASHTO HS-20 truck (325 kN). The dynamic amplification factor is 1.15. The coefficient of variation of the live load accounting for the multiple presence effects and the dynamic vibrations is assumed to be 19%. The nominal moment capacities of the sections are assumed to be 8190 kN-m, 23 400 kNm, and 19 217 kN-m for the positive moment capacity of the first span, the negative bending capacity over the middle support, and the positive bending of the second span, respectively. The bias for the moment capacities is 1.12 and the coefficient of variation is 10%. We are also assuming independence between the different moment capacities. The input data is summarized in Table 1.

The reliability analysis for each critical section produced a reliability index $\beta = 1.90$ for the positive moment of the first span, $\beta = 2.27$ for the negative moment section and $\beta = 2.17$ for the positive moment in the second span.

In general, when analyzing the safety of a bridge based on the reliability of its components and critical members, the lowest reliability index is used as the

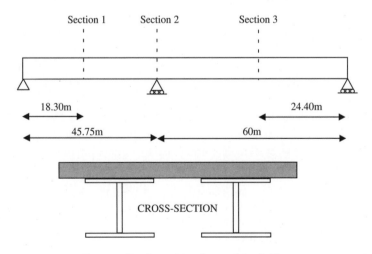

Fig. 1.   Configuration of two-girder bridge.

Table 1. Input data for reliability analysis of 2-girder bridge system.

Variable	Section 1	Section 2	Section 3
Nominal moment capacity (kN-m)	8190	23400	19217
Dead load moment (kN-m)	3640	13755	10750
Live load moment (kN-m)	1351	2183	2082
Resistance bias	1.12	1.12	1.12
Resistance COV (%)	10%	10%	10%
Distribution of resistance	Lognorm.	Lognorm.	Lognorm.
Dead load bias	1.05	1.05	1.05
Dead load COV	9%	9%	9%
Distribution of dead load	Normal	Normal	Normal
Live load bias	2.07	2.07	2.07
Dynamic amplification factor	1.15	1.15	1.15
Live load COV (%)	19%	19%	19%
Distribution of live load	Extreme I	Extreme I	Extreme I
Member reliability index, β	1.90	2.27	2.17

reliability index for the whole bridge. This assumption is very commonly used although it is not strictly correct. The error is due to the fact that the reliability of a structural system depends on the reliability of each of the system's components and the way that these components would combine to produce system collapse or the failure mode.

4. Reliability of Bridge Structural Systems

The calculation of the reliability index for bridge structures depends on the accurate formulation of the safety margin equation, which is normally based on practical methods of structural analysis. Current methods for bridge evaluation are based on the assumption that bridge structural systems behave in a linear elastic mode although member capacities are proportioned using ultimate limit states. Thus, current bridge specifications largely ignore the influence of material nonlinearity on the redistribution of forces in a bridge structural system. Although such a conservative approach is often desirable in engineering practice, it does not provide accurate estimates of the true load carrying capacity of structural systems and in turn produces erroneous estimates of the reliability indices.

Several factors affect the reliability of a structural system. These are primarily related to whether the system is formed by individual components joined together in series or whether the system is formed by a number of parallel members. In addition, the level of component ductility plays an important role in characterizing the effectiveness of the structural system. Finally, the correlation between the member capacities and/or the correlation between the loads affect the reliability of the system as compared to the reliability of the individual members. Below is a more detailed discussion on these important factors.

4.1. *Series systems*

A series system, also called weakest link system, is a system where the failure of any member will produce the failure of the complete system. A typical example is a chain where the failure of any link results in the failure of the whole chain. Examples in bridge engineering include determinate truss structures where the failure of any truss member produces the collapse of the whole bridge. Bridges that may have different possible modes of failure (say shear versus moment or several possible collapse mechanisms) would collapse if any one mode of failure takes place. It is clear that the level of member ductility does not affect the reliability of determinate truss systems (i.e. the full truss will fail whether the individual members fail in brittle or ductile modes). Thus, in general the level of member ductility does not influence the reliability of systems formed by several members in series. On the other hand, the level of member ductility affects the individual collapse mechanisms for girder bridges, which can be considered as systems with parallel members.

If a system is formed by two independent members in series, the system will survive only if both members 1 and 2 are safe. Therefore, the safe domain of the system is the intersection of the safe domains of members 1 and 2, and the system reliability, P_s, is:

$$P_s = 1 - P_f = P_{s1}P_{s2} = (1 - P_{f1})(1 - P_{f2}). \tag{7}$$

If the loads applied on each member or the resistances of the members are correlated (which is normally the case in bridge structures), then, finding the reliability of the system becomes more complicated. To help simplify the problem, several researchers have developed upper and lower bounds on the reliability of a series system. The best-known bounds currently in use are known as the Ditlevsen's bounds.[14]

4.2. *Ditlevsen's bounds for systems in series*

Let us assume that a bridge system would collapse if one of two members (or modes) fails. The safety margins of the members are Z_1 and Z_2. The bridge system will fail when either Z_1 or Z_2 is less than zero or when both Z_1 and Z_2 are less than zero. In bridge engineering it is usual to find that Z_1 and Z_2 are not independent. When the failure of a bridge system is modeled by two modes in series, the probability of failure of the whole system will be higher than the probability of failure of either member taken independently. The difference between the reliability of the system and that of the individual members (or modes) is related to the level of correlation between the different members (or modes).

The degree of correlation between Z_1 and Z_2 can be calculated if the expressions for Z_1 and Z_2 are known. The correlation coefficient between Z_1 and Z_2 is denoted as ρ_{12}. Ditlevsen's bounds on the probability of failure of the system can be calculated based on the probability of failure of each individual member and the correlation coefficient between every two modes. These bounds can be expressed in the form[14]:

$$P_f \leq \sum_{i=1}^{n} \Pr(F_i) - \sum_{i=2}^{n} \max_{j<i}[\Pr(F_j \cap F_i)], \qquad (8a)$$

$$P_f \geq \Pr(F_1) + \sum_{i=2}^{n} \max_{j<i}\left\{\left[\Pr(F_i) - \sum_{j=1}^{i-1}\Pr(F_j \cap F_i)\right];0\right\}, \qquad (8b)$$

where n is the total number of modes, $\Pr(F_i)$ denotes the probability of occurrence of failure mode i, and $\Pr(F_j \cap F_i)$ denotes the probability of occurrence of the intersection of failure modes i and j. If the system has more than two modes, the ordering of the modes may be important. Best results are obtained when all possible permutations of ordering are considered.

4.3. *Parallel systems*

A parallel system is a structure where the total failure of the system requires failure of all its components. Examples of parallel systems include statically indeterminate structures and multi-girder bridges. The failure state associated with the failure of a parallel system is known as a failure mode. A system may have several possible failure modes. The failure in any one mode would result in the failure of the system. Thus, the different failure modes would constitute the "elements" of a system in series. In general, any structural system would be composed of several subsystems in series (failure modes) where each subsystem may be composed of a single member or a number of elements in parallel.

The level of member ductility is important in parallel systems because ductility will influence the reliability of the system compared to the reliability of the individual members. For example, if two parallel members that are perfectly ductile form a system, then the failure of the system will occur only if the two members fail and the resistance of the system is equal to the sum of the member resistances. On the other hand, if brittle members form the system, then the capacity of the system is not necessarily equal to the sum of the member capacities but the system capacity depends on the level of reserve strength in each member. This situation occurs, because when a ductile member reaches its limit, the member will only stop taking any additional loads and the other members in the system will have to carry the additional load only. On the other hand, when a brittle member reaches its limiting capacity it will shed the load that it carries to the adjacent members which will have to sustain the loads that they were already carrying, the load being redistributed to them from the failed member, and the additional load.

The effect of member correlation on the reliability of systems is also different than that observed in series systems. For example, the reliability of ductile parallel systems decreases when the correlation coefficient increases; the reliability increases as the number of elements increases.[15]

4.4. *Example*

In this example, we will look at a bridge with the same configuration as the bridge illustrated in Fig. 1. The analysis will evaluate the reliability of the system rather than that of the individual members. When the moment applied at Sec. 1 reaches the ultimate capacity of that section, the section will undergo inelastic deformations and a redistribution of the load to the other sections of the bridge is effected. In fact, assuming perfectly plastic behavior, and assuming equal distribution of the applied load to the two parallel girders, the bridge system will fail following the formation of a collapse mechanism.

Two different mechanisms are possible for this bridge configuration as shown in Fig. 2 where the applied live load is represented by one concentrated load denoted as P. Each collapse mechanism can be represented by a failure function, Z_i which can be written as:

$$Z_1 = 2(M_1 - D_1) + (M_2 - D_2) - \frac{PL_1}{2}, \tag{9}$$

$$Z_2 = 2(M_3 - D_3) + (M_2 - D_2) - \frac{PL_2}{2}, \tag{10}$$

where M_i is the moment capacity at section i, D_i is the moment due to dead load at section i, P is the applied maximum lifetime truck load, and L_i is the length of span i. The assumptions are that the dead loads for the different sections considered are independent. Also, it is assumed that the portion of a section's capacity available to carry the live load is the total moment capacity minus the moment due to the dead load. The dead load moment is calculated using linear elastic analysis.

The statistical data associated with the failure functions (9) and (10) are the same as those shown in Table 1. It is herein assumed that the dead load effects represented by D_1, D_2 and D_3 are independent random variables. Although somewhat unrealistic, this assumption is maintained herein to simplify the calculations.

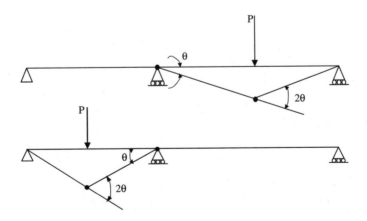

Fig. 2.   Collapse mechanisms for two-span continuous bridge.

Using a FORM algorithm, a reliability index is calculated for each of the two modes. The reliability index for mode Z_1 is obtained as 3.43. The reliability index for mode Z_2 is 3.47. Notice that these values are quite higher than the lowest member reliability index values. This observation confirms that using the member reliability index value to estimate the safety of this bridge configuration will vastly under-predict the true safety level accounting for the system's effects.

The bridge system will fail when Z_1 is less than zero, or when Z_2 is less than zero, or when both Z_1 and Z_2 are less than zero. Common variables in the equations for Z_1 and Z_2 indicate that the two modes are not independent. The correlation coefficient is calculated to be $\rho_{12} = 0.12$. Using Ditlevsen's equations, the probability of failure of the complete system is found to be bounded by: $5.524 \times 10^{-4} \le P_f \le 5.526 \times 10^{-4}$. This will produce a system reliability index $\beta = 3.26$.

4.5. *Generation of failure modes*

The reliability analysis of structural systems whether in series, parallel or combined is possible only if all the failure modes can be identified. Bridge systems are complex structures formed by hundreds of individual components. Finding all the failure modes of such systems is a very elaborate and time-consuming process. On the other hand, many failure modes have very low probabilities of failure and would not influence the reliability of the system as a whole and only a few are generally important. Several methods have been developed to calculate the reliability of structural systems using different types of approximations. Such methods include the β unzipping method and the response surface approach as well as the incremental loading technique and the branch and bound method.[16-18] Most of these are difficult to apply in practical situations as they require especially developed programs and/or make major assumptions about the behavior of the members of the system being analyzed. The response surface method was found to be most practical because of the ease of its implementation in finite element analyses using commercially available FEM packages combined with the classical FORM algorithm. Recent experimentations with Genetic Algorithms have also shown that they can provide very powerful tools for failure mode identifications.[19] Brief descriptions of these two methods and their applicability for the reliability analysis of bridge systems are discussed in the next sections.

5. Response Surface Method

An efficient numerical simulation technique that can be used to calculate the reliability of bridge systems when the failure equations cannot be explicitly formulated is the response surface method.[17,20] This method assumes that a structural analysis program that models the deterministic behavior of bridge structures is available. The program is used to obtain the capacity of the bridge system for a predetermined set of member resistances and dead loads. For the bridge example analyzed in the previous section, this means that for a given set of values for M_1, M_2, M_3, D_1, D_2, D_3 the value of P^* that will produce the collapse of the system is obtained. It should be noted that P^* is a representation of the capacity of the system to carry the live load. The applied live load P may be smaller or larger than P^*.

Several analyses are performed for different sets of data. The results of the deterministic analyses are then used to obtain a functional relationship between P^* and M_1, M_2, M_3, D_1, D_2, D_3. The functional relationship could be obtained by a multi-variable regression analysis, a Taylor series expansion, or any suitable curve fitting technique. This functional relationship is often known as the response function or the response surface, which can be used to obtain a safety margin equation and consequently perform a FORM analysis.

It should be noted that the regression analysis and the Taylor series expansion are sensitive to the points where the expansion is performed. To improve the accuracy of the results, an iterative process is proposed.[20] The iterative process consists of first performing the Taylor series expansion around the nominal values of the random variables and then repeating the expansion at points close to the failure point once the failure point is identified by FORM.

To illustrate the use of the response surface method, the same bridge example studied in the previous section is analyzed. A program is written to perform the plastic analysis of a continuous beam modeling the behavior of one of the parallel bridge girders. As a first step, the beam is analyzed assuming that the moments M_1, M_2, M_3, and the dead loads D_1, D_2, D_3 are at their nominal or design values. At these nominal values, the calculations indicate that a plastic mechanism would form at a load $P^* = 820.66\,\text{kN}$. The calculation of P^* is then repeated after each one of the six variables (moments and dead loads) is independently varied from its nominal value by $+$ or -10%. The value of P^* is noted for each one of these instances. To calculate the response function of the bridge, a first order Taylor series expansion is performed using the expression:

$$f(x_i) = f(x_i^*) + \sum \frac{\partial f}{\partial x_i}|_{x_i^*}(x_i - x_i^*), \qquad (11)$$

where x_i denotes the variables used (M_1, M_2, M_3, D_1, D_2, D_3) and x_i^* denotes the values at which the Taylor series expansion is performed. The operator $f(\ )$ represents the results of the structural analysis. In particular, $f(x_i^*)$ is the value P^* obtained when the variables M_1, M_2, M_3, D_1, D_2, D_3 are evaluated at their nominal values. In the case discussed herein $f\ (x_i^*)$ is equal to 820.66 kN.

The partial derivatives used in (11) may be calculated using a finite difference formulation such that:

$$\frac{\partial f}{\partial x_i} = \frac{f(x_i^* + 0.1x_i^*) - f(x_i^* - 0.1x_i^*)}{0.2x_i^*}. \qquad (12)$$

The safety margin becomes:

$$Z = f(x_i) - P. \qquad (13)$$

This expression for the failure surface is used in the FORM program based on the input data provided in Table 1. On the first iteration, the reliability index obtained in this analysis is 3.54.

For the first iteration, the Taylor series expansion was performed at points (nominal design values) that are in the safe region and not on the failure surface. In subsequent iterations, the calculation of the response function is performed at points around the most likely failure point identified by the form algorithm and more refined expressions of the response surface and the safety margin are obtained. The process is repeated until the reliability index converges to a stable value. For the problem at hand, the process converged after only two iterations. The convergence led to a safety margin equation, which was found to be exactly the same as that corresponding to the failure function (9). The corresponding reliability index of 3.43 was calculated.

Experience with the method has shown that the response surface approach will often converge rather quickly to the reliability index corresponding to the most critical mode. However, in order to obtain the other mode of failures, a filtering algorithm is necessary to direct the search away from the modes that have been previously identified. This can be difficult to achieve in practical applications. However, for the sample example treated in this chapter, the algorithm converged to the solution of the second failure mode after only three additional iterations. The solution included the value of the reliability index for the second mode, $\beta = 3.47$ and the expression of the second failure function (10).

Notice that the response surface method is capable of determining the critical failure modes and the reliability indices corresponding to these modes. The probability of failure of the system can be subsequently calculated using either Ditlevsen's bounds or Monte Carlo algorithms. The Monte Carlo simulation with importance sampling can be efficiently used since the most likely failure points will also be known from the response surface results.

Although the response surface method often converges to the most critical mode of failure, there is always a high chance that the method would lead to less important modes. Furthermore, in many cases, it is difficult for the method to identify other important modes. Recent advances in artificial intelligence techniques have led to the development of heuristic search methods that are quite suitable for use in structural reliability.[19] The next section illustrates the application of Genetic Algorithms for the structural reliability of bridge systems.

6. Genetic Algorithms in Bridge System Reliability

The fact that the search for the reliability index and the design point in the FORM algorithm reduce to an optimization problem makes modern day heuristic search algorithms appropriate for solving reliability problems. The Genetic Algorithm is especially suitable because of its ability not only to detect the global optimum of any search problem but also to find the local optima. This ability is important when dealing with structural problems such as bridge systems that may have several modes of failures.

6.1. *Genetic algorithm method*

Shao and Yoshisada[19] developed a method to use Genetic search algorithms to find the global reliability index of a bridge system by executing a search along several pre-set directions that lie along 45° angles in the multi-dimensional normalized random variable space. Extensions on the method were made to improve the efficiency of the search and refine it to produce more accurate results by adding new operators and using data mining techniques, which speed up the search process.[21–22]

The method begins by defining the search directions using a binary format based on the direction cosines of the search directions in the standard normal space of the random variables (see Fig. 3). These digitized directions become the chromosomes of the genetic algorithm. The fitness of a search (how good it is) is defined as $1/\beta$ where β is the reliability index. The object of the genetic search is to find the search direction (chromosome) that produces the best fitness (lowest β) which is the direction leading to the design point (optimum solution). The proposed coding scheme needed to convert the directions into chromosomes is also illustrated in Fig. 3 where each variable is represented as a gene. The gene value is a representation of the direction cosine corresponding to the associated variable.

Following this coding, a number of search directions are randomly created to form the first population. The population is also randomly divided into pairs that are mated to produce the next generation. The mating process consists of having each pair of chromosomes exchange their genes using a crossover operator, so that each of their two offsprings have parts of the genes of the parents. Additionally, one of the genes may be randomly changed in order to simulate a mutation process that introduces some new genes into the population. The mating process is illustrated as shown in Fig. 4.

Once a new chromosome is created, its fitness is obtained by following the search direction using a finite element analysis program until failure is reached. This is executed by reinverting the chromosome into the actual variable space and incrementing all the variables by the same factor and performing finite element analyses at each stage and checking whether the analysis indicates that failure has been reached or not. The factor by which the variables are multiplied to reach first failure is related to the fitness or the inverse of the reliability index. Thus, the reliability index for this particular search direction is obtained.

By checking all the search directions created through different generations, a global reliability index associated with the most critical failure mode is obtained. In addition, the algorithm leads to sub-optimum solutions that are associated with the reliability indices of the other possible failure modes. A local search algorithm and appropriate filtering operators are used to refine the results of the search and lead to accurate estimates of the reliability index values and the identification of

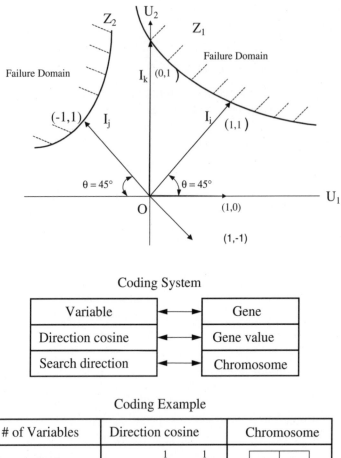

Coding System

Variable	←→	Gene
Direction cosine	←→	Gene value
Search direction	←→	Chromosome

Coding Example

# of Variables	Direction cosine	Chromosome
2 variables	$V = \dfrac{1}{\sqrt{2}}, \dfrac{1}{\sqrt{2}}$	1 | 1
3 variables	$V = \dfrac{1}{\sqrt{3}} \ \dfrac{-1}{\sqrt{3}} \ \dfrac{-1}{\sqrt{3}}$	1 | -1 | -1

Fig. 3.   Illustration of genetic coding scheme.

the corresponding design points.[22] The applicability of the genetic algorithm for solving general structural problems and bridge reliability problems has been verified by comparing the GA solution to that of other methods.[22]

6.2. *Illustrative example*

A simple example is analyzed in this section to compare the performance of the genetic algorithm with that of gradient-based and other commonly used structural reliability methods, such as FORM, Response Surface Method (RSM), Importance Sampling Method (ISM), and Neural Network techniques. The example is for a uniaxial bar with plastic behavior that is subjected to an axial load.[23] The problem

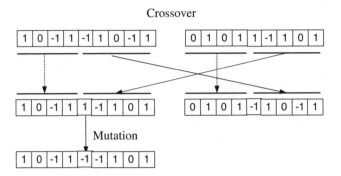

Fig. 4. Illustration of crossover and mutation operators.

Table 2. Comparison of different methods for bar example.

Technique	Reliability index β	Probability of failure P_f	Average evaluation times
Analytical solution[24]	3.049	0.001148	—
Direct MCS[23]	3.075	0.001050	100000
MCS with Importance Sampling[23]	3.050	0.001131	14000
FORM (explicit limit function)[23]	3.049	0.001147	7
Response Surface Method[23]	3.049	0.001147	50
Neural NetWork[23]	3.046	0.001141	30
Advanced Genetic Algorithm[22]	3.049	0.001148	30

has a failure function expressed by the equation:

$$Z = Af_y - P, \tag{14}$$

where the three random variables are: A = cross sectional area of the bar, f_y = yielding stress, and P = applied load.

Comparison of the results from different reliability methods for this problem is provided in Table 2. The genetic algorithm converged after a total 30 function evaluation times. The "exact" $\beta = 3.049$ was obtained. When compared to the results from other methods, it is observed that the genetic algorithm is as efficient as the Neural Network technique and is more efficient than the Response Surface Method or the Importance Sampling Method. The latter is true even if every function evaluation in the genetic algorithm would require several structural analyses to arrive at the corresponding reliability index.

The gradient-based FORM algorithm converged after only 7 iterations in this problem where the limit state function is explicitly known. However, it should be noted that the advantage of the gradient-based algorithm quickly dissipates for problems involving multiple failure modes or in the cases when the limit state function is not explicitly known and the search gradients have to be numerically evaluated using finite element analysis programs.

Table 2 shows that overall the genetic algorithm can be more efficient than the Response Surface Method and is superior to RSM because of its ability to easily solve multiple failure mode problems as previously explained.

6.3. *Analysis of cable-stayed bridge*

The reliability of a cable-stayed bridge is analyzed to demonstrate the ability of the genetic algorithm to identify the important failure modes of a large geometrically nonlinear structure. The realistic bridge model consists of one support tower and the symmetric cable stayed system shown in Fig. 5.

The main girder's sectional properties can be divided into the three categories listed in Table 3. The modulus of elasticity of the girders is $E = 2.1 * 10^5$ MPa, and the yielding stress is $\sigma_y = 345$ MPa. The total length of the bridge is 1024 meters. The height of the main tower is 269.95 meters. In this analysis, since the tower is very stiff compared to the other members, the components of the tower are modeled as rigid elements. There are a total of 35 stay cables whose design tension stress is 1860 MPa and modulus of elasticity $E = 1.95 * 10^5$ MPa. The typical span between the cables is 15 meters except for the leftmost part of the bridge, where the cables are spaced 9 meters apart.

The random variables include the live load, the dead load, the moment capacities of the three different cross sections of the main box girder, as well as the strengths of the 35 stay cables. The nominal values, biases, COV's and distribution types of all the variables are listed in Table 4. Assuming structural degradation of the stay cables, the biases of all cables' strengths are assumed to be 0.74 for the 75-year service life. The structural reliability of service life is a time-dependent problem. However, to simplify the problem, it is assumed that the biases of cable strengths

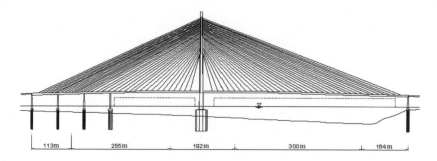

Fig. 5.   Configuration of cable-stayed bridge.

Table 3.   Geometric properties of main girder sections.

	Self weight (KN/M)	Area (m^2)	I_x (m^4)	I_y (m^4)
Section 1	215.6	2.544	10.164	261.748
Section 2	223.4	2.278	9.078	226.203
Section 3	245.0	1.743	6.755	168.786

Table 4. Random variables of cable stayed bridge.

Variable number	Variables	Nominal value	Bias	COV	Distribution type
1	Live load	129.78 KN/m	1.05	0.2	Extreme Type I
2	Dead load	See Table 3	1.03	0.08	Normal
3	Moment capacity of main girder section 1	1 353 456 KN-m	1.05	0.11	Normal
4	Moment capacity of main girder section 2	1 205 740 KN-m	1.05	0.11	Normal
5	Moment capacity of main girder section 3	889 777 KN-m	1.05	0.11	Normal
6	Yield strength cable # 1 (KN)	16 665.6	0.74	0.2	Normal
7	Yield strength cable # 2 (KN)	13 020.0	0.74	0.2	Normal
8	Yield strength cable # 3 (KN)	13 020.0	0.74	0.2	Normal
9, 10, 11, 12, 13	Yield strength cables # 4, #5, #6, #7, #8 (KN)	18 228.0	0.74	0.2	Normal
14	Yield strength cable # 9 (KN)	19 269.6	0.74	0.2	Normal
15	Yield strength cable # 10 (KN)	20 311.2	0.74	0.2	Normal
16	Yield strength cable # 11 (KN)	21 352.8	0.74	0.2	Normal
17	Yield strength cable # 12 (KN)	22 394.4	0.74	0.2	Normal
18	Yield strength cable # 13 (KN)	23 956.8	0.74	0.2	Normal
19, 20, 21, 22	Yield strength cables # 14, #15, #16, #17 (KN)	26 560.8	0.74	0.2	Normal
23, 24	Yield strength cables # 18, #19 (KN)	27 081.6	0.74	0.2	Normal
25, 26, 27	Yield strength cables # 20, #21, #22 (KN)	31 248.0	0.74	0.2	Normal
28	Yield strength cable # 23 (KN)	28 644.0	0.74	0.2	Normal
29, 30, 31, 32, 33, 34	Yield strength cables # 24, #25, #26, #27, #28, #29	31 768.8	0.74	0.2	Normal
35, 36, 37	Yield strength cables # 30, #31, #32 (KN)	31 248.0	0.74	0.2	Normal
38, 39, 40	Yield strength cables # 33, #34, #35 (KN)	30 727.2	0.74	0.2	Normal

are constant over the whole service life. All 40 random variables are assumed to be statistically independent.

In a first step of the analysis, all the resistance variables are multiplied by a reduction factor equal to 0.3. This is done to reduce the probability of failure, P_f, so that the analysis can be performed using classical Monte Carlo Simulation. The failure equation can then be roughly represented by the equation:

$$Z = 0.3R - S. \tag{15}$$

Applying the Monte Carlo Simulation direct sampling option in the structural analysis program, ANSYS, version 7.0, the probability of failure and the system's reliability index were found to be $P_f = 1.93 * 10^{-1}$ and $\beta_{\text{sys}} = 0.866$ after 300 cycles. Using the genetic algorithm for executing the search for the failure modes in association with the finite element program, ANSYS, the failure modes are identified, and the system reliability index as well as the probability of structural failure are obtained as, $P_f = 1.87 * 10^{-1}$ and $\beta_{\text{sys}} = 0.889$. In this case, the total number of structural analyses required for convergence happened to be equal to 300.

Comparing the results of the genetic algorithm with those of the Monte Carlo Simulation, it is noticed that the proposed algorithm provides reasonably accurate results. The analysis was stopped after identifying the first three failure modes. Improved accuracy can be obtained if the simulation is continued so that a larger number of modes are identified. Another reason for the difference is the assumption that all failure modes can be approximated by first-order equations. For this particular problem, this is not necessarily the case.

Ignoring the 0.3 reduction factor used in the previous test case of Eq. (15), the reliability analysis of the cable-stayed bridge is performed using the genetic algorithm with the original data of Table 4. The analysis was stopped after the three most important failure modes are identified. If L represents the live load, D represents the dead load, G_i represents the resistance of girder section i and C_j represents the resistance of cable j, the most critical linearized failure equation, Z_1, is given as:

$$
\begin{aligned}
Z_1 = {} & 3.374G_1 + 2.400G_2 + 1.700G_3 + 0.790C_1 + 0.451C_3 + 0.720C_7 \\
& + 1.026C_{10} + 1.277C_{12} + 1.082C_{14} + 1.381C_{17} + 0.473C_{21} \\
& + 0.711C_{26} + 1.284C_{30} + 1.085C_{33} + 0.850C_{34} + 0.741C_{35} \\
& - 1.421L - 10.89.
\end{aligned}
\tag{16}
$$

The corresponding reliability index is calculated to be, $\beta_{Z1} = 4.791$. Some valuable information can be obtained from Eq. (16). For example, the dead load is found not to be a critical parameter that would affect the system reliability. This is most probably due to its low COV.

From a design optimization point of view, the cables that do not appear in this failure mode can be considered to have been overdesigned. More sensitivity analyses can be performed to analyze different girder box sections, different load conditions, and cable strengths to obtain a more efficient design.

The genetic algorithm also identified the second failure mode whose linearized failure equation is obtained as:

$$
\begin{aligned}
Z_2 = {} & 2.646G_1 + 2.292G_2 + 1.392C_9 + 1.986C_{12} + 1.859C_{15} \\
& + 1.219C_{17} + 1.356C_{21} + 1.158C_{31} + 0.830C_{32} + 0.810C_{35} \\
& - 1.336L - 1.427D - 4.592.
\end{aligned}
\tag{17}
$$

The corresponding reliability index is $\beta_{Z2} = 5.379$. For the third failure mode, the linearized failure equation is:

$$
\begin{aligned}
Z_3 = {} & 1.223G_1 + 2.694G_2 + 0.450C_4 + 0.535C_7 + 0.615C_{12} \\
& + 0.445C_{13} + 1.836C_{15} + 0.743C_{16} + 1.866C_{17} + 1.347C_{20} \\
& + 1.193C_{21} + 1.018C_{22} + 0.572C_{26} + 0.648C_{28} + 1.417C_{32} \\
& - 1.089L - 6.8222.
\end{aligned}
\tag{18}
$$

The corresponding reliability index is $\beta_{Z3} = 5.931$. The system reliability when all three failure modes are considered is $\beta_{sys} = 4.7819$ producing a probability of failure $P_f = 8.68 * 10^{-7}$. When only the two first failure modes are considered, the system reliability index becomes $\beta_{sys} = 4.7820$. This shows that in this example, the first failure mode is dominant and the system's reliability can be well estimated when the first two modes are considered.

The long span cable-stayed bridge example solved in this section demonstrates the accuracy, efficiency and high flexibility of the genetic-algorithm-based reliability analysis and the ability of the method to be combined with general-purpose finite element programs to perform the reliability analysis of complex structural system.

7. Concluding Remarks

Major advances in the development and application of the theory of structural reliability have been made during the last three decades. The application of structural reliability in bridge engineering has lagged behind other structural fields such as offshore structures and nuclear power plants where reliability methods are used on a routine basis to assess the safety of individual structures. In bridge engineering, the application of reliability theory has focused on developing new design and load capacity evaluation specifications that are calibrated to produce consistent levels of reliability index values. The reliability concepts should be expanded for direct use during the safety evaluation of specific bridge structures especially when such bridges have unusual configurations. This chapter demonstrated that structural reliability methods can be easily applied to evaluate the structural safety of individual bridge components as well as complete structural systems. In particular, genetic algorithms are found to be especially useful for solving large-scale problems involving material and geometric nonlinearities. By linking the genetic algorithm to existing finite element packages, realistic models of complex bridges can be analyzed to estimate their reliability levels and identify their modes of failure.

Acknowledgments

The author would like to acknowledge the contributions of Prof. Fred Moses, Prof. Dan Frangopol, Dr. Shaowen Shao, Dr. Linzhong Deng, Dr. Jian Wang to the information provided in this chapter.

References

1. P. Thoft-Christensen and M. Baker, *Structural Reliability Theory and Its Applications*, Springer Verlag, Berlin, Germany (1982).
2. R. E. Melchers, *Structural Reliability: Analysis and Prediction*, John Wiley & Sons, New York, NY (1999).

3. A. Haldar and S. Mahadevan, *Probability, Reliability and Statistical Methods in Engineering Design*, John Wiley & Sons, New York (2000).

4. A. S. Nowak and K. Collins, *Reliability of Structures*, McGraw Hill, New York (2000).

5. A. S. Nowak, *Calibration of LRFD Bridge Design Code,* NCHRP Project 12–33, TRB, Washington DC (1993).

6. E. Aktas, F. Moses and M. Ghosn, Cost and safety optimization of structural design specifications, *Journal of Reliability Engineering and System Safety* **73**, 3 (2001) 205–212.

7. D. M. Frangopol, K. Y. Lin and A. C. Estes, Life-cycle cost design of deteriorating structures, *J. Struct. Eng.*, ASCE **123**, 10 (1997) 1390–1401.

8. American Association of State Highway & Transportation Officials, AASHTO LRFD Bridge Design Specifications, 3rd edition, Washington DC (2002).

9. American Association of State Highway & Transportation Officials, AASHTO Guide manual for Condition Evaluation and Load and Resistance Factor Rating (LRFR) of Highway Bridges, *LRFD Bridge Design Specifications* 2nd edition, Washington, DC (2003).

10. M. Ghosn and F. Moses, Reliability calibration of a bridge design code, *ASCE Journal of Structural Engineering* (April 1986).

11. M. Ghosn and F. Moses, *Redundancy in Highway Bridge Superstructures*, NCHRP Report 406, Transportation Research Board, Washington, DC (1998).

12. B. Ellingwood, T. V. Galambos, J. G. MacGregor and C. A. Comell, *Development of a Probability Based Load Criterion for American National Standard A58*, National Bureau of Standards, Washington, DC (1980).

13. M. Ghosn, F. Moses and J. Wang, *Design of Highway Bridges for Extreme Events, National Cooperative Highway Research Program*, NCHRP Report 489, Transportation Research Board, National Academy Press, Washington, DC (2003).

14. O. Ditlevsen, Narrow reliability bounds for structural systems, *J. Struct. Mech.* **7** (1979) 435–451.

15. S. Hendawi and D. M. Frangopol, System reliability and redundancy in structural design and evaluation, *Struct. Safety*, Elsevier **16**, 1 + 2 (1994) 47–71.

16. P. Thoft-Christensen and Y. Murotsu, *Application of Structural System Reliability Theory*, Springer-Verlag, Berlin (1986).

17. G. Augusti, A. Baratta and F. Casciati, *Probabilistic Methods in Structural Engineering*, Chapman, Hall, New York (1984).

18. F. Moses, System reliability development in structural engineering, *Struct. Safety* **1**, 1 (1982) 3–13.

19. S. Shao and M. Yoshisada, Approach to failure mode analysis of large structures, *Probabilistic Engineering Mechanics* **14** (1999) 169–177.

20. M. Ghosn, F. Moses and N. Khedekar, Response function and system reliability of bridges, *Probabilistic Structural Mechanics: Advances in Structural Reliability Methods*, Springer-Verlag (1994) 220–236.

21. J. Wang and M. Ghosn, Hybrid data mining/genetic shredding algorithm for reliability assessment of structural systems, *ASCE Journal of Structural Engineering* (2005).

22. J. Wang and M. Ghosn, Linkage-shredding genetic algorithm for reliability assessment of structural systems, *Journal of Structural Safety* **27**, 1 (2005) 49–72.

23. H. M. Gomes and A. M. Awruch, Comparison of response surface and neural network with other methods for structural reliability analysis, *Journal of Structural Safety* **26** (2004) 49–67.

24. A.-H. S. Ang and W. H. Tang, *Probability Concepts in Engineering Planning and Design*, 1, John Wiley & Sons, New York (1975).

CHAPTER 6

STOCHASTIC RESPONSE OF FIXED
OFFSHORE STRUCTURES

SER-TONG QUEK*, XIANG-YUAN ZHENG[†] and CHIH-YOUNG LIAW[‡]

*Department of Civil Engineering, National University of Singapore
117576, Singapore
E-mails: *cveqst@nus.edu.sg
[†]cvezxy@nus.edu.sg
[‡]cvelcy@nus.edu.sg*

This chapter focuses are on two frequency-domain methods to estimate the stochastic response of fixed offshore structures. The sources of nonlinearity including that due to drag and waves that lead to complexity in stochastic analyses are discussed. Both the Volterra series and cumulant spectral approaches are presented where the drag-related wave forces (including that due to inundation) are polynomialized into zero-memory processes. It is shown that up to fourth-order statistical moments of the response can be computed and in good agreement with values estimated from time domain simulations and the response is found to deviate significantly from Gaussianity.

1. Introduction

One main challenge in offshore engineering is the ability to precisely and efficiently predict the wave forces acting on the structures and the resulting structural responses, accounting for inherent and environmental variabilities. Fixed platforms such as jacket and jack-up rigs are mainly subjected to Morison-type wave loads acting on their slender members. The difficulty in prediction of forces and responses, and estimating their stochasticity lies in nonlinearities due to the nonlinear force-kinematics relationship and/or the nonlinear nature of waves.

1.1. *Sources of nonlinearities*

In practical engineering analysis and design, linear wave theory has been popularly accepted as a basis for modeling the short-term random wave storm, where the wave elevation can be adequately modeled as Gaussian processes. Despite this, the structural response shows deviations from Gaussianity, which can be attributed to three main nonlinear effects, two associated with drag forces and the third with wave inundation. The non-Gaussian response behavior will be more evident if nonlinearity of waves is considered. Nonlinear random wave based on Stokes theory is non-Gaussian in nature.

The first nonlinear effect of wave loading arises from the distributed nonlinear drag force which can be evaluated using the well-known Morison equation.

This is usually the predominant wave force component for the slender structural members of a fixed platform and is the main cause of the so-called superharmonic phenomenon.[1] In the absence of current, the power spectrum of the wave force exhibits sharp peaks at odd-multiples of the peak wave frequency ω_p. If current is not negligible, the spectral peak at $3\omega_p$ becomes less dominant and a peak at $2\omega_p$ becomes observable.[2,3]

The second nonlinear effect is due to the relative velocity term, which must be included in the modified Morison expression in order to account for wave-structure interaction. The superharmonic responses influenced by the relative velocity have been studied by Bouyssy and Rackwitz,[4] and Tognarelli et al.[5] If the structural displacement is relatively large, such as in the case of tension leg platforms, sub-harmonic[6–9] and even chaotic responses[10] will occur in addition to superharmonic resonance.

The third nonlinear effect is that induced by the varying water surface, known as "inundation effect",[11–13] also referred to as "the effect of wave intermittency"[14–16] or "splash-zone wave force".[17] Unlike the Morison drag force, inundation introduces only superharmonic forces at even-multiples of ω_p in the absence of current.[18–20] If the structural natural frequency ω_N of an offshore platform is close to $2\omega_p$, the nonlinear inundation effect will be significant,[21,22] especially in shallow water.[16,22]

The superharmonic phenomena can also be produced by wave nonlinearities. Many previous researches were devoted to deterministic studies of wave nonlinear effects on the structural response or the non-Gaussian distribution of wave height. In the stochastic approach, the second-order Stokes theory has been employed[14,23] to investigate the response spectrum of platform and it was concluded that wave non-linearities is not negligible. As the use of higher-order spectral analyses to account for wave nonlinearities is still not fully developed, this chapter will only concentrate on the more established stochastic analysis based on linear random waves.

1.2. *Frequency-domain analyses of wave loadings*

For structural reliability prediction, a good description of the extreme response behavior is critical, which is generally more appropriately characterized as a non-Gaussian process, especially for drag-dominant cases. The non-Gaussian probability density function (PDF) of Morison force may be approximated by the Pierson-Holmes[24] (P-H) system of frequency curves which is defined by the first four moments. Quasi-static studies[25–28] showed that structural response follows the P-H class of distribution. However, this is only true for very rigid offshore structures where $\omega_N \gg \omega_p$. In other cases, dynamic considerations are necessary to obtain accurate higher-order moment estimations for representative PDF approximations.[29]

Stochastic dynamic analyses to estimate the response higher-order statistics can be carried out in either time or frequency domain. In time-domain analyses, the spectral representation method for Monte Carlo simulation is widely used.[4,14,21,30–33] To capture the dynamic characteristics accurately, many harmonic frequency components and a large ensemble of sample functions are needed, which leads to time-consuming computational work. In addition, the physics behind the factors contributing to the statistical moments of interest is harder to discern compared to the frequency domain approach.

1.2.1. *Volterra series approach*

To a certain extent, frequency-domain approaches may partially overcome the drawbacks of computational time and the manifestation of physical behavior inherent in the time domain approach. One such approach is the use of Volterra series, a generalization of Taylor series expansion of a function.[34,35] It is a power series with memory[36] and is most suitable for a nonlinear system that is expressible as polynomials[34] to obtain frequency response functions (FRF) or transfer functions in control theory. Morison drag force has been polynomialized for employing the Volterra series approach to estimate the responses of offshore structures.[5,23,37–40] The drag cubicization for jack-up platform response analysis was conducted by Li *et al.*[39] Spectral analysis based on fourth-order polynomialization of inundation force was performed by Liaw and Zheng.[41,42] Figure 1 displays a finite-memory model[36] in the dynamic analysis of fixed offshore structure subjected to Morison and inundation forces. The first system relating the water particle kinematics to wave fluctuation is given explicitly by linear wave theory, while the last system is governed by a linear differential equation linking the structural states to the input excitation. The first system becomes nonlinear under Stokes wave theory and the resultant particle kinematics are no longer Gaussian.

For a system with strong nonlinearity, higher-degree polynomial approximation is essential. Higher-order (beyond second-order) spectral analyses based on Volterra series approach becomes quite tedious due to the complicated system input-output relationship.[36,43] In addition, because higher-order statistics involve multi-dimensional integrals, the efficiency of such an approach drastically decreases when the order of polynomials increases.[39]

1.2.2. *Cumulant spectral approach*

Another frequency-domain approach for the finite-memory model (Fig. 1) is cumulant spectral analysis, which is more efficient than the Volterra series method. It is

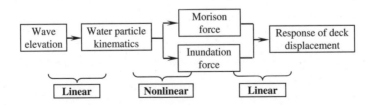

Fig. 1.   Nonlinear input-output model of stochastic analysis of fixed platform.

based on the inverse Fourier transforms of the response power, bi- and tri-spectra to obtain the desired moments, namely, variance, skewness and kurtosis.[44]

Most fixed offshore platforms can be idealized as linear time-invariant (LTI) systems, with the wave loading as input and the structural response as output. For a LTI system, the relationship of the nth-order cumulant spectra of the input and output has been well established.[45,46]

The first essential step is to formulate the cumulant spectra of the wave loading. This can be realized by applying Fast Fourier Transform (FFT) to the cumulant functions of the wave loading. The power spectral analysis of Morison force is an illustration of the basic cumulant spectral analysis. Borgman[47] analytically derived the autocorrelation function, a second-order cumulant, of the Morison drag force at a fixed location based on Price's theorem.[48,49] The corresponding power spectrum could be evaluated by 1 dimensional (1D) Fourier transform of autocorrelation. Such an idea was extended to find the spectral density of the total wave loading, the latter being the summation of Morison forces at distinct positions.[1] Further extension to the more complicated spectral analysis of inundation force has been performed[42] and its superiority in terms of numerical efficiency over Volterra-series approach was substantiated.[42]

Higher-order spectral analysis of a LTI system driven by non-Gaussian excitations can be also found in Refs. 50–54. Hu and Lutes[51] derived the fourth-order cumulant function of Morison drag force using power series expansion, which involves the product of up to four velocity correlation functions (eighth-order joint moments). The corresponding tri-spectrum was computed by frequency convolutions, instead of the more efficient 3D-FFT. The same tri-spectrum was obtained by Hermite cubicization of the drag using least squares fit.[54]

A complete and more meaningful cumulant spectral analysis of the total force on fixed platform was presented by Zheng and Liaw.[22] This analysis provides an efficient method for obtaining the power-, bi- and tri-spectra of total Morison loadings and the inundation force, with spatial correlations amongst them taken into account. The second-, third- and fourth-order cumulants of total wave force were derived based on moment decomposition and condensation algorithms developed for the purpose of reducing tedious derivations and time-consuming computations. Numerical results showed the significant contribution of inundation effects leading to larger deviation from Gaussianity in the deck response. It was also pointed that tenth and twelfth joint moments of velocities are indispensable to obtain a proper response kurtosis value; applying least square fit in drag force approximation results in unreasonable kurtosis estimations.[55]

2. Polynomial Approximation of Nonlinear Wave Forces

To facilitate either quasi-static or frequency-domain computations, one simplification often adopted is the polynomial approximation of nonlinear drag forces.[5,25–28,37–42] Statistical linearization, which is convenient for both spectral derivation and numerical computation, has long been used to simplify the Morison drag force.[5,56–59] However, linearization can only capture the response mean and variance and unable to account for non-Gaussian features. This necessitates higher-degree polynomial approximation of nonlinear drag. In this section, least squares and moment-based approximations will be discussed.

2.1. *Morison and inundation drag forces*

Morison *et al.*[56] suggested that the in-line force per unit length acting on a stationary slender vertical cylinder can be expressed as:

$$f(z,t) = f_I + f_D, \tag{1}$$

where f_I and f_D are inertia and drag forces respectively, in which:

$$f_I = C_M A_I (\partial u/\partial t) = k_I a, \tag{2}$$

$$f_D = C_D A_D u |u| = k_D u |u|, \tag{3}$$

where C_M and C_D are respectively the inertia and drag coefficients, assumed constant along the submerged cylinder; $A_I = \pi \rho D^2/4$ and $A_D = \rho D/2$; ρ is the water mass density; D is the equivalent diameter of the cylinder; $u = u(z, t)$ is the water particle velocity (to be replaced by $u + C$ if current C exists) at the position z; z is positive upwards from still water level (SWL); and a is the particle acceleration.

If fluid-structure interaction is modeled, Morison's equation needs to be modified as:

$$f = -A_I(C_M - 1)\ddot{v} + k_I \dot{u} + k_D(u - \dot{v}) |u - \dot{v}|, \tag{4}$$

where v is the displacement of the cylinder. The interaction represented by the third term on the RHS results in fluid damping effect.[60] Due to the relatively small $\dot{v}$ for fixed platforms, this effect can be included by adjusting the structural damping for analyses,[3] while retaining Eqs. (1–3).

The free-surface inundation, which constitutes one part of the modal force, can be expressed as[18]:

$$P = \int_0^\eta \Phi(z) f(z,t) dz \approx \Phi(0) \cdot f(0,t) \cdot \eta, \tag{5}$$

where $\Phi(z)$ is the mode shape of the structure under free vibration; and η is the wave elevation, usually assumed Gaussian-distributed with a zero mean. In Eq. (5), the inundation is simplified as a point load acting at the SWL,[18,20] and its drag contribution is approximated as:

$$PD = \Phi(0) C_D A_D [u(0,t) |u(0,t)| \eta]. \tag{6}$$

Note that both two drag forces, f_D and PD, are represented by nonlinear square-law term with sign (Eqs. (3) and (6)), the following normalizations are first introduced:

$$x = \eta/\sigma_\eta, \quad y = u/\sigma_u, \quad c = C/\sigma_u, \tag{7}$$

where σ_η and σ_u are the standard deviations of η and u (written concisely in place of $u(0,t)$). Based on linear wave theory, both x and y are standard Gaussian distributed. The nonlinear problem reduces to dealing with the two functional forms given by:

$$r = (y+c)\,|y+c|\,, \tag{8}$$

$$s = x(y+c)\,|y+c|\,. \tag{9}$$

2.2. Least squares approximation (LSA)

Least squares approximation (LSA) and moment-based approximation (MBA) of r and s have been discussed in detail in Ref. 61. As such, only the approximation schemes and some useful results are presented herein. LSA seeks the minimization of the mean square error between the approximate and original forcing functions. For the Morison drag term, the nth-degree polynomial approximation is

$$r = (y+c)\,|y+c| \approx r' = a_0 + a_1 y + \cdots + a_n y^n, \tag{10}$$

with the mean square error ε defined as:

$$\varepsilon = E[(r - r')^2]. \tag{11}$$

Minimization leads to n linear equations:

$$\frac{\partial \varepsilon}{\partial a_i} = 0 \quad (i = 0, 1, \ldots, n), \tag{12}$$

which can be grouped into two independent sets of equations, one for the odd-degree, the other for the even-degree coefficients.[61] In the absence of current ($c = 0$), the linear, cubic and quintic fits are[31]:

$$r' = \sqrt{\frac{8}{\pi}}y, \quad r' = \sqrt{\frac{2}{\pi}}\left(y + \frac{1}{3}y^3\right), \quad r' = \sqrt{\frac{2}{\pi}}\left(\frac{3}{4}y + \frac{1}{2}y^3 - \frac{1}{60}y^5\right). \tag{13}$$

If current is present, the cubic[61] and quartic[2] fits have also been obtained. The PDF of r and r' can then be derived using:

$$p(r) = p(y)\,|dy/dr| \quad \text{and} \quad p(r') = p(y)\,|dy/dr'|\,, \tag{14}$$

where the PDF of y is

$$p(y) = \frac{1}{\sqrt{2\pi}} \exp\left(-\frac{y^2}{2}\right). \tag{15}$$

The quartic approximation of s (representing inundation drag) takes the following form:

$$s' = a_0 + a_1 x + a_2 xy + a_3 xy^2 + a_4 xy^3. \tag{16}$$

Like r, the even-degree coefficients are solved independently of the odd-degree by minimizing the mean square error, that is:

$$\begin{cases} a_0 + a_2 E[xy] + a_4 E[xy^3] = E[s] \\ a_0 E[xy] + a_2 E[x^2 y^2] + a_4 E[x^2 y^4] = E[xys] \\ a_0 E[xy^3] + a_2 E[x^2 y^4] + a_4 E[x^2 y^6] = E[xy^3 s] \end{cases} , \tag{17}$$

and

$$\begin{cases} a_1 E[x^2] + a_3 E[x^2 y^2] = E[xs] \\ a_1 E[x^2 y^2] + a_3 E[x^2 y^4] = E[xy^2 s] \end{cases} . \tag{18}$$

The expectations in Eqs. (17–18) are estimated based on the joint probability density function (JPDF) of x and y,[61] given by:

$$p(x, y) = \frac{\exp\{-(x^2 - 2\rho_{xy}xy + y^2)/2(1 - \rho_{xy}^2)\}}{2\pi\sqrt{1 - \rho_{xy}^2}} \tag{19}$$

where the correlation coefficient $\rho_{xy} = \rho_{\eta u}$ can be evaluated from the wave spectrum $S_{\eta\eta}(\omega)$.[61] For the case $c = 0$ and $\rho_{xy} = 1.0$, the quartic fit is

$$s' = -0.199 + 1.197xy + 0.2xy^3. \tag{20}$$

2.3. *Moment-based approximation (MBA)*

Bruce[26] obtained cubicization of Morison drag term

$$r' = a_0 + a_1 y + a_2 y^2 + a_3 y^3 \tag{21}$$

by directly equating the first four central moments of r to those of r' for the four polynomial coefficients:

$$\mu_i^r = \mu_i^{r'}, \quad (i = 1, 2, 3, 4). \tag{22}$$

The central moments μ_i are related to the moments m_i:

$$\begin{cases} \mu_1 = m_1 \\ \mu_2 = \sigma^2 = m_2 - m_1^2 \\ \mu_3 = m_3 - 3m_2 m_1 + 2m_1^3 \\ \mu_4 = m_4 - 4m_3 m_1 + 6m_2 m_1^2 - 3m_1^4 \end{cases} \tag{23}$$

in which moments of r and r' are estimated based on $p(y)$ in Eq. (15). The roots of the four coupled nonlinear equations in Eq. (23) need to be solved to obtain a_i ($i = 0, 1, 2, 3$). In the absence of current, a_0 and a_2 vanish because mean and skewness equalizations are naturally satisfied and only equalizations of second- and fourth-order moments (variance and kurtosis) need to be imposed.

Note that MBA approach is valid in general only for monotonic function.[29] Considering that inundation drag term s is non-monotonic, a hybrid method, taking the respective advantages of LSA and moment equalization, was proposed.[61] In quartic approximation of s in Eq. (16), a_1 and a_3 are obtained using Eq. (18) based

Table 1. Comparisons of drag term approximations; no current.

Drag term	Variance	Kurtosis
$r = y\|y\|$	3	$8\frac{2}{3}$
LSA linear	2.5465	0
LSA cubic	2.9709	13.5918
LSA quintic	2.9921	8.8049
MBA cubic	3.0000	8.6667

on LSA approach while a_0, a_2 and a_4 are computed by equating the mean, variance and kurtosis corresponding to the first, second and fourth equations in Eq. (23). In the absence of current, this hybrid method degenerates into moment equalizations only for a_0, a_2 and a_4.

The MBA method has been found to be superior to LSA for the same-degree polynomializations of both Morison and inundation drags, in terms of their (a) curve fit, (b) PDF, and (c) first four moments.[61] As shown in Table 1 for approximations of Morison drag without current, MBA cubicization appears even better than LSA quintic approximation. A frequency-domain analysis based on fifth-degree approximation seems impractical for mathematical manipulations and numerical computations.[4] Based on the kurtosis obtained, LSA cubicization appears unsuitable.

3. Volterra-Series Based Frequency-Domain Analysis

3.1. A third-order Volterra series model

The Volterra series is a functional extension of linear (first-order) systems to bilinear (second-order), trilinear (third-order), and higher-order systems.[36] These extensions require the multi-dimensional FRF, $H(f, g)$ and $H(f, g, h)$ respectively, to describe the bilinear and trilinear systems in place of the simple one dimensional FRF $H(f)$ for a linear system.

In the time-domain, the general third-order Volterra equivalent system can be described as[36,43]:

$$y(t) = y_1(t) + y_2(t) + y_3(t)$$

$$= \int h_1(\tau)x(t - \tau)d\tau$$

$$+ \iint h_2(\tau_1, \tau_2)x(t - \tau_1)x(t - \tau_2)d\tau_1 \, d\tau_2,$$

$$+ \iiint h_3(\tau_1, \tau_2, \tau_3)x(t - \tau_1)x(t - \tau_2)x(t - \tau_3)d\tau_1 \, d\tau_2 \, d\tau_3 \qquad (24)$$

where $x(t)$ is a Gaussian input; $y_i(t)$ ($i = 1, 2, 3$) are the first-, second- and third-order response components respectively; $h_1(\tau)$, $h_2(\tau_1, \tau_2)$ and $h_3(\tau_1, \tau_2, \tau_3)$ are the linear, bilinear and trilinear kernels of the impulse response functions respectively.

Each kernel and the corresponding FRF form a Fourier transform pair, where for an nth-order model:

$$\begin{cases} h_n(\tau_1, \ldots, \tau_n) = \int \cdots \int H_n(\omega_1, \ldots, \omega_n) e^{i(\omega_1\tau_1 + \cdots + \omega_n\tau_n)} d\omega_1 \cdots d\omega_n \\ H_n(\omega_1, \ldots, \omega_n) = \int \cdots \int h_n(\tau_1, \ldots, \tau_n) e^{-i(\omega_1\tau_1 + \cdots + \omega_n\tau_n)} d\tau_1 \cdots d\tau_n \end{cases}. \quad (25)$$

For Gaussian input, the odd-order response components will be mutually correlated, but uncorrelated with even-order system outputs. The even-order system outputs will also be mutually correlated. For non-Gaussian input data, all output components are usually correlated.

3.2. *Application to spectral analysis of wave force*

The total wave modal force Q acting on a platform leg can be decomposed into that described by Morison equation up to SWL, F, plus that due to inundation effects, P. These two components can be further split into components caused by inertia and drag, where the latter is polynomialized. Hence, Q can be written as[55]:

$$\begin{aligned} Q = F + P &\cong (I + D) + (PI + PD) \\ &= (I + D1 + D3) + (PI + PD0 + PD2 + PD4)' \end{aligned} \quad (26)$$

where I, $D1$ and $D3$ correspond respectively to the inertia, linear and cubic drag parts of the total Morison force F; PI denotes the inundation inertia term; and PD the inundation drag which has three parts. The details are given as follows:

$$F : \begin{cases} I = \int_{-d}^{0} \Phi(z) f_I(z,t) dz = \int_{-d}^{0} k_I \Phi(z) \dot{u}(z,t) dz \\ D1 = \int_{-d}^{0} \Phi(z) f_{D1}(z,t) dz = \int_{-d}^{0} k_D c_1(z) \Phi(z) u(z,t) dz, \\ D3 = \int_{-d}^{0} \Phi(z) f_{D3}(z,t) dz = \int_{-d}^{0} k_D c_3(z) \Phi(z) u^3(z,t) dz \end{cases} \quad (27)$$

$$PI = \Phi(0) k_I \dot{u}_0 \eta \quad \text{and} \quad PD : \begin{cases} PD0 = k_D \Phi(0) b_0 \\ PD2 = k_D \Phi(0) b_2 u_0^2 \eta, \\ PD4 = k_D \Phi(0) b_4 u_0^3 \eta \end{cases} \quad (28)$$

where $u_0 = u(0,t)$ and the coefficients c_1, c_3 and b_0, b_2, b_4 are associated with the polynomial coefficients discussed in Sec. 2. The relationship amongst wave elevation, water particle kinematics, wave forces and structural response is depicted in the finite-memory model of Fig. 2 which comprises one LTI system with memory each before and after the zero-memory nonlinear systems. This model consists of three phases: (I) linear transformations from the wave elevation to water particle kinematics, a single-input and multi-output system; (II) nonlinear transformations from kinematics to the modal force Q, a multi-input and single-output system,

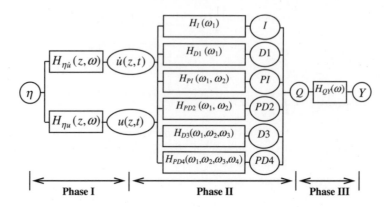

Fig. 2.   Fourth-order Volterra series input-output model.

where the input comprises six components (including two quadratic, one cubic and one quartic); (III) linear transformation from Q to modal displacement Y, a single-input and single-output system.

Such a model allows FRF of wave forces to be derived.[36] The FRF for I, $D1$ and $D3$ are respectively[39,55]:

$$H_I(\omega) = \int_{-d}^{0} k_I \Phi(z) H_{\eta\dot{u}}(z, \omega) dz, \tag{29}$$

$$H_{D1}(\omega) = \int_{d}^{0} k_D c_1(z) \Phi(z) H_{\eta u}(z, \omega) dz, \tag{30}$$

$$H_{D3}(\alpha, -\alpha, \omega) = \int_{-d}^{0} k_D c_3(z) \Phi(z) H_{D3}(z, \alpha, -\alpha, \omega) dz, \tag{31}$$

where $H_{D3}(z, \alpha, \beta, \gamma)$ is the product of linear transfer functions $H_{\eta u}(z, \omega)$ at three distinct frequencies, that is:

$$H_{D3}(z, \alpha, \beta, \gamma) = H_{\eta u}(z, \alpha) H_{\eta u}(z, \beta) H_{\eta u}(z, \gamma). \tag{32}$$

Based on linear wave theory:

$$H_{\eta u}(z, \omega) = \omega r(z), \tag{33}$$

$$H_{\eta\dot{u}}(z, \omega) = i\omega^2 r(z) = i\omega H_{\eta u}(z, \omega), \tag{34}$$

where $r(z) = \cosh k(z + d)/\sinh kd$. The FRF of asymmetric bivariate functions PI, $PD2$ and $PD4$ are obtained by an averaging method[36,55]:

$$H_{PI}(\omega_1, \omega_2) = \frac{1}{2}[H_{\eta\dot{u}_0}(\omega_1) + H_{\eta\dot{u}_0}(\omega_2)], \tag{35}$$

$$H_{PD2}(\omega_1, \omega_2) = \frac{1}{2}[H_{\eta u_0}(\omega_1) + H_{\eta u_0}(\omega_2)], \tag{36}$$

$$H_{PD4}(\omega_1, \omega_2, \omega_3, \omega_4) = \frac{1}{4}[H_{\eta u_0}(\omega_1)H_{\eta u_0}(\omega_2)H_{\eta u_0}(\omega_3)$$
$$+ H_{\eta u_0}(\omega_1)H_{\eta u_0}(\omega_2)H_{\eta u_0}(\omega_4)$$
$$+ H_{\eta u_0}(\omega_1)H_{\eta u_0}(\omega_3)H_{\eta u_0}(\omega_4)$$
$$+ H_{\eta u_0}(\omega_2)H_{\eta u_0}(\omega_3)H_{\eta u_0}(\omega_4)], \tag{37}$$

where $H_{\eta u_0}(\omega) = H_{\eta u}(0, \omega)$ and $H_{\eta \dot{u}_0}(\omega) = H_{\eta \dot{u}}(0, \omega)$.

Note that in Eqs. (27–28) F consists of only odd-degree terms of water particle kinematics, while P comprises only even-degree terms. Hence, the cross-spectrum of F and P vanishes and power-spectrum of Q is:

$$S_{QQ}(\omega) = S_{FF}(\omega) + S_{PP}(\omega). \tag{38}$$

3.2.1. *Power-spectrum of F*

According to Borgman,[1] the power spectrum of total Morison force F is:

$$S_{FF}(\omega) = S_{II}(\omega) + S_{DD}(\omega), \tag{39}$$

where the summation of the cross terms between the inertia force I and drag force D vanishes. Note the linear and cubic drag terms ($D1$ and $D3$) are correlated, which implies that:

$$S_{FF}(\omega) = S_{II}(\omega) + [S_{D1D1}(\omega) + S_{D1D3}(\omega) + S_{D3D1}(\omega) + S_{D3D3}(\omega)]. \tag{40}$$

Applying the third-order Volterra-series model, Eq. (40) is given by[39,55]:

$$S_{FF}(\omega) = |H_I(\omega)|^2 S_{\eta\eta}(\omega) + |H_{D1}(\omega)|^2 S_{\eta\eta}(\omega)$$
$$+ 6H_{D1}(\omega) \left(\int H_{D3}(\alpha, -\alpha, \omega)S_{\eta\eta}(\alpha)d\alpha \right) S_{\eta\eta}(\omega)$$
$$+ 9 \left| \int H_{D3}(\alpha, -\alpha, \omega)S_{\eta\eta}(\alpha)d\alpha \right|^2 S_{\eta\eta}(\omega)$$
$$+ 6 \int\int |H_{D3}(\alpha, \beta - \alpha, \omega - \beta)|^2 S_{\eta\eta}(\alpha)S_{\eta\eta}(\beta - \alpha)S_{\eta\eta}(\omega - \beta)d\alpha \, d\beta. \tag{41}$$

The third term on the RHS reflects the cross-effect between $D1$ and $D3$ while the last two are contributed solely by $D3$. For drag approximations based on LSA method, the preceding equation can be further reduced into a more compact form.[55]

3.2.2. *Power-spectrum of P*

It has been proven[42] that similar to F, the power-spectrum of the inundation P is the summation of the respective spectra of PI and PD. Considering the cross-correlation of $PD2$ and $PD4$, $S_{PP}(\omega)$ is written as:

$$S_{PP}(\omega) = S_{PI\,PI}(\omega) + S_{PD\,PD}(\omega)$$

$$= S_{PI\,PI}(\omega) + [S_{PD2\,PD2}(\omega) + S_{PD2\,PD4}(\omega)$$

$$+ S_{PD4\,PD2}(\omega) + S_{PD4\,PD4}(\omega)]. \tag{42}$$

The constant $PD0$ contributes to the spectrum at $\omega = 0$ only. Applying the fourth-order Volterra-series model involving only the second- and fourth-order components, it follows that[36,42,55]:

$$S_{PI\,PI}(\omega) = 2\,(k_I\Phi(0))^2 \int |H_{PI}(\alpha, \omega - \alpha)|^2\, S_{\eta\eta}(\omega - \alpha)S_{\eta\eta}(\alpha)d\alpha, \tag{43}$$

and the most complex term $S_{PD4\,PD4}(\omega)$ is:

$$S_{PD4\,PD4}(\omega) = (b_4 k_D \Phi(0))^2 \cdot [9\mu^4\sigma_{u_0}^4\delta(\omega) + 18\int |\sigma_{u_0}^2 H_{PD2}(\alpha, \omega - \alpha)$$

$$+ \mu^2 H_{\eta u_0}(\alpha)H_{\eta u_0}(\omega - \alpha)|^2 S_{\eta\eta}(\alpha)S_{\eta\eta}(\omega - \alpha)d\alpha$$

$$+ 24 \iiint |H_{PD4}(\alpha, \beta - \alpha, \gamma - \beta, \omega - \gamma)|^2$$

$$\times S_{\eta\eta}(\alpha)S_{\eta\eta}(\beta - \alpha)S_{\eta\eta}(\gamma - \beta)S_{\eta\eta}(\omega - \gamma)d\alpha\,d\beta\,d\gamma] \tag{44}$$

where $\mu^2 = E[\eta u_0] = \int H_{\eta u_0}(\omega)S_{\eta\eta}(\omega)d\omega$ and $\delta(\omega)$ is the Dirac-Delta function.

3.3. *Higher-order spectra analysis of F*

Li *et al.*[39] considered the effects of current and performed higher-order spectral estimations of response based on Volterra-series. The skewness and kurtosis were evaluated in terms of central moments, involving many multi-frequency spectra. For kurtosis evaluation, the contribution to the fourth-order central moment by the cubic drag consists of six terms each involving a six-fold integral, the second of which is illustrated here:

$$T_6^2 = 648 \int_{-\infty}^{\infty} \ldots \int H_Y(-\omega_1, \omega_1, -\omega_2)H_Y(\omega_2, -\omega_3, -\omega_4)H_Y(\omega_3, -\omega_5, \omega_5)$$

$$\times H_Y(\omega_4, -\omega_6, \omega_6)S(\omega_1)\cdots S(\omega_6)d\omega_1\ldots d\omega_6. \tag{45}$$

The numerical computation seems prohibitively excessive. If fourth- or higher-degree approximation of drag is assumed, both analytical derivation and numerical calculations become impractical.

4. Cumulant Spectral Analysis

4.1. *Input-output spectral relationship*

Figure 3 outlines the broad steps in the cumulant spectral analysis of a LTI system. The relationship of the nth-order $(n \geq 2)$ cumulant spectra of the stationary

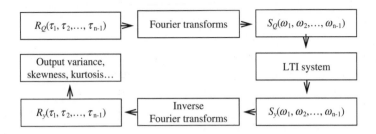

Fig. 3. Cumulant spectral analysis of a LTI system.

input Q and output y is given as[45,46]:

$$S_y(\omega_1, \omega_2, \ldots, \omega_{n-1}) = H_{Qy}(\omega_1) \cdot H_{Qy}(\omega_2) \cdots H_{Qy}(\omega_{n-1})$$

$$\times H_{Qy}^*(\omega_1 + \omega_2 + \cdots + \omega_{n-1}) S_Q(\omega_1, \omega_2, \ldots, \omega_{n-1}), \quad (46)$$

where $H_{Qy}(\omega)$ is the linear transfer function:

$$H_{Qy}(\omega) = \frac{1}{m(\omega_N^2 - \omega^2 + 2i\xi\omega_n\omega)} \quad (47)$$

for the governing equation:

$$m\ddot{y} + c\dot{y} + ky = Q. \quad (48)$$

In modal analysis of a fixed platform, m, c and k are structural modal mass, damping and stiffness, respectively. Here, the first vibration mode is considered for its largest contribution to the platform response, with the natural frequency $\omega_N = \sqrt{k/m}$. Obviously $S_y(\omega_1, \omega_2, \ldots, \omega_{n-1})$ depends on $S_Q(\omega_1, \omega_2, \ldots, \omega_{n-1})$, where the latter is available through the $(n-1)$-dimensional Fourier transform of the nth-order cumulant functions of Q[44]:

$$S_Q(\omega_1, \ldots, \omega_{n-1}) = \int \cdots \int_{-\infty}^{\infty} R_Q(\tau_1, \ldots, \tau_{n-1})$$

$$\times \exp\{-j(\omega_1\tau_1 + \cdots + \omega_{n-1}\tau_{n-1})\} d\tau_1 \cdots d\tau_{n-1}. \quad (49)$$

Similarly, if $S_y(\omega_1, \ldots, \omega_{n-1})$ is known, the nth-order cumulant functions of displacement can be estimated through the inverse Fourier transform:

$$R_y(\tau_1, \ldots, \tau_{n-1}) = \frac{1}{(2\pi)^{n-1}} \int\int \cdots \int_{-\infty}^{\infty} S_y(\omega_1, \ldots, \omega_{n-1})$$

$$\times \exp\{j(\omega_1\tau_1 + \cdots + \omega_{n-1}\tau_{n-1})\} d\omega_1 \cdots d\omega_{n-1}. \quad (50)$$

By choosing $n = 2$, 3 and 4, $S(\omega_1, \ldots, \omega_{n-1})$ represents the power-spectrum, bi-spectrum and tri-spectrum respectively and $R(\tau_1, \ldots, \tau_{n-1})$ represents the autocorrelation function, third- and fourth-order cumulant functions. Setting

$\tau_1 = \tau_2 = \tau_3 = 0$, the skewness and kurtosis are obtained in normalized form:

$$\kappa_3 = \frac{R(0,0)}{R^{3/2}(0)} \quad \kappa_4 = \frac{R(0,0,0)}{R^2(0)}. \tag{51}$$

$R(0)$ is equal to variance when the mean is zero. Another way to obtain $R(0, 0, \ldots, 0)$ is to directly integrate $S(\omega_1, \omega_2, \ldots, \omega_{n-1})$, that is,

$$R(0,0,\ldots,0) = \frac{1}{(2\pi)^{n-1}} \int_{-\infty}^{\infty} \cdots \int S_y(\omega_1, \omega_2, \ldots, \omega_{n-1}) \cdot d\omega_1 d\omega_2 \cdots d\omega_{n-1}, \tag{52}$$

which is more time-consuming than Fourier transform in Eq. (50). The critical issue here is the formulation of the input cumulant functions.

4.2. *Correlation functions of F & P*

The second-order cumulant function is the same as the correlation function. The autocorrelation function of F is derived from the double integral over the cross-correlations of the distributed Morison force[42,55]:

$$R_{FF}(\tau) = \int_{-d}^{0} \int_{-d}^{0} \Phi(z)\Phi(z')R_{ff}(z, z', \tau)dz\, dz', \tag{53}$$

where the cross-correlation of f is:

$$R_{ff}(z, z', \tau) = R_{f_I f_I}(z, z', \tau) + R_{f_D f_D}(z, z', \tau), \tag{54}$$

$$R_{f_I f_I}(z, z', \tau) = k_I(z)k_I(z')R_{\dot{u}\dot{u}}(z, z', \tau). \tag{55}$$

Based on Price theorem, $R_{f_D f_D}(z, z', \tau)$ is expressed as[1,55]:

$$R_{f_D f_D}(z, z', \tau) = k_D(z)k_D(z')R_{uu}(z, z', \tau)$$

$$\times \left[(a_1 + 3a_3)^2 \sigma_u(z)\sigma_u(z') + 6c_3(z)c_3(z')R_{uu}^2(z, z', \tau) \right]. \tag{56}$$

All cross-correlations of wave kinematics involved are obtainable from the corresponding cross-spectra:

$$\begin{cases} R_{uu}(z, z', \tau) = \dfrac{1}{2\pi} \int S_{uu}(z, z', \omega)e^{i\omega\tau}\, d\omega \\[2mm] R_{\dot{u}\dot{u}}(z, z', \tau) = \dfrac{1}{2\pi} \int S_{\dot{u}\dot{u}}(z, z', \omega)e^{i\omega\tau}\, d\omega \end{cases}, \tag{57}$$

where

$$\begin{cases} S_{uu}(z, z', \omega) = H_{\eta u}(z, \omega)H_{\eta u}(z', \omega)S_{\eta\eta}(\omega) \\[2mm] S_{\dot{u}\dot{u}}(z, z', \omega) = -H_{\eta\dot{u}}(z, \omega)H_{\eta\dot{u}}(z', \omega)S_{\eta\eta}(\omega) \end{cases}. \tag{58}$$

The autocorrelation function of the inundation load P is[55]:

$$R_{PP}(\tau) = R_{PI\,PI}(\tau) + R_{PD\,PD}(\tau), \tag{59}$$

where

$$R_{PI\,PI}(\tau) = \Phi^2(0)k_I^2 \left(R_{\eta\eta}(\tau)R_{\dot{u}_0\dot{u}_0}(\tau) - R_{\eta\dot{u}_0}^2(\tau) \right). \tag{60}$$

Utilizing Price's theorem,

$$
\begin{aligned}
R_{PD\,PD}(\tau) = \Phi^2(0)k_D^2 \Big[&\mu^4 \big(b_2 + 3b_4\sigma_{u_0}^2 \big)^2 \\
&+ \big(b_2 + 3b_4\sigma_{u_0}^2 \big)^2 \big(R_{\eta\eta}(\tau)R_{u_0 u_0}(\tau) + R_{\eta u_0}^2(\tau) \big) \\
&+ 12b_4\mu^2 \big(b_2 + 3b_4\sigma_{u_0}^2 \big) R_{\eta u_0}(\tau)R_{u_0 u_0}(\tau) \\
&+ 6b_4^2 R_{u_0 u_0}^2(\tau) \big(3\mu^4 + R_{\eta\eta}(\tau)R_{u_0 u_0}(\tau) + 3R_{\eta u_0}^2(\tau) \big) \Big].
\end{aligned}
\tag{61}
$$

The correlation functions of wave kinematics are provided by the following Fourier transform pairs:

$$
\begin{cases}
R_{\eta\eta}(\tau) \leftrightarrow S_{\eta\eta}(\omega) \\
R_{\eta\dot{u}_0}(\tau) \leftrightarrow S_{\eta\dot{u}_0}(\omega) = H_{\eta\dot{u}_0}(\omega)S_{\eta\eta}(\omega) \\
R_{\eta u_0}(\tau) \leftrightarrow S_{\eta u_0}(\omega) = H_{\eta u_0}(\omega)S_{\eta\eta}(\omega) \\
R_{u_0 u_0}(\tau) \leftrightarrow S_{u_0 u_0}(\omega) = |H_{\eta u_0}(\omega)|^2 \, S_{\eta\eta}(\omega) \\
R_{\dot{u}_0 \dot{u}_0}(\tau) \leftrightarrow S_{\dot{u}_0 \dot{u}_0}(\omega) = |H_{\eta\dot{u}_0}(\omega)|^2 \, S_{\eta\eta}(\omega)
\end{cases}
\tag{62}
$$

It follows that the power spectrum of Q is estimated using the following Fourier transform:

$$
S_{QQ}(\omega) = \int (R_{FF}(\tau) + R_{PP}(\tau))e^{-i\omega\tau}\,d\tau,
\tag{63}
$$

and the structural response spectrum is:

$$
S_{yy}(\omega) = |H(\omega)|^2 \, S_{QQ}(\omega).
\tag{64}
$$

4.3. *Fourth-order cumulant function of Q*

In the foregoing spectral analysis, F and P are uncorrelated. However, this is no longer true for higher-order spectral analysis where F and P will act together to form new joint moments. Considering that the drag terms (D and PD) are of stronger nonlinearities than the inertia terms (I and PI), and that the magnitudes of the latter are usually smaller than the former, it may be justified to only consider the correlation of dominant drag terms in the cumulant spectral analysis of Q. Such a simplification, in view of Eq. (26), leads to[44]:

$$
\begin{aligned}
R_Q(\tau_1, \ldots, \tau_n) &= R_F(\tau_1, \ldots, \tau_n) + R_P(\tau_1, \ldots, \tau_n) \\
&\cong R_Z(\tau_1, \ldots, \tau_n)
\end{aligned}
\qquad \text{for } n \geq 3,
\tag{65}
$$

where $Z = (D1 + D3) + (PD2 + PD4)$, being the assembly of linear, quadratic, cubic and quartic drag-related forces. Denoting $D = D1 + D3$ and $X = PD2 + PD4$, one can write

$$
R_F(\tau_1, \ldots, \tau_n) \cong R_D(\tau_1, \ldots, \tau_n),
\tag{66}
$$

and

$$
R_P(\tau_1, \ldots, \tau_n) \cong R_X(\tau_1, \ldots, \tau_n).
\tag{67}
$$

Rewrite $PD2$ and $PD4$ in Eq. (28) as:

$$PD2 = B_2 \eta u_0, \quad PD4 = B_4 \eta u_0^3, \tag{68}$$

where $B_2 = b_2 k_D \Phi(0)$ and $B_4 = b_4 k_D \Phi(0)$. The non-zero means are:

$$\overline{PD2} = B_2 \mu^2 \quad \overline{PD4} = 3B_4 \sigma_{u_0}^2 \mu^2. \tag{69}$$

The fourth-order cumulant function of Z can be expressed in terms of the following moment functions[44]:

$$
\begin{aligned}
R_Z(\tau_1, \tau_2, \tau_3) = {} & m_4^Z(\tau_1, \tau_2, \tau_3) - m_2^Z(\tau_1) m_2^Z(\tau_3 - \tau_2) - m_2^Z(\tau_2) m_2^Z(\tau_3 - \tau_1) \\
& - m_2^Z(\tau_3) m_2^Z(\tau_2 - \tau_1) - m_1^Z[m_3^Z(\tau_2 - \tau_1, \tau_3 - \tau_1) \\
& + m_3^Z(\tau_2, \tau_3) + m_3^Z(\tau_2, \tau_4) + m_3^Z(\tau_1, \tau_2)] + 2(m_1^Z)^2 \Big[m_2^Z(\tau_1) \\
& + m_2^Z(\tau_2) + m_2^Z(\tau_3) + m_2^Z(\tau_3 - \tau_1) + m_2^Z(\tau_3 - \tau_2) \\
& + m_2^Z(\tau_2 - \tau_1) \Big] - 6(m_1^Z)^4,
\end{aligned}
\tag{70}
$$

in which the mean is:

$$m_1^Z = \overline{PD2} + \overline{PD4}, \tag{71}$$

and the second-order moment function is:

$$m_2^Z(\tau) = R_D(\tau) + R_X(\tau), \tag{72}$$

where $R_D(\tau)$ is the double integral of $R_{f_D f_D}(z, z', \tau)$ in Eq. (56). $R_X(\tau)$ is exactly the correlation function shown in Eq. (61). The third-order moment function $m_3^Z(\tau_1, \tau_2)$ is for estimating the third-order cumulant function and skewness of Q. It is simpler than the fourth-order moment function which is given by:

$$
\begin{aligned}
m_4^Z(\tau_1, \tau_2, \tau_3) = {} & E\Big[(D(t) + X(t)) \, (D(t + \tau_1) + X(t + \tau_1)) \\
& \times (D(t + \tau_2) + X(t + \tau_2)) \, (D(t + \tau_3) + X(t + \tau_3)) \Big], \\
= {} & m_4^D(\tau_1, \tau_2, \tau_3) + m_4^X(\tau_1, \tau_2, \tau_3) + m_4^{D,X}(\tau_1, \tau_2, \tau_3)
\end{aligned}
\tag{73}
$$

where $E[\cdot]$ denotes expectation. The last term on the RHS, a cross moment function, reflects the correlation between drag terms D and X:

$$
\begin{aligned}
m_4^{D,X}(\tau_1, \tau_2, \tau_3) = {} & E\left[D(t)D(t + \tau_1)X(t + \tau_2)X(t + \tau_3) \right] \\
& + E\left[D(t)X(t + \tau_1)D(t + \tau_2)X(t + \tau_3) \right] \\
& + E\left[D(t)X(t + \tau_1)X(t + \tau_2)D(t + \tau_3) \right] \\
& + E\left[X(t)D(t + \tau_1)D(t + \tau_2)X(t + \tau_3) \right] \\
& + E\left[X(t)D(t + \tau_1)X(t + \tau_2)D(t + \tau_3) \right] \\
& + E\left[X(t)X(t + \tau_1)D(t + \tau_2)D(t + \tau_3) \right].
\end{aligned}
\tag{74}
$$

4.4. *Fourth-order moment function of D*

In numerical computations, a platform leg is discretized into N elements, with element lengths Δz_i $(i = 1, 2, \ldots, N)$. The linear and cubic modal Morison drags $D1$ and $D3$ are then approximated as:

$$\begin{cases} D1(t) \approx k_D \sum_{i=1}^{N} \Delta z_i c_1(z_i)\Phi(z_i)u(z_i, t) \\ D3(t) \approx k_D \sum_{i=1}^{N} \Delta z_i c_3(z_i)\Phi(z_i)u^3(z_i, t) \end{cases} \tag{75}$$

Let $A_i = k_D \Delta z_i c_3(z_i)\Phi(z_i)$ and $u_i(t) = u(z_i,t)$, $D3$ takes a simpler form:

$$D3 = \sum_{i=1}^{N} A_i u_i^3(t). \tag{76}$$

Note that $D1$ is a linear combination of Gaussian water particle velocities u at N distinct positions. Hence, $D1$ is also Gaussian. However, $D3$ is non-Gaussian; it is the combination of N trilinear finite-memory functions.[36] The fourth-order moment function of D is:

$$\begin{aligned} m_4^D(\tau_1, \tau_2, \tau_3) = E\Bigg[&\left\{ D1(t) + \sum_{i=1}^{N} A_i u_i^3(t) \right\} \left\{ D1(t + \tau_1) \right. \\ &+ \left. \sum_{j=1}^{N} A_j u_j^3(t + \tau_1) \right\} \left\{ D1(t + \tau_2) + \sum_{k=1}^{N} A_k u_k^3(t + \tau_2) \right\} \\ &\times \left\{ D1(t + \tau_3) + \sum_{l=1}^{N} A_l u_l^3(t + \tau_3) \right\} \Bigg]. \end{aligned} \tag{77}$$

This term contains joint moments of even-orders from fourth to twelfth. The fourth-order moment comes only from the linear process $D1$, that is:

$$E4 = E[D1(t)D1(t + \tau_1)D1(t + \tau_2)D1(t + \tau_3)]. \tag{78}$$

The twelfth-order moment is from the cubic drag terms with four distinct time shifts, given by:

$$E12 = \sum_{i=1}^{N}\sum_{j=1}^{N}\sum_{k=1}^{N}\sum_{l=1}^{N} A_i A_j A_k A_l E[u_i^3(t)u_j^3(t + \tau_1)u_k^3(t + \tau_2)u_l^3(t + \tau_3)]. \tag{79}$$

The sixth-, eighth- and tenth-order joint moments ($E6$, $E8$, $E10$) are from the cross effect between linear and the cubic drag terms.[55] For instance, $E6$ have four different forms and one of them is:

$$E6_A = \sum_{i=1}^{N} A_i E[D1(t)D1(t + \tau_1)D1(t + \tau_2)u_i^3(t + \tau_3)]. \tag{80}$$

The spatial correlation of the distributed Morison forces is represented by the summation along the vertical coordinate, from element $i = 1$ to N. Observe that the summation kernels of all joint moments involve only the Gaussian $D1(t)$ and $u_i(t)$. Thus, Price's theorem can be used to decompose these higher-order moments into the algebraic combinations of the following three correlation functions that are obtainable from the corresponding spectra, that is:

$$\begin{cases} R_{D1}(\tau) \leftrightarrow S_{D1}(\omega) = S_{\eta\eta}(\omega) |H_{D1}(\omega)|^2 \\ R_{D1u_i}(\tau) \leftrightarrow S_{D1u_i}(\omega) = H_{D1}(\omega)H_{\eta u}(z_i, \omega)S_{\eta\eta}(\omega) \\ S_{u_i u_j}(\omega) \leftrightarrow S_{u_i u_j}(\omega) = H_{\eta u}(z_i, \omega)H_{\eta u}(z_j, \omega)S_{\eta\eta}(\omega) \end{cases} \quad (81)$$

The number of combinations is as many as $11 \cdot 9 \cdot 7 \cdot 5 \cdot 3 \cdot 1 = 10\,395$ for the $E12$ kernel $E[u_i^3(t)u_j^3(t+\tau_1)u_k^3(t+\tau_2)u_l^3(t+\tau_3)]$, which makes the derivation manually tedious and prone to error. Fortunately, for every joint moment, only four distinct sets of Gaussian data are involved. Hence, there are many repetitive combinations which can be condensed. Two efficient algorithms[55] have been developed to perform the moment decomposition and condensation. It was found that $E12$ actually has only 47 basic non-repeating combinations; each combination is a product of six correlation functions. Also note that $E12$ is a four-fold summation along the height of cylinder of elements from $i, j, k, l = 1$ to N. Evidently, $E12$ is the most complicated joint moment and its numerical formulation contains all the necessary procedures required for $E4$, $E6$, $E8$ and $E10$.

In the actual formulation of $E12$, the four-fold summation due to the spatial correlation and the aforementioned 47 combinations are two critical bottlenecks. To facilitate computation, two schemes are introduced.[55] One is a "page-by-page" scheme: The 3D kernel (about τ_1, τ_2 and τ_3) is formed by performing the loop only about τ_3. For each loop, a 2D "page" is generated using matrix multiplications for the product of correlations with respect to τ_1 and τ_2. The symmetry of τ_3 allows nearly half of the corresponding number of loops to be saved. The other scheme takes advantage of four inherent symmetries associated with the permutations of integrating variables (i, j, k, l), such that the minimum number of loops can be realized for the four-fold summation of $E12$:

$$i = 1, 2, \ldots, N; \quad j = 1, 2, \ldots, i; \quad k = 1, 2, \ldots, j; \quad l = 1, 2, \ldots, k. \quad (82)$$

The reduction of the number of loops is therefore rather significant. However, $E12$ still accounts for over 90% of the total numerical effort in 4th-order cumulant analysis.

The above procedures are also applicable for formulating the fourth-order moment function of X and cross-moment function of X and D (Eq. (74)) that have joint moments of the order higher than 12, but less spatial correlations represented by summations, as detailed in Ref. 55.

5. Time-Domain Simulations

5.1. *Linear wave*

In time-domain simulations, the standard spectral representation method has been widely utilized in the dynamic analysis of offshore platforms. Given the wave spectrum $S_{\eta\eta}(\omega)$, the time history of the linear random wave elevation can be generated, that is:

$$\eta(x,t) = \sum_{i=1}^{M} \eta_i(t) \approx \sum_{i=1}^{M} a_i \cos(\phi_i), \qquad (83)$$

and the horizontal water particle kinematics:

$$\begin{cases} u(x,z,t) = \displaystyle\sum_{i=1}^{M} \omega_i \frac{\cosh k_i(d+z)}{\sinh k_i d} \eta_i(x,t) \\ \dot{u}(x,z,t) = \displaystyle\sum_{i=1}^{M} \omega_i^2 \frac{\cosh k_i(d+z)}{\sinh k_i d} [a_i \sin(\phi_i)] \end{cases}, \qquad (84)$$

where $\phi_i = k_i x - \omega_i t + \theta_i$ and the amplitude of frequency component is $a_i = \sqrt{2S_{\eta\eta}(\omega_i)\Delta\omega}$; the circular frequency and wave number are related by the dispersion function $\omega^2 = kg\tanh kd$. Each harmonic component $\eta_i(t)$ is assumed to be independent of the others, with the phase angle θ_i being randomly and uniformly distributed between 0 and 2π. For accurate representation, the number of frequencies M should be large.

Once the time histories of wave kinematics are available, Morison forces and inundation force can be obtained without polynomial approximations for the modal force $Q(t)$. The modal displacement can be computed using Newmark time-integration scheme.

For a LTI system driven by local Morison force, it has been suggested[4] that an ensemble of 120 or more sample functions be simulated to obtain a reasonably stable fourth-order response moment. Each simulation, as recommended in SNAME,[62] corresponds to a wave storm of 3 hours' duration. To investigate the effect of number of frequency discretizations on simulation results, Fig. 4 gives the variation of drag kurtosis, response kurtosis and response variance versus $\log_2 M$. The value of variance is normalized using the values computed with $M = 2^{14} = 16\,384$. It can be seen that the variance is stable even with $M = 2^7 = 128$. However, both drag and response kurtosis increase with M. The drag kurtosis approaches the exact value 8.6667 and the response kurtosis is almost stable when $16\,384$ harmonic components are employed. Therefore, the value of 8.6667 may act as a criterion to test the effectiveness of simulations: A reasonable response kurtosis relies on the kurtosis of Morison drag force that should be close to 8.6667. The large value of M indicates that simulation is time-consuming and can become impractical when modal force is treated, because Morison forces at various positions need to be simulated before being integrated. Fortunately, the computing efforts can be greatly reduced if FFT technique, instead of the direct summation in Eq. (84), is used.[63]

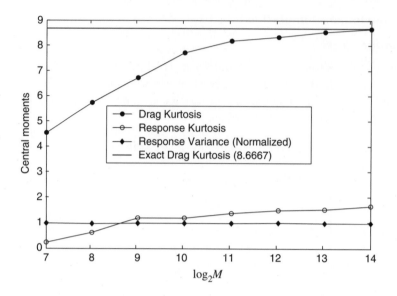

Fig. 4.   Number of frequency discretizations on simulated moments.

5.2. *Nonlinear wave*

Based on second-order Stokes theory, nonlinear random wave kinematics can be simulated using the method proposed by Hasselmann.[64] In deep waters, the second-order random water particle kinematics are:

$$
\begin{cases}
u_2(x, z, t) = -\displaystyle\sum_{n=1}^{M}\sum_{m=n}^{M} a_n a_m \omega_m (k_m - k_n) e^{(k_m - k_n)z} \cos(\phi_m - \phi_n) \\
\dot{u}_2(x, z, t) = -\displaystyle\sum_{n=1}^{M}\sum_{m=n}^{M} a_n a_m \omega_m (k_m - k_n) e^{(k_m - k_n)z} \sin(\phi_m - \phi_n)
\end{cases}, \tag{85}
$$

which can be reduced into single-fold summations to expedite computations. That is, for velocity:

$$
u_2(x, z, t) = -R\left[\sum_{n=1}^{M} (A_n B_n - k_n A_n C_n)\right], \tag{86}
$$

where the operator $R[\ ]$ signifies taking the real part in the bracket and:

$$
A_n = a_n e^{-k_n z} e^{-i\phi_n}, \quad B_n = \sum_{m=n}^{M} d_m k_m,
$$

$$
C_n = \sum_{m=n}^{M} d_m, \quad d_m = a_m \omega_m e^{k_m z} e^{-i\phi_m}. \tag{87}
$$

5.3. *Numerical example*

To illustrate the nonlinear effects of wave forces, an idealized jack-up platform with $\omega_N = 0.848\,\text{rad/s}$ and $\xi = 0.07$ is considered. The equivalent cylinder diameter D_{eq} is 1.97 m for calculating the water mass occupied by the leg and the inertia and drag coefficients are taken as 1.60 and 3.25. Wave conditions (Table 2), typically for North Sea condition, are considered with the JONSWAP wave spectrum.[62]

Figure 5 displays the response power spectrum. The frequency-domain results are based on LSA approximation of drag forces: (1) Linear approximation of Morison drag and quadratic approximation of inundation drag (i.e. second-order); and (2) cubic approximation of Morison drag and quartic approximation of inundation drag (i.e. fourth-order).

It can be seen that the superharmonic responses at $2\omega_p$ are significant and the fourth-order results agree well with that obtained from time domain simulation. Lower-order drag approximations led to underestimation of the superharmonic response.

Figure 6 presents the response spectrum, with and without inundation effects. Without inundation, the spectral amplitude at $2\omega_p$ is underestimated by about 50%.

Tables 3 and 4 show the mean, variance, skewness and kurtosis of the modal wave forces, F, P and Q, and the corresponding induced modal displacements. Results

Table 2. Wave conditions.

Parameters	
Water depth, d (m)	75
Peak wave frequency, ω_p (rad/s)	0.417
Peak enhancement factor, γ	3.3
Significant wave height, H_s (m)	12.9

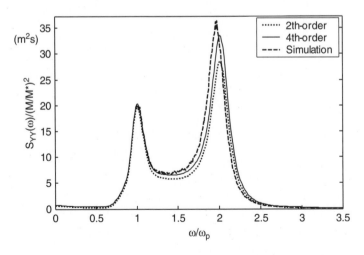

Fig. 5.   Power spectra of structural modal response.

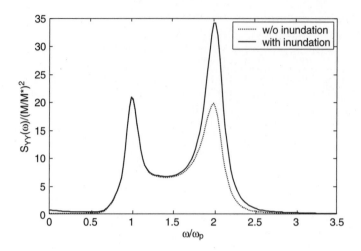

Fig. 6.   Inundation effects on response spectrum.

Table 3.   Cumulants of modal wave forces.

		F	P	Q
Mean (N)	TD	−1.2861e−02	5.1856e+01	5.1843e+01
	FD	0	5.1343e+01	5.1343e+01
Variance (N^2)	TD	1.1275e+05	1.4199e+04	1.2689e+05
	FD	1.1342e+05	1.3722e+04	1.2714e+05
Skewness	TD	−2.0636e−03	5.3046	2.4657
	FD	0	5.3278	2.2100
Kurtosis	TD	6.2697 (8.2635)	4.8619e+01	1.5137e+01
	FD	6.0990 (8.0199)	5.0111e+001	1.5961e+01

Table 4.   Cumulants of modal displacements.

		Y_F	Y_P	Y_Q
Mean (m)	TD	−8.3284e−05	3.1544e−01	3.1536e−01
	FD	0	3.1233e−01	3.1233e−01
Variance (m^2)	TD	1.5822e+01	3.7729	1.9579e+01
	FD	1.5690e+01	3.9749	1.9665e+01
Skewness	TD	−7.4168e−04	1.1049	1.5057e−01
	FD	0	1.0410	3.5072e−01
Kurtosis	TD	2.3868	1.4560e+01	5.0497
	FD	2.2750	1.3690e+01	5.2622

from frequency-domain (FD) cumulant spectral analysis and time-domain (TD) simulation are compared. Two and three dimensional FFT techniques are applied in bi- and tri-spectral analyses. The number of FFT points is 128 and the cylinder is discretized into 15 elements. Cubic approximation of the Morison drag force and quartic approximation of the inundation drag based on MBA approach are employed. Computational time for cumulant spectral analysis is about 25 times less than that for time history simulations.

For both wave forces and structural displacements, the cumulants estimated in the frequency-domain are close to the simulation results. The importance of including inundation is illustrated where the kurtosis of Y_Q is two times larger than that of Y_F. The kurtosis of total Morison drag (shown in parenthesis) is smaller than that of the local drag (8.6667), which can be attributed to spatial correlations. Due to the linear filtering effect, the kurtosis of modal displacement is smaller than that of modal force, indicating lesser deviation from Gaussianity. This can be also seen from the perceptible differences in the spectral surface variation exhibited in the tri-spectra (for $\omega_3 = 0$) of the modal force Q and modal displacement Y_Q as shown in Figs. 7 and 8.

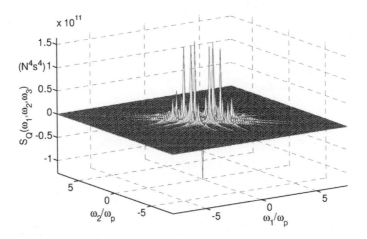

Fig. 7. Trispectrum of Q.

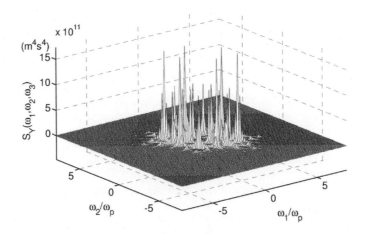

Fig. 8. Trispectrum of Y_Q.

Table 5.   Effects of wave nonlinearities.

	Mean	Variance	Skewness	Kurtosis
F	$-7.9515e-01$	$1.2457e+05$	-1.1082	8.3828 (11.4201)
Y_F	$-3.6895e-03$	$1.7611e+01$	$-1.0759e-01$	2.8049

The above results are based on the case of linear random waves. To investigate the effects incurred by wave nonlinearities, Table 5 lists the first four moments of the total Morison force F and induced modal displacement Y_F, based on second-order Stokes wave theory (Sec. 5.2).

Compared with the data in Tables 3 and 4, stronger non-Gaussianities are observed: (a) For both F and Y_F, the skewness no longer tends to be zero, implying the emergence of second-order superharmonic phenomena that is not existent in the linear random wave case (b) kurtosis values of F and drag D (shown in brackets) increase for around 30%; and kurtosis of Y_F increases for more than 15%. The increases in variance are also observable.

6. Concluding Remarks

It is shown in this chapter the complexity in performing stochastic analysis of fixed offshore structures. The frequency-domain treatment of two types of drag nonlinearities arising from wave loadings is presented. Volterra series and cumulant spectral analysis approaches for higher-order spectral analyses of Morison force, inundation force and the resultant total force are discussed. Least squares and moment-based polynomial approximations of drag nonlinearities are presented for the frequency-domain manipulations.

For the example illustrated, the response power spectra and cumulants obtained using cumulant spectral analysis are in good agreement with simulation results, but with less computational time. Inundation has the effect of increasing the response skewness and kurtosis. The need for including wave nonlinearities is confirmed by the numerical results.

References

1. L. E. Borgman, *ASCE J. Waterways and Harbors* **93**, 2 (1967) 129.
2. O. T. Gudmestad and J. J. Connor, *Appl. Ocean Res.* **5**, 4 (1983) 184.
3. A. Naess and C. S. S. Yim, *ASCE J. Eng. Mech.* **122**, 5 (1996) 442.
4. V. Bouyssy and R. Rackwitz, *ASME J. Offshore Mech. & Arctic Eng.* **119** (1997) 30.
5. M. A. Tognarelli, J. R. Zhao and A. Kareem, *ASCE J. Eng. Mech.* **123**, 5 (1997) 512.
6. C. Y. Liaw, *ASCE J. Eng. Mech.* **113**, 3 (1987) 366.
7. C. Y. Liaw, *ASME J. Offshore Mech. & Arctic Eng.* **111** (1989) 29.
8. I. Paik and J. M. Rosset, *Int. J. Offshore & Polar Eng.* **7**, 4 (1997) 301.
9. C. Y. Liaw, *Eng. Struc.* **10** (1988) 117.
10. A. G. L. Borthwick and D. M. Herbert, *J. Fluids & Struc.* **2** (1988) 479.
11. J. Lighthill, *Behavior of Offshore Struc.*, London **1** (1979) 1.

12. R. G. Tickell and J. R. Bishop, *4th Int. Offshore Mech. & Arctic Eng. Symp. (OMAE),* Dallas, Texas **1** (1985) 142.
13. M. Isaacson and J. Baldwin, *ASCE J. Waterways, Port, Coastal & Ocean Eng.* **116**, 2 (1990) 232.
14. J. S. Hu, M. Gupta and D. Zhao, *ASCE J. Eng. Mech.* **121**, 7 (1995) 819.
15. C. C. Tung, *ASCE J. Eng. Mech. Div.* **101**, 1 (1975) 1.
16. C. C. Tung, A*SCE J. Eng. Mech.* **121**, 2 (1995) 274.
17. Y. S. Li and A. Kareem, *ASCE J. Eng. Mech.* **119**, 1 (1993) 91.
18. C. Y. Liaw, T. Naik and C. G. Koh, *ASCE J. of Waterways, Port, Coastal & Ocean Eng.* **127**, 6 (2001) 319.
19. O. T. Gudmestad and G. A. Poumbouras, *Appl. Ocean Res.* **10**, 1 (1988) 43.
20. C. Y. Liaw and X. Y. Zheng, *Int. J. Offshore & Polar Eng.* **11**, 2 (2001) 87.
21. H. Kjeoy, N. G. Boe and T. Hysing, *The Jack-Up Drilling Platform Design, Construction & Operation,* Elsevier Appl. Sci., London **125** (1990).
22. X. Y. Zheng and C. Y. Liaw, *Int. J. Non-Linear Mech.* **39**, 9 (2004) 1519.
23. A. Kareem, C. C. Hsieh and M. A. Tognarelli, *ASCE J. Eng. Mech.* **124**, 6 (1998) 668.
24. W. J. Pierson and P. Holmes, *ASCE J. of Water. & Harbors* **91**, 4 (1965) 1.
25. R. Burrows, *Applied Math. Modeling* **7** (1983) 317.
26. R. L. Bruce, *Behavior of Offshore Structures,* Delft, The Netherlands, July 1–5, **899** (1985).
27. G. Najafian and R. Burrows, *Appl. Ocean Res.* **16** (1994) 205.
28. G. Najafian, R. Burrows and R. G. Tickell, *J. Fluids & Struc.* **9** (1995) 585.
29. S. R. Winterstein, *ASCE J. Eng. Mech.* **114**, 10 (1988) 1772.
30. S. O. Rice, In *Selected Papers on Noise and Stochastic Process,* Edited by N. Wax, NY (1954) 180.
31. L. E. Borgman, *ASCE J. Water & Harbors* **95**, 4 (1969) 557.
32. M. Shinozuka, *Computers and Structures* **2**, 6 (1972) 855.
33. S. R. Winterstein and R. Torhaug, *J. Offshore Mech. & Arctic Eng., Trans. of the ASME* **118**, 2 (1996) 103.
34. M. Schetzen, *The Volterra and Wiener Theories of Nonlinear Systems,* John Wiley & Sons, NY (1980).
35. W. J. Rugh, *Nonlinear System Theory — The Volterra/Wiener Approach,* The Johns Hopkins Univ. Press, Baltimore (1981).
36. J. S. Bendat, *Nonlinear System Analysis and Identification from Random Data,* John Wiley & Sons, New York (1998).
37. M. Olagnon, M. Prevosto and P. Joubert, *ASME J. Offshore Mech. & Arctic Eng.* **110** (1988) 278.
38. S. T. Quek, X. M. Li and C. G. Koh, *Appl. Ocean Res.* **16** (1994) 113.
39. X. M. Li, S. T. Quek and C. G. Koh, *ASCE J. Eng. Mech.* **121**, 10 (1995) 1056.
40. O. M. Boaghe, S. A. Billings and P. K. Stansby, *Appl. Ocean Res.* **20**, 4 (1998) 199.
41. C. Y. Liaw and X. Y. Zheng, *Proc. Fifth Int. Conf. on Stochastic Struc. Dynamics.* August 11–13, Hangzhou, China. Editors: W. Q. Zhu, G. Q. Cai and R. C. Zhang, CRC Press, Boca Raton, FL, USA (2003) 345.
42. X. Y. Zheng and C. Y. Liaw, Resp. spect. estim. for fixed offshore struc. with inund. effect included, *J. of Offshore Mech. & Arctic Eng.,* in press.
43. A. Kareem, J. Zhao, M. A. Tognarelli and K. R. Gurley, *Uncertainty Modeling in Vibr. & Fuzzy Anal. of Struc. Systems.,* Edited by B. M. Ayyub, A. Guran and A. Haldar, World Scientific Publishing Company, Singapore (1998) 101.
44. C. L. Nikias and A. P. Petropulu, *Higher-Order Spectra Analysis,* PTR Prentice Hall, NJ (1993).

45. D. R. Brillinger and M. Rosenblatt, In *Spectral Analysis of Time Series*, Edited by B. Harris, Wiley, NY (1967) 153.

46. D. R. Brillinger and M. Rosenblatt, In *Spectral Analysis of Time Series*, Edited by B. Harris, Wiley, NY (1967) 189.

47. L. E. Borgman, *Annals of Mathematical Statistics* **38**, 1 (1967) 37.

48. A. Papoulis, *Probability, Random Variables, and Stochastic Processes*, 3rd Edition, McGraw-Hill, New York (1991).

49. R. Price, *IRE Transactions on Inform. Theory*, Vol. IT-4 (1958) 69.

50. J. S. Hu, *ASCE J. Eng. Mech.* **116**, 1 (1990) 107.

51. J. S. Hu and L. D. Lutes, *ASCE J. Eng. Mech.* **113**, 2 (1987) 196.

52. G. I. Schueller and C. G. Bucher, *Stochastic Structural Dynamics: Progress in Theory and Applications*, Edited by S. T. Ariaratnam, G. I. Schueller and I. Elishakoff, Elsevier Appl. Sci. Publishers LTD, NY (1988) 219.

53. G. Q. Cai and Y. K. Lin, *ASCE Probabilistic Mech. & Struc. and Geotech. Reliability, Proc. of the Specialty Conf.* (1996) 732.

54. J. S. Hu and S. Dixit, *ASME 7th Int. Conf. on Offshore Mech. and Arctic Eng.*, Houston, Texas (1988) 109.

55. X. Y. Zheng, *Nonlinear Frequency-domain Anal. of Fixed Struc. Subjected to Morison-type Wave Forces*, PhD thesis, Nat'l Univ. of Singapore (2003).

56. J. R. Morison, M. P. O'Brien, J. W. Johnson and S. A. Schaaf, *J. of Petroleum Tech.* **189** (1950) 149.

57. P. D. Spanos and T. W. Chen, *J. Eng. Mech. Div.* **107**, 6 (1981) 1173.

58. R. S. Langley, *Appl. Ocean Res.* **6**, 3 (1984) 126.

59. B. J. Leira, *Appl. Ocean Res.* **9**, 3 (1987) 150.

60. J. Penzien, *Behavior of Offshore Struc.*, Trondheim, Norway **1** (1976).

61. C. Y. Liaw and X. Y. Zheng, *ASCE J. Eng. Mech.* **130**, 3 (2004) 294.

62. SNAME Recommended practice for site specific assessment of mobile jack-up units (1st edition). HSE, editor; *Soc. of Naval Arch. & Marine Eng.* (1994) 16.

63. A. Preumont, *Random Vibration & Spectral Analysis*, Kluwer Academic Publishers, London (1994).

64. K. Hasselmann, *J. Fluid Mech.* **12** (1962) 481.

CHAPTER 7

APPLICATION OF RELIABILITY METHODS TO FATIGUE ANALYSIS AND DESIGN

PAUL H. WIRSCHING

Department of Aerospace and Mechanical Engineering
The University of Arizona, Tucson, AZ, USA 85721
E-mail: phw@u.arizona.edu

Fatigue is generally considered the most important failure mode in mechanical and structure systems. Modern reliability methods are available for managing the relatively large uncertainties in fatigue design factors. This chapter provides a review of fatigue reliability methods.

1. The Fatigue Process

The physical process of fatigue is described in Fig. 1. Under the action of a long-term oscillatory stress, a fatigue crack initiates typically (but not always) on the surface at a point of stress concentration. In the second phase, the crack grows in a direction orthogonal to the stress. In the final phase, the crack reaches a point where the remaining material cannot support the quasi-static peak load.

Fatigue strength can be defined by an S-N curve as shown in Fig. 1. Fatigue data is characterized by relatively large scatter as measured by the coefficient of variation, C_N, which can range from 25% to 150%.

It has been reported that fatigue accounts for 80% to 90% of all observed service failures in mechanical and structural systems. Fatigue failures are often catastrophic; many have come without warning and have caused significant property damage as well as loss of life. And many fractures have been observed in applications where failures previously had not been encountered.

There are several reasons for the dominance of this failure mode and the difficulties in trying to design to avoid it. First, the process is inherently unpredictable as evidenced by the statistical scatter in laboratory data (see below). Fatigue is a very subtle process and it is very difficult to translate laboratory data of material behavior to field predictions. Environments (corrosion, thermal) play a very important role as does geometry and the nature of the loading (torsion, axial, bending). It is extremely difficult to accurately model the mechanical environments to which the system is exposed over its design lifetime. In summary, the fatigue problem involves very complicated interactions of several processes.

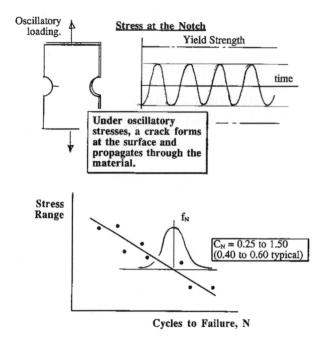

Fig. 1.   The physical process of fatigue.

2. Engineering Descriptions of Fatigue Strength

Texts on fatigue by Dowling,[1] Frost *et al.*,[2] Stephens *et al.*,[3] Almar-Ness,[4] and Gurney[5] are but a sample of definitive references on fatigue. These texts describe the three fundamental models of fatigue strength that are commonly used by engineers.

 (i) Characteristic S-N curve. Examples are provided in Figs. 1 and 2. Plotted on the ordinate is the stress level (typically amplitude or range of a constant amplitude stress process). Cycles to failure are plotted on the abscissa. This is the most commonly used model for design for high cycle fatigue (more than roughly 10^4 cycles)

 (ii) The general strain life model. This is an extension of the basic S-N model to include "low cycle" fatigue in which the material experiences cyclic plasticity. Total strain amplitude is plotted versus life. Life predictions are made using what is called "local strain analysis".

(iii) Fracture mechanic fatigue model. It is assumed that there is a pre-existing crack (or flaw) that grows under each load cycle. The crack grows until it reaches a critical crack length when fracture occurs. Fatigue strength is defined by a da/dN (crack growth rate) versus ΔK (stress intensity factor range) curve.

The focus of this chapter will be on the characteristic S-N approach. Reliability methods that are presented here for the S-N model apply as well to the fracture mechanics and strain-life models as well.

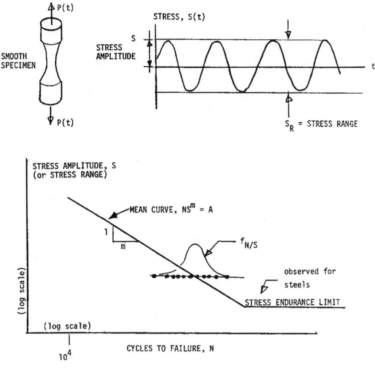

Fig. 2. A summary of fatigue testing.

The S-N curves representing fatigue strength can be defined in laboratory tests in which a smooth "dogbone" specimen is subjected to a constant amplitude stress S (amplitude or range) as shown in Fig. 2.

The testing machine runs until the specimen breaks, cycles to failure N are recorded. S is plotted as a function of N in log-log space as shown.

If the S-N data tend to follow a linear trend in log-log space, the following relationship is implied:

$$NS^m = A, \tag{1}$$

where m and A are empirical constants, called the fatigue strength exponent and the fatigue strength coefficient respectively. This model is very useful because of its simplicity. It provides a reasonable level of accuracy for strength of welded joints and those components for which life is dominated by the crack propagation phase.

3. Miner's Rule

Two examples of one common form of variable amplitude loading is shown in Fig. 3. These random processes would be typical of stresses in components of ground, flight, and marine vehicles. Wind, ocean wave, and earthquake environments produce

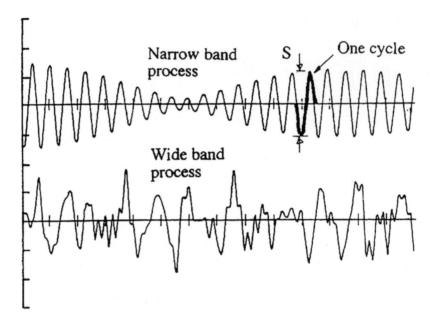

Fig. 3.   Examples of variable amplitude loading.

stresses of this general nature. Other variable amplitude processes such as vehicle loading on a bridge, loads on cranes, etc. would have a different signature.

Narrow band stress as shown is characterized by a quasi-constant period whereas several frequencies seem to participate in wide band stress.

The basic goal of analysis is to predict fatigue life under variable amplitude stress.

Almost all available fatigue date for design purposes is based on constant amplitude tests. In practice however, fatigue stresses are typically variable amplitude or random. The key issue is how to use the mountains of available constant amplitude data to predict fatigue in a component. While many models have been proposed for life prediction under variable amplitude loading, the original model, a linear damage rule, originally suggested by Palmgren[6] and developed by Miner[7] maintains its popularity because of its simplicity, and because no other model seems to consistently perform any better. As a result, Miner's rule is enshrined in almost every design code worldwide.

Miner's rule is illustrated in Fig. 4. A random process is decomposed into block loading. Consider the first block. At stress level, S_1, there are n_1 cycles applied. But the S-N curve indicates that it would take N_1 cycles to failure at S_1. Thus n_1/N_1 can be thought of as fractional damage and it follows that failure will occur when the sum of all of the fractional damages exceed one.

Miner's rule is popular because of its simplicity. Any realization of a long-term stress process can be decomposed into blocks of constant amplitude stress such as shown in Fig. 4. Subsequent application of Miner's rule is straightforward.

Block Loading to Simulate Random Process

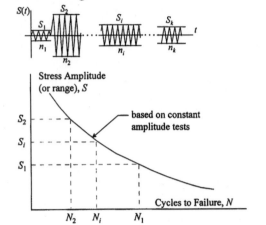

- Fractional damage at stress,

$$D_i = \frac{n_i}{N_i}$$

- Total damage is a sum of fractional damages

$$D = \sum_{i=1}^{k} \frac{n_i}{N_i}$$

- Failure $\longrightarrow D \geq 1.0$

Fig. 4.   Overview of Miner's rule.

In the special case where fatigue strength is defined by Eq. (1), it can be shown[8] that fatigue damage can be expressed as

$$D = \frac{n}{A} E(S^m), \tag{2}$$

where n is the number of cycles of application, D is the damage accumulated from n cycles, and S is a random variable denoting the stress range.

It is often convenient to express damage in terms of an equivalent constant Miner's stress, denoted as S_e where

$$D = \frac{n}{A} S_e^m, \tag{3}$$

and

$$S_e = \left[E(S^m) \right]^{1/m}. \tag{4}$$

For many applications it has been found that the Weibull distribution provides a good fit to the long term distribution of stress ranges. The distribution function of the Weibull is,

$$F_S(s) = 1 - \exp\left[-\left(\frac{s}{\delta}\right)^\xi \right] \quad \text{for } s > 0, \tag{5}$$

where ξ is the Weibull shape parameter and δ is the Weibull scale parameter. It has been found for both marine and aerospace application that ξ typically is in the range 0.70 to 1.5.

For the Weibull, the general expression for $E(S^m)$ is,

$$E(S^m) = \delta^m \Gamma\left(\frac{m}{\xi} + 1\right). \tag{6}$$

Damage D is then computed by substituting Eq. (6) into (2).

Computation of fatigue damage as described above is based on the assumption that the stress process is narrow band (see Fig. 3) and that all stress cycles are clearly identified. However for irregular stress histories, e.g. wide band stress, it is not so easy to define stress cycles. The rainflow algorithm [Dowling[1]] provides a method for counting stress cycles. Wirsching and Light[9] and Lutes and Larsen[10] have presented simplified methods, based on rainflow counting, for quantifying wide band damage.

4. Uncertainties in the Fatigue Process

The total uncertainty in the fatigue process has both physical (evidenced by scatter in data) and modeling uncertainty (analysis assumptions for models used) making major contributions. A brief summary of these uncertainties follows.

4.1. Strength modeling error; The quality of Miner's rule

Because Miner's linear damage rule has a very simple form, it should be expected to have significant modeling error. Damage, D, at failure is denoted as Δ. A brief summary of tests on the quality of Miner's rule, i.e. observed median and coefficient of variation (COV) of Δ is provided in Table 1.

As a default value, this author[13] has used Median(Δ) = 1.0, COV(Δ) = 0.30.

4.2. Strength uncertainty

Fatigue tests to define the S-N curve have are subject to considerable scatter. Typically welded joint data show a COV $\approx$ 0.50 on life, N, but pooled data can be as high as 1.50. For smooth specimen data, the COV of N at high stresses can be approximately 0.10 and for low stress something like 0.70 to 1.00. The COV of stress, S, given N is typically 0.10. It should be emphasized that these are only approximate values as the scatter is highly dependent upon the material and the type of test.

4.3. Stress uncertainty

In general the total environment to which a system is subjected over its intended service life will be a random process. This is physical uncertainty. Then there will be modeling error associated with the analysis process: (1) Translating environment to loads on the system, (2) performing the static/dynamic analysis typically with a

Table 1.   Example data on Δ.

	Median(Δ)	COV(Δ)
Miner,[7] the original study	0.95	0.26
SAE,[11] comprehensive fatigue study	1.09	0.90
Shin and Lukens,[12] extensive survey	0.90	0.67
Gurney,[5] survey of welded joint data	0.85	0.28

finite element computer code, and (3) computing the fatigue process at the fatigue sensitive point. While the uncertainty analysis should be done on a case by case basis, it is usually impractical to invest in a large experimental program.

In the first approximation, the simplest model is to use a blanket random variable, B, that includes all stress uncertainty. For example when Miner's stress is used, B is defined by the stress equation,

$$E[S^m] = BS_e, \tag{7}$$

where S_e would be defined as the best estimate of Miner's stress. Typical values of B are 0.10 to 0.15 for reasonably well defined loads in general mechanical systems, 0.20 to 0.25 for general civil structures subject to environmental loads.

5. Managing Uncertainty; Structural Reliability

Methods of structural reliability are used to manage uncertainties, i.e. probabilities of failure, or indices thereof, are estimated relative to the intended service life. Given the stress level, life N can be expressed as a function of random design factors, X. Given the intended service life, N_S, the probability of failure is,

$$p_f = P[N(X) \le N_S]. \tag{8}$$

In general this will be a very difficult calculation and numerical methods are used. There are several excellent references that discuss how to estimate probabilities of functions of random variables, e.g. Madsen, Krenk, and Lind,[14] Melchers,[15] Haldar and Mahadeven.[16] Methods of probability estimates are both analytical and experimental (i.e. Monte Carlo).

There are also several computer codes that automate the probabilistic analyses. For example, NESSUS (Southwest Research Institute) has the following methods, (1) the mean value method, (2) first order reliability method (FORM), (3) advanced first order reliability method (FPI), (4) second order reliability method (SORM), (5) Monte Carlo, (6) plane based adaptive importance sampling, (7) curvature based adaptive importance sampling, (8) radius based importance sampling, (9) advanced mean value method (AMV), (10) latin hypercube, and (11) response surface method. These include what is generally considered the most popular numerical approximation methods. Any or all of these techniques can be employed in the fatigue reliability problem and the selection depends in part upon the form of $N(X)$.

6. Example. The Lognormal Format

This is one example where direct probabilistic analysis can be performed. First define the limit state. From Eq. (3), at failure, $n = N$ cycles to failure, and $D = \Delta$, damage at failure. Introducing the variable B (see Eq. (7)) the expression for damage can be written in terms of N.

$$N = \frac{A\Delta}{B^m S^m}. \tag{9}$$

Note that this is an explicit function, i.e. it can be written in closed form. It is also a multiplicative function. And if it can be assumed that B, A, and Δ all have lognormal distributions, then N also will be lognormal. Invoking the properties of the lognormal distribution along with Eq. (8), a closed form expression for the probability of failure results. The exact form of the probability of failure is

$$p_f = \Phi(-\beta), \tag{10}$$

where Φ is the standard normal distribution function and β is the safety index defined for this limit state as

$$\beta = \frac{\ln(\tilde{N}/N_S)}{\sigma_{\ln N}}, \tag{11}$$

where the tilde denotes median values, and

$$\tilde{N} = \frac{\tilde{A}\tilde{\Delta}}{\tilde{B}^m S_e^m}, \tag{12}$$

$$\sigma_{\ln N} = \sqrt{\ln\left[\left(1 + C_\Delta^2\right)\left(1 + C_A^2\right)\left(1 + C_B^2\right)^{m^2}\right]}. \tag{13}$$

The C's denote coefficients of variation.

This author in Refs. 13, 17, and 18 has discussed the lognormal format applied to fatigue.

The lognormal format has the following features:

(i) It provides an exact and easy to use closed form expression for the safety index.
(ii) In general, lognormal models are known to provide a good fit to most design variables.
(iii) Extensive (unpublished) studies by the author have shown that the lognormal provides the best fit (relative to other widely used two-parameter models) for cycles to failure data in fatigue.

Example: The best estimate Miner's stress range in a butt welded joint is $S_e = 8.0$ ksi. Uncertainty in the stress range is quantified by the random variable B, assumed to be lognormal with median and COV.

$$\tilde{B} = 1.0, \quad C_B = 0.15.$$

The fatigue strength coefficient A is assumed to have a lognormal distribution with median and COV, (UK DEn/HSE, D-curve[19,20])

$$\tilde{A} = 1.21E10 \text{ (ksi units)}, \quad C_A = 0.51.$$

The slope of the S-N curve is $m = 3.0$.

Uncertainty in Miner's rule is modeled with the random variable, D, also assumed to have a lognormal distribution with median and COV,

$$\tilde{\Delta} = 1.0, \quad C_D = 0.30.$$

Goal of analysis. Estimate the *safety index* and the *probability of failure* for a service life of $N_S = 2 \times 10^6$ cycles.

Solution: First compute the median life by Eq. (12):

$$\tilde{N} = 2.36E7 \text{ cycles}$$

Then the standard deviation of $\ln N$ by Eq. (13):

$$\sigma_{\ln N} = 0.72$$

The safety index is computed from Eq. (11).

$$\beta = 3.43.$$

The estimated probability of failure is,

$$p_f = \Phi(-\beta) = 3E - 4.$$

7. Same Example ... Different Statistical Distributions

For the above example, an exact solution from probability theory was available. This was possible because the life was an explicit multiplicative function and all design factors had lognormal distributions. In the more general case where the life function is continuous, but not multiplicative or has all lognormal variates, a numerical solution is necessary. (See Sec. 5.)

Consider the above example with a change of distributions, Δ is normal, B is Type I EVD, and A is Weibull. While the median and COV is useful for the lognormal, the means μ_X and standard deviations σ_X are necessary for the other distributions. For the lognormal, the relationship between the parameters is given as

$$\mu_X = \tilde{X}\sqrt{1 + C_X^2} \quad C_X = \frac{\sigma_X}{\mu_X} \tag{14}$$

Using these forms, the parameters are given in Table 2.

The limit state function is,

$$\begin{aligned} g &= N - N_S \\ &= \frac{\Delta A}{B^m S_e^m} - N_S. \end{aligned} \tag{15}$$

Upon execution of Prorgram NESSUS (SORM option), the estimated safety index and failure probability is,

$$\beta = 2.61, \quad p_f = 0.0045.$$

Table 2. Definition of distributions.

	Distribution	Mean, μ_X	Standard deviation, σ_X
D	normal	1.04	0.313
B	Type I EVD	1.01	0.152
A	Weibull	1.36E10	6.92E9

8. Reliability Analysis when Life is an Implicit Function

An implicit function is loosely defined as a function in which the relationship between the response (dependent) variable and the independent variables does not have a simple closed-form expression. Such a function may be defined by a computer code (e.g. a local strain analysis code, a finite element code) in which life N is computed as a function of several variables, X_i, $i = 1, n$.

$$N(X) = N(X_1 \cdots X_n). \tag{16}$$

The problem is that the code has to be run whenever a function evaluation is required and a FORM/SORM might require up to 1000 function evaluations. Now when the limit is an explicit function (see above examples), this operation may take a very few seconds on a PC. But when a large computer code, taking significant clock time, is required for one function evaluation, FORM/SORM becomes impractical.

The advanced mean value (AMV) method developed by Wu *et al.*[21] is an efficient algorithm for performing reliability analysis, using a minimum number of function evaluations, when the limit state function is implicit.

Steps in the execution of the AMV method are:

 (i) Approximate $N(X)$ by a linear form,

$$N(X) = a_0 + \sum_{i=1}^{n} a_i(X_i - \mu_i), \tag{17}$$

where μ_i is the mean value of each of the n random design factors, X_i. Noting that the coefficients, a_i through a_n are just the partial derivatives of the variables, estimates can be made by function evaluations that perturb each variable individually about the mean (all other variables at their mean values) a small amount, i.e. $a_i \approx \Delta N/\Delta X_i$.

 (ii) A FORM or SORM is run using the linear form to approximate the cumulative distribution function (cdf) of life, N, at specific discrete points in the sample space of N. (The number of points is arbitrary, but four points on either side of the mean going to ± 3 or 4 sigma generally will provide a reasonable estimate of the cdf.) This gives a "first-order" estimate of the cdf of N. In general, this is not expected to be an "accurate" solution because the limit state is being approximated by a linear function.

(iii) A second-order improvement in the cdf is made by re-evaluating N using the design point from the SORM analysis at each of the discrete probability levels defined in step (ii).

It has been shown in unpublished studies that for a well behaved limit state a reasonably accurate cdf can be constructed with as few as 14 function evaluations. For the same level of accuracy, the response surface method requires about 80 function evaluations. Direct Monte Carlo may require well over 100 000 depending on the requirement on the extent of the tails.

9. Comments; A More General Case Where System Reliability is a Consideration and Maintenance is Performed

The fracture mechanics model is useful for structural health maintenance. Consider a member having a growing fatigue crack. The crack growth rate is random. Periodic inspection is performed. The member is repaired/replaced when a crack is discovered that is deemed to be "dangerous". But discovery is uncertain. Probability of detection curves are used to assess the chances of finding a crack of a given length. But the restored member now has a new random crack growth rate. Clearly the process is discontinuous. In order to estimate reliability over the service life, special techniques must be employed, but generally simulation is required. This is particularly the case when there is a structural system of components involved.

Wirsching and Mansour[22] provide a more detailed overview of the maintenance problem.

10. Fatigue Design Criteria

Fatigue design criteria can be based on a maximum allowable failure probability or minimum allowable safety index in which the designer is required to perform a probabilistic analysis. But reliability technology can be used to translate a maximum allowable failure probability into user-friendly safety factors. The following methods, all used by various agencies and companies, are demonstrated in the following using a very simple example.

The methods presented are: (1) The design stress approach, (2) target damage level, (3) factor of safety on life, and (4) partial safety factors.

The basic design requirement, stated in terms of both a target (minimum allowable) safety index and target (maximum allowable) probability of failure.

$$\beta_0 = 2.0, \quad p_0 = \Phi(-\beta) = 0.0228.$$

The intended service life (design life), $N_S = 2E6$ cycles.

The UK Department of Energy D-Curve is specified as the S-N curve to be used. Values of the fatigue strength exponent, m, and the fatigue strength coefficient, A, and design value A_0, are given in Table 3.

Table 3. Parameters for examples.

	Median	COV
Δ	1.0	0.30
B	1.0	0.15
A	1.21E10 (ksi)	0.51
A_0	4.64E9 (ksi)	
m	3.0	

These are the same parameters that were used in Example, Sec. 6. The design factors, Δ, B, and A are assumed to have lognormal distributions so that the lognormal format can be used.

10.1. *Fatigue design criteria; The design stress approach*

Using the lognormal format described in Sec. 6, the safety index can be written as

$$\beta = \frac{\ln(\tilde{N}/N_S)}{\sigma_{\ln N}} = \frac{\ln\left[\frac{\tilde{\Delta}\tilde{A}}{\tilde{B}^m N_S S_e^m}\right]}{\sigma_{\ln N}}. \tag{18}$$

For safe design $\beta > \beta_0 = 2.0$.
    Solve Eq. (18) for S_e:

$$S_e < \left[\frac{\tilde{\Delta}\tilde{A}}{\tilde{B}^m N_S \exp(\beta_0 \sigma_{\ln N})}\right]. \tag{19}$$

Here $\sigma_{\ln N} = 0.72$ (see Example, Sec. 6). Substituting the parameters into Eq. (19),

$$S_e < 11.3\,\text{ksi}.$$

The design stress approach is used by ABS in its Rules for Steel Vessels.[23]

10.2. *Target damage level*

This is a common approach used by designers. First compute damage at the end of the service life, N_S:

$$D = \frac{N_S S_e^m}{A_0}, \tag{20}$$

where A_0 is the design curve. This will be a curve on the safe (lower) side of the S-N data. For the HSE data, it is shifted to the left from the median curve, two standard deviations on a log basis.

Design is safe if $D \le \Delta_0$, and the goal of analysis is to derive Δ_0.
    Typical values in codes

$$\Delta_0 = 0.50 \text{ general}$$
$$= 0.10 \text{ serious failure consequences.}$$

Given: Parameters of the random design factors and β_0, the target safety index, a probability based Δ_0 can be derived:

$$\Delta_0 = \frac{\lambda\tilde{\Delta}}{\tilde{B}^m \exp(\beta_0 \sigma_{\ln N})}, \tag{21}$$

$$\lambda = \frac{\tilde{A}}{A_0}. \tag{22}$$

From the previous example,

$$\beta_0 = 2.0, \quad \sigma_{\ln N} = 0.72.$$

Also,

$$\lambda = \frac{\tilde{A}}{A_0} = \frac{1.21\text{E}10}{4.64\text{E}9} = 2.61.$$

Then, from Eq. (21),

$$\Delta_0 = 0.62.$$

10.3. *Fatigue design criteria; Factor of safety on life*

The safety check expression for life is

$$N > (FS_{life})N_S, \tag{23}$$

where N is the life that a designer would use, i.e. based on the design S-N curve.

$$N = \frac{A_0}{S_e}. \tag{24}$$

The factor of safety on life is just the inverse of Δ_0:

$$FS_{life} = \frac{1}{\Delta_0}. \tag{25}$$

Continuing the above example,

$$\Delta_0 = 0.62.$$

And,

$$FS_{life} = 1.61,$$

used by API RP2A and other marine structural design criteria.

10.4. *Partial safety factor format*

Reliability technology can be employed to derive partial safety factors (PSF) ... a factor applied to the nominal value of each random design factor. The fatigue limit state is defined in Eq. (15).

$$g = \frac{\Delta A}{B^m S_e^m} - N_S. \tag{26}$$

The three random design factors are, Δ, A, and B. A deterministic analog of the limit state that expresses a safe condition is

$$\frac{(\gamma_\Delta \Delta_n)(\gamma_A A_n)}{(\gamma_B B_n)^m S_e^m} \geq N_S, \tag{27}$$

where γ_Δ, γ_A, and γ_B are partial safety factors and Δ_n, A_n, and B_n are nominal (characteristic) values for Δ, A, and B respectively. The goal is to derive γ_Δ, γ_A, and γ_B given the target safety index, β_0. Thus, when the safety check expression of Eq. (27) is satisfied, a reliability exceeding β_0 is ensured.

The partial safety factor concept is described in Mansour, Wirsching *et al.* and Mansour, Wirsching *et al.* [2000]. Application to the fatigue limit state is described in the following.

The goal of analysis is to derive a safety check expression of the form

$$(\gamma_1 \tilde{\Delta})(\gamma_2 \tilde{A}) > (\gamma_B \tilde{B})^3 S_e^3 N_S. \tag{28}$$

The process of deriving partial safety factors has been automated by Program PSF (developed under contract with the Materials Properties Council). Here, after some reduction,

$$\gamma_1 = 0.789,$$
$$\gamma_2 = 0.524,$$
$$\gamma_3 = 1.20.$$

For convenience in design, the safety check expression should be written in terms of the design curve, A_0:

$$\frac{\tilde{A}}{A} = \frac{1.21\text{E}10}{4.64\text{E}9} = 2.61.$$

Upon substitution into Eq. (28), it follows that

$$(0.789 \, \tilde{\Delta})(1.367 A_0) > (1.202 \, \tilde{B})^3 S_e^3 N_S.$$

Thus the three partial safety factors are defined:

$$\gamma_\Delta = 0.789,$$
$$\gamma_A = 1.367,$$
$$\gamma_B = 1.202.$$

It has generally been assumed that

$$\tilde{\Delta} = \tilde{B} = 1.0.$$

Thus the safety check expression becomes

$$1.079 A_0 > (1.202 \, S_e)^3 N_S.$$

It is seen that the strength term is on the left side of the inequality; stress is on the right side. Taking this one step farther:

$$A_0 = S_e^3 (1.61 \, N_S).$$

The requirement reduces to a factor of safety on life of 1.61, in agreement with the results of the example of Sec. 10.3.

11. Concluding Remarks

(i) In general, why are reliability methods employed? Ans: (1) Probabilistic methods are convenient tools to describe physical phenomena too complex to treat with the current level of scientific knowledge, (2) probabilistic design procedure promise to improve the quality of engineered systems.

(ii) Fatigue design factors are subject to relatively very large uncertainties.

(iii) Reliability technology provides the analytical tools for managing the uncertainties, i.e. producing cost effective designs having structural integrity.

(iv) Reliability technology can be used to derive probability based safety factors for conventional design.

(v) A continuing problem ... quantifying stress and model uncertainty.

References

1. N. E. Dowling, *Mechanical Behavior of Materials*, 2nd edition, Prentice-Hall, NJ (1999).
2. N. E. Frost, K. J. Marsh and L. P. Pook, *Metal Fatigue*, Dover Publications (2000).
3. R. I. Stephens, A. Fatemi, R. R. Stephens and H. O. Fuchs, *Metal Fatigue in Engineering*, John Wiley and Sons, New York (2001).
4. A. Almar-Ness, *Fatigue, Handbook: Offshore steel Structures*, Tapir Publishers, Trondheim, Norway (1985).
5. T. R. Gurney, *Fatigue of Welded Structures*, 2nd edition, Cambridge University Press (1979).
6. A. Palmgren, in Zeitschrift des Vereines Deutscher Ingenieure **68**, 14 (1924) 339.
7. M. A. Miner, *Transactions of the ASME* **67** (1945) 159–164.
8. P. H. Wirsching, T. L. Paez and K. Ortiz, *Random Vibratons, Theory and Practice*, Wiley Interscience, New York (1995).
9. P. H. Wirsching and M. C. Light, *Journal of the Structural Division*, ASCE **106** (1980) 1593–1607.
10. L. Lutes and M. L. Larson, *Journal of the Structural Division*, ASCE **116** (1990) 1149–1164.
11. Society of Automotive Engineers, *Fatigue Under Complex Loading*, AE-6, SAE, Warrensville, PA (1977).
12. Y. S. Shin and R. W. Lukens, *Random Fatigue Life Prediction* **72**, ASME, New York (1983).
13. P. H. Wirsching, *Journal of the Structural Division*, ASCE **110**, 10 (1984).
14. H. O. Madsen, S. Krenk and N. C. Lind, *Methods of Structural Safety*, Prentice-Hall, NJ (1986).
15. R. E. Melchers, *Structural Reliability Analysis and Prediction*, John Wiley, New York (1999).
16. A. Haldar and S. Mahadeven, *Probability, Reliability, and Statistical Methods in Engineering Design*, John Wiley, New York (1999).
17. P. H. Wirsching, *Probabilistic Structural Mechanics Handbook*, Chapman and Hall (1995).
18. P. H. Wirsching, *Progress in Structural Engineering and Materials* **1**, 2 (1999) 200–206.
19. HSE, Fatigue Background Guidance Document, Health and Safety Executive Report OTH 92 390, BSI, London (1992).

20. HSE, Offshore Installations Guidance on Design, Construction and Certification, Health and Safety Executive, BSI, London (1995).

21. Y. T. Wu, H. R. Millwater and T. A. Cruse, *AIAA Journal* **28** (1990) 1663–1669.

22. P. H. Wirsching and A. E. Mansour, *Prevention of Fracture in Ship Structure*, The National Research Council, NTIS #PB97-141568 (1997).

23. ABS, Rules for Building and Classing Steel Vessels, American Bureau of Shipping, Houston, TX (2001).

CHAPTER 8

PROBABILISTIC MODELS FOR CORROSION IN STRUCTURAL RELIABILITY ASSESSMENT

ROBERT E. MELCHERS

Centre for Infrastructure Performance and Reliability
The University of Newcastle, Australia, 2308
E-mail: rob.melchers@newcastle.edu.au

The basic concepts in structural reliability analysis using probability theory are overviewed and the important effect of structural deterioration is noted. For representation of material loss due to corrosion and for pitting corrosion appropriately accurate models are required. These should account for the many variables involved in natural environments. Recently introduced probabilistic phenomenological models for marine immersion corrosion are then described both for general and for pitting corrosion. These show that the nature of the corrosion process changes from being controlled by oxygen diffusion and later by anaerobic bacterial action. This change has important implications for longer-term prediction of general corrosion loss and for maximum pit depth. A summary is given also of the effects of steel composition, water velocity, salinity, pollution and season of first immersion.

1. Introduction

In the design of new structures and the assessment of existing structures, long-term durability increasingly is becoming an issue of interest. This is the case both for offshore structures and shipping and for land-based structural systems. Because of the uncertainties involved it is appropriate to employ structural reliability methods. In addition to models for the applied loadings $\mathbf{Q}$ these require probabilistic models for the structural strength properties $\mathbf{R}(t)$, here shown as a function of time since typically structural strength and capacity tend to deteriorate with time. This must be considered in any reliability assessment. Thus models are required also to describe the deterioration that occurs as a function of time. Structural reliability assessment is also an important component of risk based inspection, maintenance planning and execution and in 'whole-of-life' project assessment.[1]

Although protective measures such as cathodic protection and surface coatings are almost invariably applied in practice, they are known to be not always wholly effective or sufficiently robust, particularly in the splash zone and at structural details. Evidence of this is widespread. As a result, for the design of new structures subject to corrosive conditions the usual approach is to make an allowance for future loss of material. For example, for ships a 10% plate thickness allowance for 'wastage' is typical in Classification Society rules for commercial vessels. Once this is reached in practice the wasted element must be replaced, usually at considerable cost. Evidently, wastage allowances do not provide for localized corrosion, for example, as might occur at welds or other locations of non-uniform materials.

Monitoring of corrosion loss can be through thickness measurements (e.g. using ultrasonics) or through indirect methods such as the measurement of 'corrosion potentials'. However, indirect methods provide only the potential for corrosion to occur but do not provide an indication of the actual rate of corrosion. Other methods, such as impedance measurements, are very difficult to use in field conditions and are of variable reliability. Moreover, they are expensive to use and require access to the region of interest. In addition, there is the problem of extrapolating from highly localized data points to larger structural elements (such as ships and off-shore structures).

None of these approaches are helpful in estimating the remaining safe life of an existing structure. Typically a much higher level of accuracy is required, both for predictions of structural safety and for prediction of likely future corrosion. This requires some level of mathematical modelling of structural corrosion. The topics of main interest to structural engineers are (i) the amount of strength (and stiffness) loss due to corrosion that is likely to occur with time and (ii) the likely time to penetration of steel plate or pipeline wall under given exposure conditions. Currently available corrosion handbooks for designers typically deal with corrosion 'rates', suggesting that corrosion occurs at a constant rate, even though it has been known for a long time from field observations that this is not usually the case. While the corrosion literature contains a wealth of information derived from laboratory observations, mainly electro-chemical, little of this is of direct use to structural engineers.[2] Until recently, the information available to structural engineers about marine corrosion tended to be anecdotal, not well organized and of limited in use to simple applications. It was also generally insufficient for reliability analyses purposes.[1] This situation is changing, as will be described in this chapter. In the process, greater understanding of the corrosion process is being obtained, and is in a form suitable for use by structural engineers. This is quite distinct from the information typically of interest to corrosion scientists.

Most of this chapter will deal with the representation of the corrosion of mild steel in seawater using models based on probability theory. The discussion will focus on immersion corrosion, both general and pitting. Probabilistic models for tidal, splash zone and atmospheric corrosion are currently under development and are expected to follow the same general approach, even though the number of environmental variables involved is much greater and understanding of the fine detail of the corrosion processes is still incomplete.

This chapter is not a complete summary of all aspects of marine immersion corrosion. It deals with the corrosion of small steel coupons as (reasonably good) surrogates for larger steel surface areas. Readers interested in a summary of the state of the art relevant to ships and offshore structures might consult Melchers.[3,4]

2. Factors in Marine Corrosion

Corrosion of metals requires the presence of moisture. For oxidation of a metal, oxygen must have access to the corroding surface. Inhibition or lack of either will reduce the rate of corrosion or stop it occurring. However, increasingly it has been recognized that under certain conditions corrosion may occur without the presence of oxygen. As will be described below, this involves anaerobic bacterial action.

The corrosion process involves electro-chemical reactions. This recognition has been very significant for the development of corrosion science and has had a profound influence on laboratory-based experimentation. However, owing to obvious practical difficulties it has been less helpful for *in situ* or field observations. The fundamentals of the electro-chemical theory of corrosion are available in a number of texts.[2,5] There are also compendia of practical marine corrosion observations.[6-8] As will become evident, these fundamental pieces of information are important for the proper development of probabilistic models for metal corrosion loss.

Figure 1 shows the forms in which corrosion can occur for structural steels.[3] So-called 'uniform' corrosion is the nominal loss of thickness of material as derived from weight loss measurements. No account is taken of limited pitting corrosion, even though it occurs to a greater or lesser degree in all marine corrosion. Hence 'uniform' corrosion is a convenient but not necessarily accurate description, although for mild and low alloy steels corrosion usually approximates a near-uniform loss. Uniform corrosion is of most interest for overall degradation of structural strength, as in plates and structural members.

Pitting corrosion is important for loss of containment, such as for pressure vessels and to a lesser extent shipping and offshore platforms. Crevice corrosion is relevant mainly to stainless steels at fittings, bolts, etc. Galvanic corrosion caused by differences in material properties around the heat affected zone of welds may be important for strength considerations of local details and for overall strength of stiffened plates.

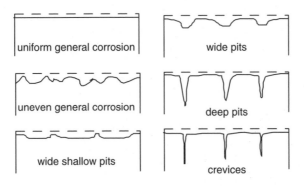

Fig. 1.   Corrosion types.

It has been shown that the material properties themselves, such as elastic moduli and yield strength are not influenced by the corrosion of adjacent material. Table 1 sets out the main types of corrosion, summarizes some characteristics and indicates where it may be an issue for ships, offshore structures and coastal structures. The different marine corrosion environments are shown in Table 2.

Since seawater is remarkably uniform from location to location[9] it might be expected that for 'at-sea' conditions there would be a general similarity between observations. However, a superficial comparison (Fig. 2) of data for general corrosion of mild steel coupons under immersion conditions obtained from all available sources to 1994 suggests that this is not so, and has been used in the past to argue that it is not possible to make sense of the various field data. Moreover, some have argued that only field-testing can determine local corrosion rates.

Table 1.   Main types of corrosion.[1,3]

Type of corrosion	Material	Feature	Effect
General (or 'uniform')	mild steels, high tensile steels	roughly uniform over extended areas	reduction in plate thickness, structural capacity
Pitting	limited for mild steels, mainly high tensile and stainless steels	highly localized penetration, often with clusters of pits	local reduction in thickness and stress intensification, possibly leakage
Crevice	mainly stainless and some high tensile steels	highly localized elongated penetration	localized stress intensifcation, local failure
Galvanic	dissimilar metals in contact	localized	localized severe material loss

Table 2.   Corrosion environments ($\mathbf{E}$ = vector of environmental and material parameters).[1]

Factor E_i	Importance
bacteria	very important, long-term
biomass	likely of low importance
oxygen supply	very important, shorter term
carbon dioxide	little importance
salinity	not important by itself
pH	important
carbonate solubility	not very important
pollutants	importance varies
temperature	very important
pressure	not important
suspended solids	not important
wave action	important
water velocity	important

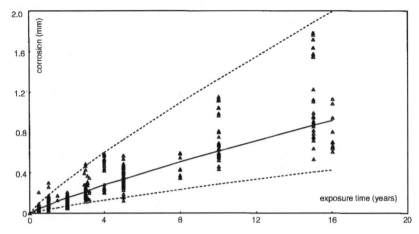

Fig. 2. Compilation of general immersion corrosion data from all available field data sources to 1994. The 5 and 95 percentile confidence bands are also shown.[3,10]

The data in Fig. 2 shows a large degree of scatter. This is due to (i) not all environmental and material variables having been properly isolated, (ii) a preponderance of short-term data, often from laboratory trials using artificial seawater and (iii) a lack of an appropriate theoretical model on which to base data comparisons. Although simplistic models[11,12] can be developed from Fig. 2, it is possible to do much better, as will now be described.

3. Probabilistic Model for Marine General Corrosion

3.1. *General form*

As seen in Fig. 2, actual field observations, being obtained under imperfect conditions, typically display variability. Since there are many factors (Table 2) that are unlikely to be known precisely, unlikely to be measured accurately or cannot be controlled, it is essential that a comprehensive model for corrosion loss as a function of time take an account of uncertainty. Such a model has been proposed earlier for immersion corrosion loss as a function of time t^{11-14}:

$$c(t, \mathbf{E}) = b(t, \mathbf{E}) \cdot fn(t, \mathbf{E}) + \varepsilon(t, \mathbf{E}), \tag{1}$$

where $c(t, \mathbf{E})$ is the weight loss of material, $fn(t, \mathbf{E})$ is a mean valued function, $b(t, \mathbf{E})$ is a bias function, $\varepsilon(t, \mathbf{E})$ is a zero mean error function (a probability 'bucket') and $\mathbf{E}$ is a vector of environmental (and material) parameters.

To use (1), models must be developed for each of its functions. The key element is the mean value function $fn(t, \mathbf{E})$. A model (Fig. 3) for it for 'at-sea' conditions has been developed[13] based on theoretical corrosion science concepts and suitable laboratory observations.[15]

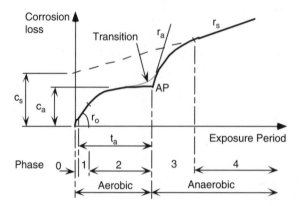

Fig. 3. Immersion corrosion loss as a function of exposure period, showing changing phases of corrosion process — schematic.[13]

3.2. *Mean value model*

The mean-value function $fn(t, \mathbf{E})$ has a number of sequential phases. Each corresponds with a (different) process controlling the (instantaneous) rate of corrosion. By restricting attention mainly to 'at-sea' conditions, the waters are on average fully aerated, have low velocity and low wave action, unpolluted and are of average ocean seawater salinity. The phases of the model may be summarized as follows.[13]

Phase 0

This phase is a very short time period during which corrosion is under 'activation' or kinetic control and is governed by the rate at which local chemical reactions can occur unhindered by external diffusion or transportation limitations.

When the steel surface is first exposed to seawater, it will be subject to invasion marine organisms. Also, its corrosion behavior will be governed by cathodic and anodic polarization, involving dissolution of the Fe ions and their diffusion away from the metal surface into the electrolyte according to

$$Fe \Rightarrow Fe^{++} + 2e^-, \tag{2}$$

and oxygen supply to the corroding surface and the uptake of electrons from (2) according to:

$$O_2 + 2H_2O + 4e^- \rightarrow 4OH^-. \tag{3}$$

The development of anodic and cathodic polarization will be a transient and very short-term process before equilibrium is established. Although either of (2) or (3) could control the corrosion process, it has been established that the diffusion of metal ions is not normally the limiting condition.[16] The consumption of oxygen at or near the corroding surface reduces the local oxygen concentration and leads to an oxygen flux towards the corroding surface. In practice this process often soon takes over as the process controlling the rate of the corrosion.

Phase 1

This phase is the period of 'oxygen concentration' control, in which the corrosion rate is governed by the rate of arrival of oxygen through the water and corrosion product layer to the corroding surface.[2] While the diffusion processes involved are not (quite) linear it is sufficient for practical application to model it as a linear function r_0.[13] As would be expected from theoretic considerations the initial corrosion rate r_0 was found to correlate well with Arrhenius equation.

The rate of corrosion r_0 under concentration control is proportional to the limiting discharge current i for oxygen diffusion (or 'corrosion current'), given approximately by[2]:

$$r_0 \propto i = \frac{nFDC}{\delta},$$ (4)

where δ is the thickness of the diffusion layer immediately adjacent to the corroding surface and which represents the region in which oxygen concentration draw-down occurs, n is the number of electrons taking part in the discharge of the oxygen molecules according to (3), F is the Faraday constant, D is the diffusion coefficient for oxygen and C is the concentration of oxygen in the bulk seawater (i.e. away from the corroding surface).

Laboratory observations have shown that the diffusion layer thickness δ is reduced by water velocity and turbulence. Hence i and thus the rate of corrosion, is increased. This has been verified for field conditions.[17] It is also likely to be the case for 'at-sea' conditions due to wave and current action. Due to turbulence and mixing, fully aerated conditions are usually assumed for 'at-sea' conditions.

In stagnant waters the water in the region near the corroding surface may become so depleted of oxygen that the transport of oxygen from the air-water interface and through the water to the corroding material becomes the limiting criterion. Typically this results in very low rates of corrosion.

Phase 2

The corrosion rate in this phase is controlled by the rate of oxygen diffusion through the increasing thickness of corrosion product to the corroding surface. It is therefore a non-linear function in time. The theoretical description[18] follows from the fact that the rate of oxygen supply depends on the rate at which oxygen can permeate through the corrosion products and any marine growth, modelled as:

$$\frac{dO_2}{dt} = \frac{k(C_0 - C_i)}{D_{cp}},$$ (5)

where O_2 is the transfer mass of oxygen, C_0 is the concentration of oxygen in the bulk seawater, C_i is the concentration of oxygen at the corrosion interface, D_{cp} is the diffusion coefficient for the corrosion products and k is a constant. If the reaction is controlled by the rate of arrival of oxygen, it will be consumed as soon as it arrives at the corrosion interface, thus C_i is zero.

Theory and field data show that there is a linear relationship between oxygen consumption and corrosion[2,7] and hence the corrosion rate should be directly proportional to C_0. This allows the mass-loss with time to be derived.[18] In the simplest case of a uniform density corrosion product, the amount of (general) corrosion $c(t)$ is given by:

$$c(t) = \sqrt{a \cdot t + b}, \tag{6}$$

where a and b are constants. In the (theoretically impossible) case of diffusion control from the very beginning of exposure, (6) becomes (with b small) the well-known 'square root' function:

$$c = d \cdot t^{0.5} + e, \tag{7}$$

where d and e are constants. More generally, allowance must be made for non-uniformity of corrosion product and for phase 1 behavior (see above). As a result the exponent will be less than 0.5. In addition, calibration is required to empirical data.[18]

With the build-up of the corrosion product layer with time, the rate of supply of oxygen to the corroding surface will decline and eventually widespread anaerobic conditions will set in between the corrosion product and the corroding metal. This is modeled as occurring at point AP in Fig. 3. It has been recognized for many years that long-term corrosion behavior departs from (7) and involves sulphate-reducing bacteria.[19,20] These are ubiquitous in nature.

Phase 3

In this phase the corrosion process is controlled by sulphate-reducing bacteria (SRB) under anaerobic conditions, assumed established more or less uniformly over the corroding surface at the end of phase 2 (i.e. at AP). In reality there will be some degree of non-homogeneity and the resulting transition is shown schematically in Fig. 3. Field observations for mild and low alloy steels[21] and microbiological considerations[19,20] indicate that during phase 3 there is a period rapid growth of sulphate reducing bacteria (SRB) with high corrosion. A mathematical model for this is currently being developed.

Phase 4

This phase is currently assumed to be linear in time, supported by (limited) long term observations for a variety of steels. These also suggest that the corrosion rate is largely independent of the actual activity levels of the SRB provided their activity remains sufficiently stimulated by replenishment of nutrients.[22]

3.3. Calibration of mean value model

As noted, detailed mathematical modelling for $fn(t, \mathbf{E})$ is at the present time limited to phases 1 and 2. However, this has not been necessary for model calibration using the parameters shown in Fig. 3.[13] Limiting to 'at-sea' conditions, allows most of the factors in Table 2 to be eliminated. The most important remaining factor is mean seawater temperature, shown earlier[23] to be a significant influence not only for structural steels but also for copper-nickel and aluminium.[24,25] Thus $fn(t, \mathbf{E})$ becomes $fn(t, T)$ and similarly for $b(t, \mathbf{E})$ and $\varepsilon(t, \mathbf{E})$. The functions fitted to the parameters t_q, c_a, r_0, r_a, c_s and r_s in Fig. 3 have been calibrated both for structural steels[13] and for copper-bearing steels[26] and are summarized as functions of average seawater temperature T in Table 3.[27]

For many practical engineering applications only relatively short-term corrosion is of interest since repair or maintenance action will normally be taken before corrosion becomes too severe. The initial corrosion rate r_0 as a function of average seawater temperature is shown in Fig. 4.

Table 3.   Calibrated parameters for $fn(t, T)$ as a function of T.

Parameter	Functional relationship	Correlation coefficient
t_a	$t_a = 6.61 \exp(-0.088T)$	$R = 0.99$
c_a	$c_a = 0.32 \exp(-0.038T)$	$R = 0.944$
r_0	$r_0 = 0.076 \exp(0.054T)$	$R = 0.963$
r_a	$r_a = 0.066 \exp(0.061T)$	$R = 0.97$
c_s	$c_s = 0.075 + 5678T^{-4}$	—
r_s	$r_s = 0.045 \exp(0.017T)$	$R = 0.71$

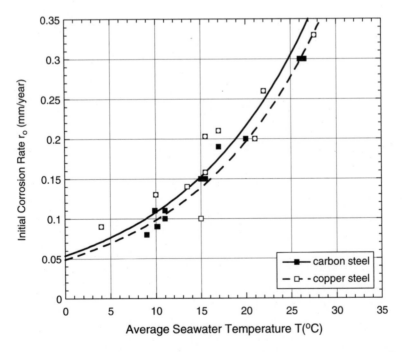

Fig. 4.   Initial corrosion rate r_0 as a function of average seawater temperature T.

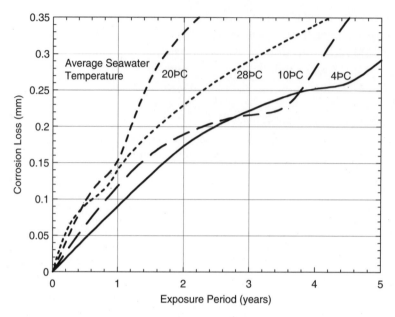

Fig. 5. Corrosion loss — exposure period plots for mild steel as a function of average seawater temperature.[26]

However, it can be quite misleading to use this rate for extended time periods since, as indicated in Fig. 3, the corrosion process changes with time and the changes are functions of seawater temperature. This is illustrated in Fig. 5. It shows that the total amount of corrosion loss departs considerably from linearity and in some cases produces a higher corrosion loss for lower average temperatures, a matter that has been viewed as caused primarily by hard marine growth in other observations.[28]

When there is a considerable annual variation in seawater temperature, the seasonal timing of the exposure of the steel to corrosion can have an influence on the form of the corrosion loss — time curve.[4,6] Importantly, in cases where variation is likely to be important standard practice is to time exposure to produce the worst effect — thus exposure in springtime is typical. The above relationships are based on this assumption. Where first exposure coincides with lower seawater temperature conditions for an extended period, it is a simple matter to make an appropriate allowance.

3.4. Bias and uncertainty functions $b(t, T)$ and $\varepsilon(t, T)$

The bias function $b(t, T)$ can be estimated by comparing the predictions of the mean-value model $fn(t, T)$ with the mean values of the corrosion loss $c(t)$ for given T obtained from field trials.[14] This showed that $b(t, T)$ is in the range 0.9–1.1 for the aerobic phases 1 and 2 (Fig. 6). At some sites serious excursions were found only in the anaerobic phases. Since the SRB are very sensitive to nutrient levels,

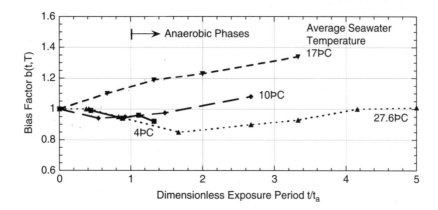

Fig. 6. Typical bias function $b(t, T)$ as a function of dimensionless exposure period.

the possibility of water pollution was investigated for a number of sites and this appears to correlate well (see below).

Since $\mathbf{E}$ is a multi-variate random vector, it would be expected that in the ideal case $\varepsilon(t, \mathbf{E})$ would be represented by a multi-dimensional joint probability density function, dependent on t. In practice there is insufficient information for such a complete description and it has been proposed to represent $\varepsilon(t, \mathbf{E})$ simply as

$$\varepsilon(t, \mathbf{E}) = \varepsilon(t, T) \approx N(0, \sigma_T), \tag{8}$$

where $N(0, \sigma)$ is the (multi-variate) standard normal probability density function. The standard deviation $\sigma_T(t, T)$ is then

$$\sigma_T^2 = \sum_{i=1}^{n} \sigma_i^2, \tag{9}$$

where σ_i is the standard deviation of variable i. Thus the total uncertainty σ_T is defined completely by (9) once the individual uncertainties σ_i are determined.

There are very few suitable data available to estimate $\varepsilon(t, T)$. Typically, only the mean of (usually only) two coupon weight-loss (i.e. corrosion loss) results are reported in the literature. Figure 7 shows some results including specially commissioned observations at Taylors Beach and Coffs Harbour, Australia.[29]

Figure 7 shows that the CoV for coupon variability is around 0.03–0.07 for all t provided $t/t_a < 1.5$. This range includes aerobic and early anaerobic conditions. Considerably higher values can occur in the later anaerobic region. For unpolluted waters the CoV appears reasonably constant with time. Also, variability will be a function of seawater temperature and the following preliminary function has been proposed[14]:

$$\sigma_c(t, T) = (0.006 + 0.0003\, T)(t/t_a), \qquad \text{for } t/t_a < 1.5. \tag{10}$$

Variability due to differences in metal composition and in site location have been described in the literature.[14]

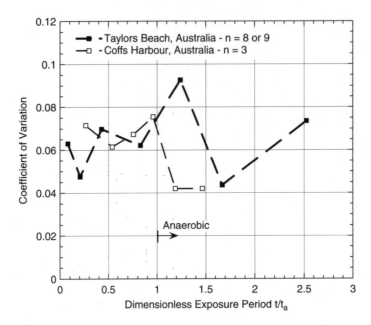

Fig. 7.   Point estimates of CoVs for coupon variability as a function of t/t_a for average seawater temperatures of about $20°C$.

3.5. *Example application*

As an example, consider the expected corrosion loss at $15°C$ average seawater temperature.[1] From Table 3 the corrosion model parameters are $t_a = 1.7$ years, $c_a = 0.18$ mm, $r_0 = 0.18$ mm/year, $r_a = 0.175$, $c_s = 0.175$ mm and $r_s = 0.06$ mm/year. These allow the mean-value function to be constructed. Also the curves representing one standard deviation may be constructed similarly as shown in Fig. 8. For simplicity the bias function was assumed unity throughout.[14]

4. Probabilistic Model for Marine Pitting Corrosion

4.1. *Background*

Pitting corrosion of steels used for structural purposes (i.e. mild and low alloy steels) may be important for applications involving containment systems such as a tanks, pipelines or (nuclear) waste containers and in this case pit growth as a function of time of exposure is primarily of interest. Laboratory observations show that some corrosion pits initiate, grow a limited amount and then stop, while others will grow rapidly in much the same environment.[30,31] Typically it is assumed that pits are hemispherical and grow in depth (and diameter) according to a power law relationship given by

$$x = at^b, \tag{11}$$

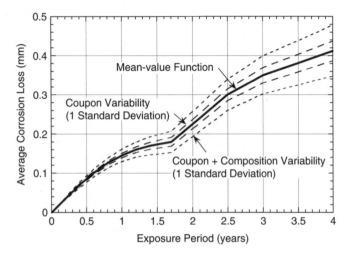

Fig. 8. Example mean-value corrosion-time curve and ± one standard deviation for coupon variability and for combined composition and coupon variability for 15°C average seawater temperature.

with b in the range 0.33–0.5, estimated by calibration to experimental observations.[32] This model, although entirely empirical and widely used to represent pit depth growth, usually can be calibrated only to relatively short-term test observations. Nevertheless, it has been used to estimate penetration times for long-term exposures, such as for nuclear containments.[33]

Detailed laboratory observations of pit growth show, however, that for structural steels pits may have much more complex shapes and in particular are often deeper than suggested by (11).[30] Moreover, it has been shown recently that field observations, particularly for longer exposure periods, do not correlate well with (11).[34] As a result, a new multi-phase phenomenological model for pit-depth growth with time has been proposed (Fig. 9). Evidently, it has the same general form as the model (Fig. 3) for general corrosion but with an allowance for very early pitting behavior.

The model for pit depth growth shows that after rapid pit depth growth in phase 0, pitting depth will increase relatively slowly under the predominantly aerobic conditions prevailing under phases 1 and 2. However, it will increase rapidly when widespread anaerobic conditions are established (Phase 3 of Fig. 9). The model is consistent with data obtained from a variety of sources when appropriate account was taken of average seawater temperature and when the waters are equivalent to normal unpolluted seawater. Figure 10 shows a typical application.[34]

For the following discussion, let the model in Fig. 9 be represented by the generic function $f(t, A, \mathbf{E})$, where t is the period of exposure, A is the exposed surface area and $\mathbf{E}$ represents a vector of factors characterizing the exposure environment, steel composition and surface finish factors.

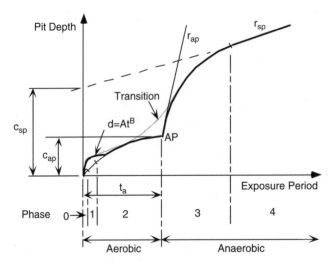

Fig. 9.   Multiphase phenomenological model for pitting corrosion loss as a function of exposure period (with parameterization).

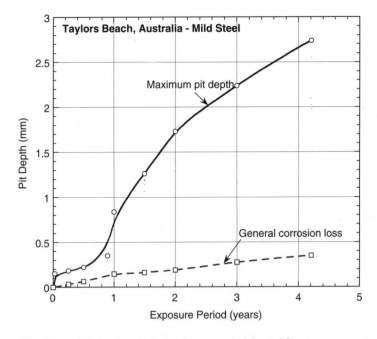

Fig. 10.   Pit depth as a function of exposure period for 20°C average seawaters.

4.2. *Probabilistic pit growth model*

Figure 9 may be considered a mean value function for pit depth. It is reasonable to suppose that, like most natural random phenomena, the depth distribution of all pits that will exist on a surface at any point in time can be described by the Normal distribution. [However, because of the practical difficulty of measuring all pits (or a sufficiently large sample) on a surface, neither the variance nor the distribution is usually available.] The exceedence probability for a given value of pit depth d_q can be given by:

$$P[d \leq d_q] = q, \tag{12}$$

where $d_q(t, A, \mathbf{E})$ represents the qth percentile of the pit depth, i.e. This means that there is a probability $1 - q$ that a pit having a depth greater than d_q will occur. The pit depth d_q for an exposed surface area A can then be given as

$$d_q(t, A, \mathbf{E}) = b(t, A, \mathbf{E})f(t, A, \mathbf{E})|_q. \tag{13}$$

Here $b(t, A, \mathbf{E})$ is a bias function which should be unity when $f(t, A, \mathbf{E})$ represents $d_q(t, A, \mathbf{E})$ exactly under all conditions. The term $f(t, A, \mathbf{E})|_q$ indicates that $f(t, A, \mathbf{E})$ is evaluated at the qth percentile.

For practical applications in which only the maximum pit depth is of interest, extreme value theory indicates that for probability distributions with exponential tails (such as the normal distribution) the maximum can be described by an extreme value distribution, such as the Gumbel distribution.[35] This has been used for nearly 50 years in pitting corrosion studies.[36,37] Of particular interest is the statistical distribution of maximum pit depth as the pits evolve over time. An important assumption underlying this approach is that the extreme pit depths are all statistically independent events.

Figure 11 shows measurements of maximum pit depth for coupons recovered from Taylors Beach on the Eastern Australian seaboard plotted on the Gumbel

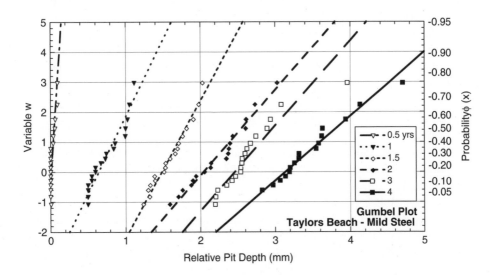

Fig. 11.   Maximum (relative) pit depths plotted on Gumbel paper.

paper. Evidently, the data could be interpreted as consistent with a straight line, indicating consistency with the Gumbel distribution and allowing, in principle, extrapolation to longer exposure times and estimation of the probability of occurrence of greater pit depth.[38] This is the most common approach and, as noted, the one that has been used for estimating, for example, the probability of nuclear containment perforation over hundreds of years.[33]

4.3. *Dependence and homogeneity for maximum pit depths*

A different view of the data in Fig. 11 is that it more closely can be modeled by a bi-modal distribution consisting of an extreme value distribution for the smaller extreme-depth pits and approximately a (narrow) Normal distribution for the deeper extreme-depth pits.[32] The proportion between these partial distributions moves away from the exponential with time. Moreover, this bi-modal distribution is similar to the distribution for the complete population of pit depths. For example, the change in distribution can be seen clearly in Fig. 11 for the pits for exposure times of up to about one year, which, for Taylors Beach, corresponds to the time when phase 3 becomes active.

Systems reliability theory indicates that similarity between the probability distribution for the extreme pit depths and that for the underlying pit depth distribution can exist only if the extreme pit depths are highly correlated.[39,40] In this case, it follows that the probability distribution for extreme pit depth should also be normal. This conclusion is a major departure from the accepted approach to representing the probability distribution for extreme pit depths, but interestingly, both the data recent data for pitting of mild steel and much literature data support it.[39] Moreover, it has led to a re-assessment of the basis for the common assumption of independence between maximum depth pits, considering that in marine immersion conditions:

(i) the metal surfaces of the coupons may be modeled as essentially spatially homogeneous (since the coupons typically are cut from the same sheet of metal),
(ii) the metal surfaces are all subject to a very similar local (i.e. immersion) environment, and
(iii) the physical, electro-chemical and thermodynamic laws are highly likely to govern all pitting behaviour similarly for similar metals and environmental conditions.

These points indicate that the growth pattern for extreme pits should be very similar and that therefore the extreme pit depths are likely to be highly dependent.[39] Moreover, it has also been suggested that the distribution of all pits is not homogeneous, as is required for analyses based on conventional extreme value theory.[35,41] The reason for the non-homogeneity arises from the well-established observation that pit growth may be divided into two types — 'metastable' pitting, involving delayed pit initiation or intermittent pit growth and 'stable' pitting, which some pits evolve into eventually. Pitting behavior characterized as 'superstable' pitting has been advanced to account for pits that will initiate immediately on first immersion and then continue to grow in depth in the stable-pitting mode, that is, without interruption. Only these types of pits, having maximal stable and minimal meta-stable pit growth behaviour, could become the extreme depth pits — the others simply could not 'catch-up'.[40] These remarks are confined to pit depths within phases 0–2 of Fig. 11.

4.4. *Comparison*

Figure 12 shows the different interpretation of pit data using a normal distribution for the larger pit depths in the aerobic range (phases 0–2). It is evident from comparison to Fig. 11 that there are large differences in exceedence probabilities for pits in the aerobic phases between the two probability models.

At the present time the statistical behavior for pitting under anaerobic conditions has not been investigated in detail since the mechanisms involved in pit growth and pit extreme behavior under anaerobic conditions are still not sufficiently well-understood. Accordingly, the probability distribution for these more advanced pit depths currently is still best be represented by the Gumbel distribution (Fig. 11). However, in view of the above observations, it is appropriate to consider just the deeper extreme valued pits rather than the complete extreme value population. This technique is a recognized approach when changes in extreme value behavior are known.[41]

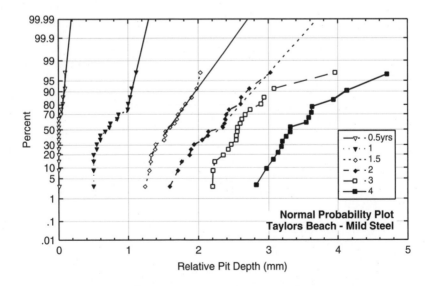

Fig. 12.   Pit data for larger pits represented using a normal distribution.

5. Other Factors

For completeness some brief observations will now be made about the effects of steel composition, water velocity, salinity, pollution and the season of first immersion season on general corrosion and on pitting corrosion. These factors have not so far been incorporated in the above probabilistic models. Most effort has been focussed on developing the necessary mean value functions and probabilistic models have yet to be developed.

5.1. *Steel composition*

The chemical composition of some steels makes them more resistant to immersion corrosion than others. Until recently efforts to use correlation studies with empirical field data produced inconsistent and sometimes conflicting results.[42] Much of this is eliminated by invoking Figs. 3 or 9 together with very careful reassessment of the data sources and their supplementation with additional environmental data.[43,44] Figure 13 summarizes schematically the effect of particular alloys relative to mild steel for general corrosion. A similar effect has been observed for pitting corrosion. At this time the data is insufficient to report percentage changes.

Fundamental corrosion science considerations indicate that alloys should have little effect on the initial corrosion rate r_0 since in phases 1 and 2 the rate of corrosion is controlled by the rate of oxygen diffusion.[2,5] This means that conventional high strength steels corrode at about the same rate as mild structural grade steels, rendering the use of high strength steels problematic, such as in their use for ship hulls.

The effect of alloys may be different in different phases of the corrosion model. A good example is the effect of chromium as an alloy. For short-term corrosion it is beneficial, but in the longer term it can be very detrimental owing to its proneness to attack by SRB. Importantly, the timing involved is temperature dependent, which

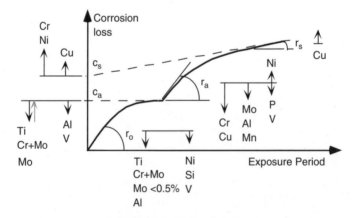

Fig. 13. The nominal effect of different alloys on the parameters of the model of Fig. 3. The notation for the alloying elements follows conventional chemistry notation. The length of the arrows indicates, schematically, the contribution of the alloying effect to each of the parameters.

explains, for example, why only beneficial effects of chromium were observed in the 7-year North Sea (10°C) corrosion trials.[45] Whilst those in the tropical Panama Canal Zone (27.6°C) showed detrimental effects already after about 4 years.[7,21]

5.2. *Water velocity*

The calibration of the model of Fig. 1 employed the field data for which there typically was some wave action and sometimes low water currents. Under really quiescent conditions the early corrosion rate is reduced, consistent with a lower rate of oxygen supply to the coupons.[46] Conversely, it is well-known that higher water velocities increase the corrosion rate.[2,5] A widely quoted relationship between corrosion rate and velocity is due to LaQue.[47] Unfortunately, it was obtained in a short term (36 day) laboratory experiment using a closed recirculating seawater system, casting doubt on its veracity since such a system would not provide proper aeration and would inhibit the influence of marine growth and bacteria in the corrosion processes. A rather different relationship was obtained in recent field trials.[48] Figure 14 shows that after a relatively short period of time (<2 months) the effect of (tidal) water velocity effectively is to shift the corrosion — time curve upwards. When this is reworked into average corrosion rates, the results are very different from those of LaQue.[47]

5.3. *Saline and brackish waters*

It is often thought that waters of lower salinity are less corrosive. However, laboratory observations nearly 100 years ago already showed that chloride concentration

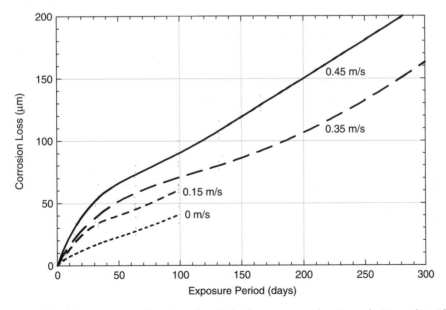

Fig. 14. Corrosion loss as a function of period of exposure and water velocity under tidal conditions.

itself has little direct effect on corrosion rates when the rate of corrosion is controlled by the rate of diffusion of oxygen to the corroding surface. More recent studies add confirmation even under mild water velocity conditions.[15]

The corrosivity of water is closely tied to the calcium and magnesium carbonate balance (hardness) in the seawater and to pH.[5] This is because calcium and magnesium carbonate deposits form part of the corrosion product layer and thus regulate the supply of oxygen to the corrosion interface. Increased pH will increase calcite deposition owing to the changed solubility,[5] providing, in the longer term, greater protection against aerobic corrosion (phases 1–2) through limiting oxygen diffusion.

The pH of seawater is remarkably uniform across the oceans and typically has a daily fluctuation in the range 8.0–8.2. In coastal regions dilution by river-flows or stormwater runoff changes (usually lowers) the pH and hence offers less corrosion product protection against corrosion. For example, clean, neutral freshwater has a pH of 7.0 while laboratory grade triple distilled tap water is in the range 5.5–6.0. In some cases the diluting waters may increase the pH owing to their hardness, lowering net corrosivity.

These remarks are relevant for the aerobic phases 0–2. There is some laboratory evidence[49] that the calcium carbonate content of water may influence SRB activity and hence corrosion in phases 3 and 4.

5.4. *Water pollution*

Water pollution of coastal waters, harbors and river estuaries is often associated with excessive corrosion; however the reasons for this to occur are seldom reported in sufficient detail for correlation with corrosion studies. Although the chemical composition of the seawater can influence corrosion behavior,[5] in practice the concentrations required for this would be so high that the waters normally would be unacceptable on environmental and other grounds.

Probably the most common pollution effect is a reduction in the dissolved oxygen (DO) level of the water. Often this is due to the biological oxygen demand (BOD) of the waters discharged to the seawater. This leads to a (perhaps localized) lowering of the rate of oxygen supply to the corroding surface and in turn should lower t_a (see Fig. 3), the amount depending nonlinearly on the level of dissolved oxygen.[17] A lower t_a will produce an earlier onset of phase 3 and, because of the usually high corrosion rate at the beginning of phase 3, much increased overall levels of corrosion.

Water pollution may be associated also with nutrient discharges such as from untreated or poorly treated sewage or storm-water run-off or, increasingly more common, nutrient-loaded run-off from poor fertilization practices on agricultural areas draining to the sea. It has been found in recent tests that the effect is mainly on the anaerobic phases 3 and 4 of the model of Fig. 3, typically with much higher levels of corrosion in phase 3. Earlier corrosion phases are not significantly affected.[50]

At the present time there is insufficient quantitative information to prescribe mathematically the relationship between the level of nutrient pollution and corrosion.

5.5. *Season of first immersion*

The season (and hence the water temperature) of first immersion can have an effect on early corrosion.[4,6] Typically steel exposed at the beginning of the winter months and during the winter shows a period of relatively low level of corrosion. The rate increases during spring and summer as the waters increase in temperature. The models of Figs. 3 and 9 are based on data for spring immersions.

6. Conclusion

The present chapter has indicated that good quality probabilistic models can be developed for representing corrosion loss and for pitting corrosion. Such models are necessary in structural reliability analysis such as might be applied to structural systems, offshore platforms, pipelines and shipping. Herein only models for immersion corrosion were discussed. Similar models are currently under development for tidal and for coastal atmospheric corrosion situations although for these applications the number of environmental variables involved is very much greater and this adds considerable complexity and difficulty to model development.

Simple linear models as implied by the 'corrosion rate' can be very misleading and are unable to untangle the effects of metal composition and velocity, for example. It is necessary to recognize that there are changes in the mechanisms controlling the (instantaneous) rate of corrosion as corrosion products build-up. Importantly, longer-term corrosion in seawater is not controlled by oxidation but is controlled by bacterial activity. This recognition allows also the effects of pollution to be considered in a rational manner.

The models described herein bridge the gap between field testing and laboratory-based corrosion science. Exponents of the former have seldom been able to develop models with adequate predictive power, while the latter has been focussed only on short-term behavior with little validity for predicting longer-term effects.

Acknowledgments

The financial support of the Australian Research Council is gratefully acknowledged.

References

1. R. E. Melchers, A probabilistic model for marine corrosion of steel for structural reliability assessment, *Journal of Structural Engineering, ASCE* **129**, 11 (2003) 1484–1493.
2. D. Jones, *Principles and Prevention of Corrosion*, Second Edition, Prentice-Hall, Upper Saddle River, NJ (1996).

3. R. E. Melchers and M. Ahammed, Nonlinear modelling of corrosion of steel in marine environments, Research Report 106.09.1994, Department of Civil Engineering and Surveying, The University of Newcastle, Australia (1994).
4. R. E. Melchers, Probabilistic models for corrosion in structural reliability assessment — Part 1: Empirical models, *Journal of Offshore Mechanics and Arctic Engineering* **125**, 4 (2003) 264–271. Part 2: Models based on mechanics, *Journal of Offshore Mechanics and Arctic Engineering* **125**, 4 (2003) 272–280.
5. H. H. Uhlig, *Corrosion and Corrosion Control*, John Wiley & Sons (1963).
6. F. L. LaQue, *Marine Corrosion*, Wiley, New York (1975).
7. M. Schumacher, Editor, *Seawater Corrosion Handbook*, Noyes Data Corporation, NJ (1979).
8. G. Kreysa and R. Eckermann, *Dechema Corrosion Handbook*, Vol. 11, VCH Publishers, New York (1992).
9. F. P. Ijsseling, Guidelines for corrosion testing for marine applications, *British Corrosion Journal* **24**, 1 (1989) 55–78.
10. R. E. Melchers, Probabilistic modelling of marine corrosion of steel specimens, *Conference Proceedings, ISOPE-95*, The Hague, The Netherlands, June 12–15 (1995) 205–210.
11. R. E. Melchers, Modeling of marine corrosion of steel specimens, *Corrosion Testing in Natural Waters*, R. M. Kain and W. T. Young, Editors, ASTM STP 1300, Philadelphia (1997) 20–33.
12. R. E. Melchers, Corrosion uncertainty modelling for steel structures, *Journal of Constructional Steel Research* **52**, 1 (1999) 3–20.
13. R. E. Melchers, Modeling of marine immersion corrosion for mild and low alloy steels — Part 1: Phenomenological model, *Corrosion (NACE)* **59**, 4 (2003) 319–334.
14. R. E. Melchers, Modeling of marine immersion corrosion for mild and low alloy steels — Part 2: Uncertainty estimation, *Corrosion (NACE)* **59**, 4 (2003) 335–344.
15. A. D. Mercer and E. A. Lumbard, Corrosion of mild steel in water, *British Corrosion Journal* **30**, 1 (1995) 43–55.
16. J. L. Crolet, Mechanisms of uniform corrosion under corrosion deposits, *J. Materials Science* **28** (1993) 2589–2606.
17. R. E. Melchers, *Early Corrosion of Mild Steel in Seawater*, submitted for publication.
18. R. E. Melchers, Mathematical modelling of the diffusion controlled phase in marine immersion corrosion of mild steel, *Corrosion Science* **45**, 5 (2003) 923–940.
19. B. J. Little, R. I. Ray and R. K. Pope, Relationship between corrosion and the biological sulfur cycle: A review, *Corrosion (NACE)* **56**, 4 (2000) 433–443.
20. J.-L. Crolet, From biology and corrosion to biocorrosion, *Microbialm Corrosion*, C. A. C. Sequeira and A. K. Tiller, Editors, *Proceedings of the 2nd EFC Workshop*, Portugal, 1991, The Institute of Metals, pp. 50–60.
21. C. R. Southwell, J. D. Bultman and C. W. Hummer, Estimating service life of steel in Seawater, M. Schumacher, Editors, Seawater Corrosion Handbook, Noyes Data Corporation, NJ (1979) 374–387.
22. R. G. J. Edyvean, The effects of microbiologically generated hydrogen sulfide in marine corrosion, *MTS Journal* **24**, 3 (1990) 5–9.
23. R. E. Melchers, Effect of temperature on the marine immersion corrosion of carbon steels, *Corrosion (NACE)* **58**, 9 (2002) 768–782.
24. R. E. Melchers, Temperature effect on seawater immersion corrosion of 90:10 copper-nickel alloy, *Corrosion (NACE)* **57**, 4 (2001) 440–451.
25. R. E. Melchers, Influence of temperature on seawater immersion corrosion of Aluminium (UNS A95086), *British Corrosion Journal* **36**, 3 (2001) 201–204.

26. R. E. Melchers, Modelling of marine immersion corrosion for copper-bearing steels, *Corrosion Science* **45**, 10 (2003) 2307–2323.

27. R. E. Melchers, A new model for marine immersion sorrosion in structural reliability, *Applications of Statistics and Probability in Civil Engineering*, A. Der Kiureghian, S. Madanat and J. M. Pestana, Editors, Millpress, Rotterdam (2003) 599–604.

28. H. Y. Sun, S. D. Ma, B. R. Hou, J. L. Zhang, G. Q. Huang and X. R. Zhu, Mathematical analysis of long term fouling corrosion in seawater, *Corrosion Engineering Science and Technology* **38**, 3 (2003) 223–227.

29. R. Jeffrey and R. E. Melchers, *Corrosion Tests of Mild Steel in Temperate Seawater*, Research Report, Department of Civil, Survey and Environment Engineering, The University of Newcastle, Australia (2001).

30. G. Butler, P. Stretton and J. G. Beynon, Initiation and growth of pits on high-purity iron and its alloys with chromium and copper in neutral chloride solutions, *British Corrosion Journal* **7**, 7 (1972) 168–173.

31. G. Wranglen, Pitting and sulphide inclusions in steel, *Corrosion Science* **14** (1974) 331–349.

32. Z. Szklarska-Smialowska, *Pitting Corrosion of Metals*, Houston, NACE (1986).

33. G. P. Marsh, A. H. Harker and K. J. Taylor, Corrosion of carbon steel nuclear waste containers in marine sediment, *Corrosion (NACE)* **45**, 7 (1989) 579–589.

34. R. E. Melchers, Pitting corrosion of mild steel in marine immersion environment — 1: Maximum pit depth, *Corrosion (NACE)* **60**, 9 (2004).

35. J. Galambos, *The Asymptotic Theory of Extreme Order Statistics*, Second edition, Krieger, Malabar, FL (1987).

36. P. M. Aziz, Application of the statistical theory of extreme values to the analysis of maximum pit depth data for aluminum, *Corrosion (NACE)* **12**, 10 (1956) 495t–506t.

37. J. W. Provan and E. S. Rodriguez, Development of a Markov description of pitting corrosion, *Corrosion (NACE)* **45**, 3 (1989) 178–192.

38. R. E. Melchers, Pitting corrosion of mild steel in marine immersion environment — 2: Variability of maximum pit depth, *Corrosion (NACE)* **60**, 9 (2004b).

39. R. E. Melchers, *Statistical Characterization of Pitting Corrosion — 1: Data analysis*, Research Report, Centre for Infrastructure Performance and Reliability, The University of Newcastle, Australia (2004).

40. R. E. Melchers, *Statistical Characterization of Pitting Corrosion — 2: Probabilistic Modelling for Maximum Pit Depth*, Research Report, Centre for Infrastructure Performance and Reliability, The University of Newcastle, Australia (2004).

41. E. Castillo, Extremes in engineering applications, *Extreme Value Theory and Applications*, J. Galambos, J. Lechner, E. Simiu and N. Macri, Editors, Kluwer Academic Publishers, Dordecht, (1994) 15–42.

42. W. A. Schultze and C. J. van der Wekken, Influence of alloying elements on the marine corrosion of low alloy steels determined by statistical analysis of published literature data, *British Corrosion Journal* **11**, 1 (1976) 18–24.

43. R. E. Melchers, Effect on marine immersion corrosion of carbon content of low alloy steels, *Corrosion Science* **45**, 11 (2003) 2609–2625.

44. R. E. Melchers, Effect of small compositional changes on marine immersion corrosion of low alloy steel, *Corrosion Science* **46**, 7 (2004) 1669–1691.

45. F. Blekkenhorst, G. M. Ferrari, C. J. van der Wekken and F. P. Ijsseling, Development of high strength low alloy steels for marine applications, Part 1: Results of long term exposure tests on commercially available and experimental steels, *British Corrosion Journal* **21**, 3 (1986) 163–176.

46. M. H. Peterson and T. J. Lennox, The effects of exposure conditions on the corrosion of mild steel, copper, and zinc in seawater, *Materials Performance* **23**, 3 (1984) 15–18.

47. F. L. LaQue, Behavior of metals and alloys in sea water, The *Corrosion Handbook*, H. H. Uhlig, Editor, John Wiley & Sons, New York (1948) 391.

48. R. E. Melchers and R. Jeffrey, Influence of water velocity on marine corrosion of mild steel, *Corrosion (NACE)* **60**, 1 (2004) 84–94.

49. C.-G. Peng and J. K. Park, Principal factors affecting microbiologically influenced corrosion of carbon steel, *Corrosion* **50**, 9 (1994) 669–675.

50. R. E. Melchers, Effect of nutrient-based water pollution on the corrosion of mild steel in marine immersion conditions, *Corrosion (NACE)*, in press.

CHAPTER 9

SEISMIC RISK ASSESSMENT OF REALISTIC FRAME STRUCTURES USING A HYBRID RELIABILITY METHOD

JUNGWON HUH

Ocean Engineering Program, Yosu National University
Yeosu, 550-749, Korea
E-mail: jwonhuh@yosu.ac.kr

ACHINTYA HALDAR

Department of Civil Engineering and Engineering Mechanics
University of Arizona, Tucson, AZ85721, USA
E-mail: haldar@u.arizona.edu

Reliability analysis techniques have matured in recent years, but they are not yet widely accepted by the deterministic community. One of the major objections is that the available techniques are not adequate to evaluate the risk of real complex structures. To overcome this concern, a new, efficient and accurate hybrid finite element-based reliability method is presented to evaluate the reliability of real nonlinear structures excited by short duration dynamic loadings applied in the time domain, including earthquake loading. It intelligently integrates the response surface method, the finite element method, the first-order reliability method, and an iterative linear interpolation scheme. The method explicitly considers nonlinearities due to geometry and material characteristics, boundary or support conditions, connection conditions and the uncertainty in them. Time domain reliability of any structures that can be represented by finite elements can be evaluated with this approach, thus removing one of the major concerns of the deterministic community. The applicability of the method is demonstrated with the help of several examples.

1. Introduction

Despite significant recent progresses in the risk and reliability analysis techniques, a large segment of the engineering profession is not familiar with them and thus fails to use them in everyday practices. One of the major factors is that the available reliability methods fail to capture the realistic structural behavior. They are not similar or parallel to the deterministic methods commonly used to study the behavior of real structures. It is believed that reliability-based design and analysis concepts will be accepted by the deterministic community if the applications of probabilistic methods in estimating the reliability of real structures can be demonstrated. The lack of progress so far indicates that the task is not simple. In developing an attractive reliability evaluation method, it is necessary to first improve the deterministic algorithms used to capture the realistic behavior of a structure. The estimation of the failure probability implies that the structural behavior just before failure needs to be captured as accurately as possible. Major sources of nonlinearity including geometric and material, boundary or support, and the connection conditions must be modeled realistically. The most rigorous deterministic dynamic analysis requires that the dynamic loadings must be applied in time domain. Thus, the reliability analysis also needs to be conducted under similar conditions.

The finite element method (FEM) is commonly used to study the behavior of real structures under static and dynamic loadings. The use of FEM adds two different challenges to the reliability community. The first challenge is how to improve the computational efficiency of the deterministic algorithm. If standard procedures are followed, the probabilistic algorithm could be inefficient or impractical. The second challenge is how to incorporate uncertainty into the improved deterministic algorithm. For the seismic reliability evaluation, since the time history of a future earthquake is unknown, it introduces a major source of uncertainty. The representation of the dynamic load in the form of the power spectral density function may not be appropriate, particularly for highly nonlinear structures. Also, since most practicing engineers are not familiar with this approach, they will not use it. To comprehensively address some of the deficiencies in the currently available reliability methods, the authors recently proposed a hybrid approach consisting of the response surface method, the nonlinear finite element method, and the first-order reliability method to estimate the reliability of nonlinear structures where dynamic loading can be applied in time domain. It is presented in this chapter.

2. A Unified Time-Domain Reliability Assessment of Real Structures

The limit state functions are essential in estimating reliability using the commonly used first-order or second-order reliability method (FORM/SORM).[1] For nonlinear dynamic problems, limit state functions may not be available in an explicit form or they may change with time. The proposed hybrid method attempts to address issues associated with the reliability evaluation when the limit state functions are implicit. Using the FEM approach and the response surface method (RSM), the proposed method approximately generates the limit state function under a short duration dynamic loading. However, the basic RSM is not expected to be efficient if the failure region is unknown. To efficiently locate the failure region, it may need to be integrated with the FORM. Since FORM is an iterative algorithm, an efficient interpolation scheme is necessary to make the procedure attractive. The integration of these logics forms the foundation of the proposed hybrid approach.

2.1. *Deterministic finite element method for dynamic analysis*

The commonly used displacement-based FEM may not be efficient for tracking highly nonlinear behavior of real structures considered in this study. The assumed stress-based FEM[2-4] was found to be very efficient and accurate for this purpose, particularly for frame structures and is used in this study. In this approach, the tangent stiffness can be expressed in an explicit form without performing any integration and fewer elements are required to describe a large deformation configuration. It is not necessary to discuss it in detail here due to lack of space. However, the detail information on the subject is widely available in the literature.[2,4,5] Only very few of its essential features, in the context of the dynamic analysis, are presented here.

The nonlinear dynamic equilibrium equation can be expressed at time $t + \Delta t$ as:

$$\mathbf{M}^{t+\Delta t}\ddot{\mathbf{D}}^{(n)} + {}^t\mathbf{C}^{t+\Delta t}\dot{\mathbf{D}}^{(n)} + {}^t\mathbf{K}^{(n)}{}^{t+\Delta t}\Delta\mathbf{D}^{(n)} = {}^{t+\Delta t}\mathbf{F}^{(n)} - {}^{t+\Delta t}\mathbf{R}^{(n-1)}$$
$$- \mathbf{M}^{t+\Delta t}\ddot{\mathbf{D}}_g^{(n)}, \qquad (1)$$

where $\mathbf{M}$ is the mass matrix, ${}^t\mathbf{C}$ is the viscous damping coefficient matrix at time t, ${}^t\mathbf{K}^{(n)}$ is the global tangent stiffness matrix of the nth iteration at time t, ${}^{t+\Delta t}\Delta\mathbf{D}^{(n)}$ is the incremental displacement vector of the nth iteration at time $t + \Delta t$, ${}^{t+\Delta t}\mathbf{F}^{(n)}$ is the external load vector of the nth iteration at time $t + \Delta t$, ${}^{t+\Delta t}\mathbf{R}^{(n-1)}$ is the internal force vector of the $(n-1)$th iteration at time $t + \Delta t$, and ${}^{t+\Delta t}\ddot{\mathbf{D}}_g^{(n)}$ is the seismic ground acceleration vector of the nth iteration at time $t + \Delta t$.

The matrix ${}^t\mathbf{C}$ in Eq. (1) represents viscous damping. In a realistic seismic analysis of steel frames, the amount of damping energy to be dissipated will depend on the non-yielding and yielding state of the material, and the hysteretic behavior if the material yields. For mathematical simplicity, the effect of non-yielding energy dissipation is generally represented by equivalent viscous damping varying between 0.1% and 7% of the critical damping.[6] In the context of nonlinear finite element analysis, the Rayleigh-type damping, i.e. the damping is proportional to mass and stiffness is very common. Since the stiffness is expected to change in the nonlinear range, by assuming the damping is proportional to the stiffness is a realistic representation of the damping characteristics. The Rayleigh-type damping is used in this study. It can be represented as:

$$ {}^t\mathbf{C} = \alpha\mathbf{M} + \gamma\,{}^t\mathbf{K}, \qquad (2)$$

where ${}^t\mathbf{K}$ is the tangent stiffness matrix, and α and γ are proportional constants. They can be evaluated from the first two natural frequencies of the structure.

2.2. *Systematic response surface method*

The major purpose of using RSM in the proposed algorithm is to generate an approximate explicit expression for a performance function.[7,8] At least a second order polynomial is necessary for the nonlinear dynamic problem under consideration in this study. The following two types of second order polynomial are considered:

$$\hat{g}(\mathbf{X}) = b_0 + \sum_{i=1}^{k} b_i X_i + \sum_{i=1}^{k} b_{ii} X_i^2, \tag{3}$$

$$\hat{g}(\mathbf{X}) = b_0 + \sum_{i=1}^{k} b_i X_i + \sum_{i=1}^{k} b_{ii} X_i^2 + \sum_{i=1}^{k-1} \sum_{j>i}^{k} b_{ij} X_i X_j, \tag{4}$$

where X_i $(i = 1, 2, \ldots, k)$ is the ith random variable, and b_0, b_i, b_{ii}, and b_{ij} are coefficients of the second-order polynomial. The polynomial can be fully defined using regression analysis or by solving a set of simultaneous equations using information on responses obtained at specific data points called sampling points. The selection of sampling points is a crucial factor in establishing the efficiency and accuracy of RSM. Saturated design (SD) and central composite design (CCD) are the two most promising techniques that can be used for this purpose. SD is less accurate but more efficient. CCD is more accurate but less efficient. By considering the two design methods and the form of the polynomial, the three promising response surface models suggested by Huh and Haldar[9] are: Model (1) — saturated design using a second-order polynomial without cross terms, Model (2) — saturated design using a full second-order polynomial, and Model (3) — CCD using a full second-order polynomial. They are discussed in more detail elsewhere.[4,9,10]

Since the proposed algorithm is iterative, it is necessary to improve on the selection of the location of the center point around which the sampling points are generated in the subsequent iterations. Bucher and Bourgund[7] suggested an iterative linear interpolation scheme as shown in Fig. 1 and is used in this study. It

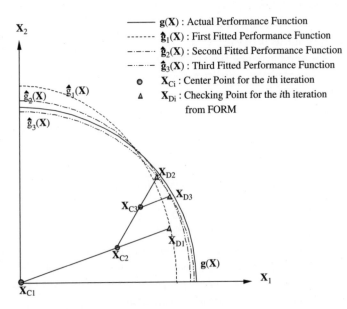

Fig. 1.   Iterative linear interpolation scheme (similar to Rajashekhar and Ellingwood[8]).

can be mathematically represented as:

$$\mathbf{x}_{C_2} = \mathbf{x}_{C_1} + (\mathbf{x}_{D_1} - \mathbf{x}_{C_1})\frac{g(\mathbf{x}_{C_1})}{g(\mathbf{x}_{C_1}) - g(\mathbf{x}_{D_1})} \quad \text{if } g(\mathbf{x}_{D_1}) \geq g(\mathbf{x}_{C_1}), \tag{5}$$

$$\mathbf{x}_{C_2} = \mathbf{x}_{D_1} + (\mathbf{x}_{C_1} - \mathbf{x}_{D_1})\frac{g(\mathbf{x}_{D_1})}{g(\mathbf{x}_{D_1}) - g(\mathbf{x}_{C_1})} \quad \text{if } g(\mathbf{x}_{D_1}) < g(\mathbf{x}_{C_1}), \tag{6}$$

where $\mathbf{x}_{C_1}$ and $\mathbf{x}_{D_1}$ are the coordinates of the center point and the checking point for the first iteration, and $g(\mathbf{x}_{C_1})$ and $g(\mathbf{x}_{D_1})$ are the actual responses of the limit state function estimated from the dynamic FEM analysis at $\mathbf{x}_{C_1}$ and $\mathbf{x}_{D_1}$, respectively. The point $\mathbf{x}_{C_2}$ can be used as a new center point for the next iteration. This iteration scheme needs to be continued until a preselected convergence criterion is satisfied. The convergence criterion of $(\mathbf{x}_{C_{i+1}} - \mathbf{x}_{C_i})/\mathbf{x}_{C_i} \leq |0.05|$ is used in the numerical examples of this study.

Considering the three models identified earlier, Huh and Haldar[9] suggested two promising schemes to improve the computational efficiency without compromising the accuracy. They are: Scheme 1 — saturated design using a second-order polynomial without cross terms [Model (1)] for the intermediate iterations and saturated design using a full second-order polynomial for [Model (2)] the final iteration, and Scheme 2 — saturated design using a second-order polynomial without cross terms [Model (1)] for the intermediate iterations and CCD using a full second-order polynomial [Model (3)] for the final iteration. Both schemes are considered here.

2.3. Consideration of uncertainty

Most of the load and resistance-related parameters in the finite element formulation are uncertain in nature. The randomness in them needs to be identified at this stage. The probabilistic characteristics of the variables are extensively studied and are widely available in the literature.[1] Additional discussions on the topic are unnecessary. For steel structures considered in this study, the resistance-related random variables are the Young's modulus (E), the yield stress (F_y), the cross sectional area (A), the moment of inertia (I), and the plastic section modulus (Z) of frames. For the concrete shear walls, the Young's modulus (E_C) and the Poisson's ratio (ν) are also considered to be random variables.

The consideration of uncertainty in the time domain seismic loading is a very complicated subject. No guideline is available on how to consider the uncertainties in both the amplitude and frequency content of the seismic loading. The uncertainty in the amplitude of the earthquake is considered in this study by treating it as a random variable. It is conceptually shown in Fig. 2. A parameter g_e is introduced to incorporate the uncertainty in the amplitude. The uncertainty in the frequency content of an earthquake is considered indirectly. The large number of time histories recorded in close proximity of each other during a specific earthquake can be used for this purpose. They have different frequency content. The estimated reliability of a given structure excited by them will indicate the effect of uncertainty in the frequency content. The uncertainty in the frequency content can also be simulated but it is beyond the scope of this study.

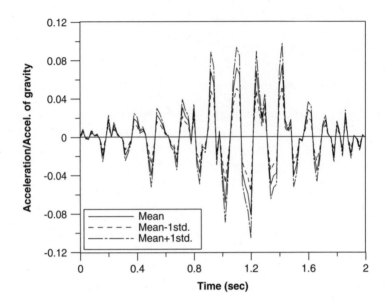

Fig. 2.   Consideration of uncertainty in the amplitude of an earthquake.

2.4. *Limit state function for risk assessment*

Risk is always estimated corresponding to a performance or limit state function.[1] Commonly used limit state functions can be broadly divided into two groups: The serviceability and strength limit states. Each limit state needs to be considered separately since a structure can fail due to excessive lateral or inter-story deflection, or due to failure of several components in strength forming a local or global mechanism. The proposed algorithm is capable of calculating risk for both types of limit states, as discussed next.

2.4.1. *Strength limit state*

The limit state function for structural strength mainly depends on the failure mode to be considered. Most of the elements in the structural system considered in this study are beam-columns, i.e. they are subjected to both axial load and bending moment at the same time. For design purposes, interaction equations are generally used to consider the combined effect of axial load and bending moment. Thus, the interaction equations suggested by the American Institute of Steel Construction's (AISC's) *Load and Resistance Factor Design*[11] manual for two-dimensional structures are given below for the discussion purpose. They are:

$$\frac{P_u}{\phi P_n} + \frac{8}{9}\left(\frac{M_{ux}}{\phi_b M_{nx}}\right) \leq 1.0; \quad \text{if } \frac{P_u}{\phi P_n} \geq 0.2, \tag{7}$$

$$\frac{P_u}{2\phi P_n} + \left(\frac{M_{ux}}{\phi_b M_{nx}}\right) \leq 1.0; \quad \text{if } \frac{P_u}{\phi P_n} < 0.2, \tag{8}$$

where ϕ and ϕ_b are the resistance factors, P_u is the required tensile or compressive strength, P_n is the nominal tensile/compressive strength, M_{ux} is the required flexural strength and M_{nx} is the nominal flexural strength. P_n and M_{nx} can be calculated using AISC's LRFD code.

Since P_u and M_{ux} are functions of time in dynamic analysis, the ratio of P_u and M_{ux} is also a function of time. Furthermore, the axial and bending strength of the members, P_n and M_{nx}, are also functions of time since the load effects change from compression to tension and vice versa in dynamic loading. The interaction equation used for static problems therefore may not be applicable for the dynamic problems. In order to follow the intent of the design guidelines, the effects of axial load and bending moment are represented separately as $\alpha_1(P_u/P_n)$ and $\alpha_2(M_{ux}/M_{nx})$, respectively. The coefficients α_1 and α_2 represent the time-variant aspects of P_u/P_n and M_{ux}/M_{nx}. These modifications do not affect the efficiency or generality of the algorithm since they are obtained simultaneously from an identical design. Only an estimation of the coefficients has to be made for both polynomials.

In general, the maximum values of both P_u and M_{ux} do not occur at the same time. It is also difficult to predict which interaction equation needs to be used for a given experimental sampling point since the P_u/P_n ratio is unknown. To address this issue, the responses for each experimental sampling point are tracked, i.e. in a deterministic dynamic analysis, recording the time when the interaction equation [left hand side of Eq. (7) or (8)] reaches the maximum value. This provides the necessary information on which interaction equation is to be used. Then the contributions of $\alpha_1(P_u/P_n)$ and $\alpha_2(M_{ux}/M_{nx})$ are evaluated for an experimental sampling point. For other experimental sampling points the contributions of axial load and bending moment can be similarly evaluated. However, these contributions are expected to be different, indicating the dynamic nature of the analysis. Interestingly, these values can be used to formulate the necessary response surface required for reliability analysis using FORM.

2.4.2. *Serviceability limit state*

For seismic loading, the design may be controlled by the serviceability, e.g. inter-story drift or the overall lateral displacement. Limit states corresponding to inter-story drift or overall lateral displacement can be formulated using the recommendations given in design codes. The general form of a serviceability limit state can be represented as:

$$g(\mathbf{X}) = \delta_{\text{allow}} - y_{\text{max}}(\mathbf{X}), \tag{9}$$

where δ_{allow} is the allowable inter-story drift or the overall lateral displacement specified in codes or design guidelines and $y_{\text{max}}(\mathbf{X})$ is the corresponding maximum inter-story drift or overall lateral displacement estimated from analysis.

2.5. Solution strategy

The solution strategy for the proposed algorithm[9,12] can be stated as follows. The initial center point is first assumed to be the mean values of the random variables for the first iteration for both schemes. Similar assumption is also made for FORM. The responses are calculated by conducting nonlinear FEM at the experimental sampling points for the response surface model being considered. A limit state function is thus generated in terms of k basic random variables. Using the explicit expression for the limit state function and FORM, the reliability index β and the corresponding coordinates of the checking point and direction cosines for each random variable are obtained. The coordinates of the new center point are obtained by applying the linear interpolation scheme. The updating of the location of the center point continues until it converges to a predetermined tolerance level. In the final iteration, the information on the most recent center point is used to formulate the final response surface using either Scheme 1 or Scheme 2. FORM is then used to calculate the reliability index and the corresponding coordinates of the most probable failure point. The graphical representation of solution strategy is provided in the form of a flowchart as shown in Fig. 3.

3. Reliability Estimation of Frames with PR Connections

Connections in steel frames are essentially partially restrained (PR) with different rigidities. They add another major source of nonlinearity but also dissipate energy at a much higher level. The commonly used analytical procedures developed for frames with fully restrained (FR) connections are not applicable for the PR connections. The PR connections not only change the dynamic properties (such as the natural frequency, stiffness, damping, etc.) of structures, but they also add a major source of uncertainty in the reliability analysis during loading, unloading, and reloading stages of dynamic excitation. The behavior of a steel frame will depend on how the PR connections are modeled in any nonlinear algorithm.[12]

3.1. Modeling of PR connections

The flexibility in the connections can be represented by a relationship between the moment M, transmitted by the connection, and the relative rotation angle θ. The relationship is generally represented by $M - \theta$ curves. Several mathematical models are available in the literature for this purpose.[10] Due to many advantages over other models, the Richard four-parameter moment-rotation model[13] is chosen here to represent the flexible behavior as shown in Fig. 4. The relationship can be expressed as:

$$M = \frac{(k - k_p)\theta}{\left(1 + \left|\frac{(k-k_p)\theta}{M_0}\right|^N\right)^{\frac{1}{N}}} + k_p\theta, \qquad (10)$$

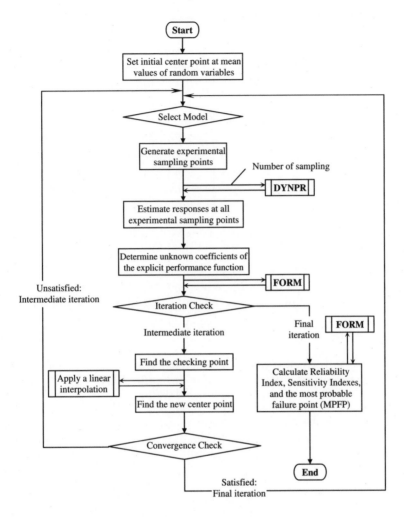

Fig. 3.   Flowchart of the proposed algorithm.

where M is the moment, θ is the relative rotation between the connecting elements, k is the initial stiffness, k_p is the plastic stiffness, M_0 is the reference moment, and N is the curve shape parameter.

Equation (10) represents only the monotonically increasing loading portion of the M-θ curves. Colson[14] and El-Salti[15] theoretically developed the unloading and reloading parts of the M-θ curves using the Masing rule. The unloading and reloading relationships of a PR connection, used in this study, can be represented as:

$$M = M_a - \frac{(k - k_p)(\theta_a - \theta)}{\left(1 + \left|\frac{(k-k_p)(\theta_a-\theta)}{2M_0}\right|^N\right)^{\frac{1}{N}}} - k_p(\theta_a - \theta). \tag{11}$$

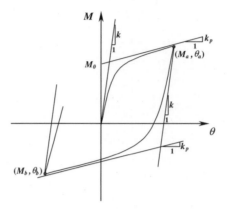

Fig. 4.   Loading, unloading, and reverse loading model of PR connections.

If (M_b, θ_b) is the next load reversal point, the reloading relationship between M and θ can be obtained by simply replacing (M_a, θ_a) with (M_b, θ_b) in Eq. (11). Therefore, the proposed method uses Eq. (10) when the connection is loading and Eq. (11) when the connection is unloading and reloading. This represents hysteretic behavior of the PR connections.

3.2. *Incorporation of PR connections into the FEM*

The assumed stress-based FEM approach used in this study uses a beam-column element to model both regular structural elements and flexible connections. One element is added for each PR connection. However, the stiffness of PR connections needs to be updated at every iteration since it depends on θ. This can be accomplished by updating the Young's modulus as:

$$E_C(\theta) = \frac{l_C}{I_C} K_C(\theta) = \frac{l_C}{I_C} \frac{\partial M(\theta)}{\partial \theta}, \tag{12}$$

where l_C, I_C, and $K_C(\theta)$ are the length, the moment of inertia, and the tangent stiffness of the connection element, respectively. When the element is loading, using Eq. (10) $K_C(\theta)$ is evaluated as:

$$K_C(\theta) = \frac{dM}{d\theta} = \frac{(k - k_p)}{\left(1 + \left|\frac{(k-k_p)\theta}{M_0}\right|^N\right)^{\frac{N+1}{N}}} + k_p. \tag{13}$$

When the element is unloading and reloading, using Eq. (11) $K_C(\theta)$ can be calculated as:

$$K_C(\theta) = \frac{dM}{d\theta} = \frac{(k - k_p)}{\left(1 + \left|\frac{(k-k_p)(\theta_a-\theta)}{2M_0}\right|^N\right)^{\frac{N+1}{N}}} + k_p. \tag{14}$$

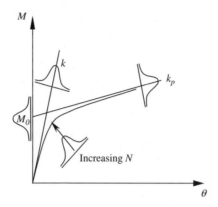

Fig. 5.   Random parameters in Richard model.

The basic FEM formulation of the structure remains unchanged, except for adding a few more elements to represent PR connections.

3.3. *Uncertainties in the connection model*

The uncertainties in the connection behavior come from the manufacturing and assembly processes and also from mathematical modeling.[1] To consider uncertainty in modeling the behavior of PR connections, the four parameters in the Richard model are considered to be the basic random variables as shown in Fig. 5.[12]

4. Reliability Estimation of In-Filled Frames

Steel frames in the presence of PR connections may not be able to transfer horizontal loads (e.g. wind and earthquake) effectively. To increase their lateral stiffness, the use of shear walls is very common. It is not an easy task to capture the realistic behavior of a combined system consisting of steel frames represented by beam-column elements and shear walls represented by plane stress elements. In addition, the associated uncertainties in modeling the combined system will make the task more challenging. The algorithm presented in the previous sections is extended to evaluate the reliability of the combined system.[16]

4.1. *Modeling of shear walls*

The basic frame is represented by two-dimensional (2D) beam-column elements and the shear walls are represented by 4-node plane stress elements. The shape of the shear wall is restricted to be rectangular. Two displacement (horizontal and vertical) dynamic degrees of freedom (DDOFs) are used at each node point. The rotation of the combined system at the node point is expected to be very small and can be neglected. This simplification was independently verified using a commercially available computer program.

To bring the shear wall stiffness into the frame structure, the components of the shear wall stiffness are added to the corresponding frame stiffness components. The explicit form of a stiffness matrix of a 4-node plane stress element can be obtained as[17]:

$$K_{sh} = \frac{t}{4\gamma} \mathbf{A}^t \mathbf{E} \mathbf{A} + \frac{t}{12\gamma} \mathbf{B}^t \mathbf{E} \mathbf{B} + \frac{t\gamma}{12} \mathbf{C}^t \mathbf{E} \mathbf{C}, \tag{15}$$

where t is thickness of the wall, γ is the ratio of b and a; i.e. $\gamma = b/a$, and $2a$ and $2b$ are long and short dimensions of the rectangular shear wall, respectively. The matrixes $\mathbf{A}$, $\mathbf{B}$, $\mathbf{C}$, and $\mathbf{E}$ in Eq. (15) can be represented as:

$$\mathbf{A} = \begin{bmatrix} -\gamma & 0 & \gamma & 0 & \gamma & 0 & -\gamma & 0 \\ 0 & -1 & 0 & -1 & 0 & 1 & 0 & 1 \\ -1 & -\gamma & -1 & \gamma & 1 & \gamma & 1 & -\gamma \end{bmatrix}, \tag{16}$$

$$\mathbf{B} = \begin{bmatrix} 0 & 0 & 0 & 0 & 0 & 0 & 0 & 0 \\ 0 & 1 & 0 & -1 & 0 & 1 & 0 & -1 \\ 1 & 0 & -1 & 0 & 1 & 0 & -1 & 0 \end{bmatrix}, \tag{17}$$

$$\mathbf{C} = \begin{bmatrix} 1 & 0 & -1 & 0 & 1 & 0 & -1 & 0 \\ 0 & 0 & 0 & 0 & 0 & 0 & 0 & 0 \\ 0 & 1 & 0 & -1 & 0 & 1 & 0 & -1 \end{bmatrix}, \tag{18}$$

and

$$\mathbf{E} = \frac{E_C}{1-\nu^2} \begin{bmatrix} 1 & \nu & 0 \\ \nu & 1 & 0 \\ 0 & 0 & \frac{1-\nu}{2} \end{bmatrix}, \tag{19}$$

where E_C is modulus of elasticity and ν is Poisson's ratio of shear walls.

The reinforced concrete (RC) shear wall is the most frequently used and is considered in this study. To consider the presence of concrete shear walls, two additional parameters, namely, the modulus of elasticity and the Poisson ratio of concrete are necessary in the deterministic formulation, as can be seen in Eq. (19). The modeling of the stiffness of shear walls is not easy. There has been extensive research on cracking in RC panels.[18–22] It was observed that the degradation of the stiffness of the shear walls occurs after cracking and can be considered effectively by reducing the modulus of elasticity of the shear walls. Lefas *et al.*[21] reported that the degradation of the stiffness after cracking could vary from 40% to 70% of the original stiffness depending on the amount of reinforcement and the intensity of axial loads. The same concept is used in this study. The shear wall is assumed to develop cracks when the tensile stress in concrete exceeds the prescribed value. The rupture strength of concrete, f_r, is assumed to be $f_r = 7.5 \times \sqrt{f'_c}$, where f'_c is the compressive strength of concrete.

4.2. *Incorporation of shear walls into the FEM*

Once the explicit form of the stiffness matrix of shear walls is obtained using Eq. (15), it can be combined with the stiffness of the steel frame to develop stiffness matrix for the combined system. In general the nonlinear governing equation for the combined system can be represented in the incremental form as:

$$\mathbf{K}_T^{(n)}\Delta\mathbf{D}^{(n)} = \mathbf{F}^{(n)} - \left[\mathbf{R}^{(n-1)} + \mathbf{K}_{sh}^{(n-1)}\mathbf{D}^{(n-1)}\right], \tag{20}$$

where $\mathbf{K}_T^{(n)} = \mathbf{K}^{(n)} + \mathbf{K}_{sh}^{(n)}$, $\mathbf{K}_{sh}^{(n)}$ is the global tangent stiffness matrix of the shear walls at the nth iteration, $\mathbf{K}_{sh}^{(n-1)}\mathbf{D}^{(n-1)}$ is the internal force vector of the shear walls at the $(n-1)$th iteration, and $\mathbf{K}^{(n)}$, $\Delta\mathbf{D}^{(n)}$, $\mathbf{F}^{(n)}$, and $\mathbf{R}^{(n-1)}$ are defined earlier. For dynamic problems, Eq. (1) needs to be modified to consider shear wall effects. Finally, the equilibrium equation of the combined system for the dynamic can be expressed as:

$$\mathbf{M}^{t+\Delta t}\ddot{\mathbf{D}}^{(n)} + {}^t\mathbf{C}^{t+\Delta t}\dot{\mathbf{D}}^{(n)} + {}^t\mathbf{K}_T^{(n)\ t+\Delta t}\Delta\mathbf{D}^{(n)}$$
$$= {}^{t+\Delta t}\mathbf{F}^{(n)} - \left[{}^{t+\Delta t}\mathbf{R}^{(n-1)} + {}^t\mathbf{K}_{sh}^{(n-1)\ t+\Delta t}\Delta\mathbf{D}^{(n-1)}\right] - \mathbf{M}^{t+\Delta t}\ddot{\mathbf{D}}_g^{(n)}. \tag{21}$$

The governing equation of the combined system consisting of steel frame and RC shear walls, i.e. Eq. (21), can be solved using the modified Newton-Raphson method with the arc-length procedure. The deterministic formulation of the problem discussed above is expected to be very accurate and efficient. The finite element representation of the RC shear walls is kept simple in order to minimize the number of basic random variables present in the formulation. More sophisticated methods can be attempted in future studies, if desired. Reliability evaluation procedures are emphasized in this study.

To consider the presence of shear walls in the context of reliability analysis, two additional variables, the modulus of elasticity (E_c) and Poisson's ratio (ν) are considered to be random.[17]

5. Numerical Examples

To demonstrate the application potential of the proposed method in the reliability estimation of real structures, the following two examples are considered. The first example considers the reliability evaluation of steel frames in the presence of PR connections. The second example considers the reliability evaluation of steel frames reinforced with concrete shear walls.

5.1. *Example 1 — Seismic reliability of steel frame structures with FR and PR connections*

5.1.1. *Seismic risk of the frame with FR connections*

A two-story steel frame structure shown in Fig. 6 is considered first. The frame consists of W27 × 84 for all beams and W14 × 426 for all columns. A36 steel is used. The frame is excited for 15 seconds by the actual acceleration time history recorded at Canoga Park during the Northridge earthquake of 1994 (N-S component) as shown in Fig. 7. Both the serviceability and strength limit states are considered.

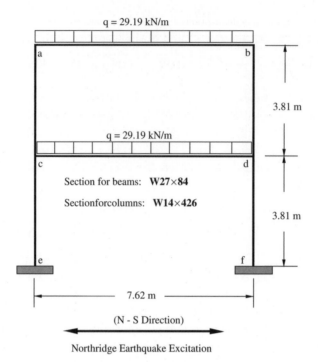

Fig. 6.   Two-story steel frame structure.

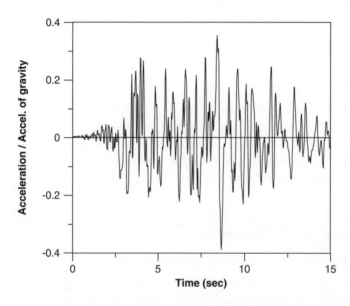

Fig. 7.   Northridge earthquake (N-S) time history.

Table 1. Statistical description of random variables (b: beam, c: column).

Random variables	Mean value	Serviceability limit state		Strength limit state	
		COV	Distribution	COV	Distribution
E (kN/m^2)	1.9994×10^8	0.06	Lognormal	0.06	Lognormal
A^b (m^2)	1.600×10^{-2}	0.05	Lognormal	—	—
I_x^b (m^4)	1.186×10^{-3}	0.05	Lognormal	0.05	Lognormal
Z_x^b (m^3)	3.998×10^{-3}	—	—	0.05	Lognormal
A^c (m^2)	8.065×10^{-2}	0.05	Lognormal	—	—
I_x^c (m^4)	2.747×10^{-3}	0.05	Lognormal	0.05	Lognormal
F_y (kN/m^2)	2.4822×10^5	—	—	0.10	Lognormal
ξ	0.05	0.15	Lognormal	0.15	Lognormal
g_e	1.00	0.20	Type I	0.20	Type I

For the serviceability limit state, the permissible lateral displacement at the top of the frame is assumed not to exceed $h/400$, where h is the height of the frame. Thus, δ_{allow} is 1.905 cm for this example and the corresponding limit state is:

$$g(\mathbf{X}) = \delta_{\text{allow}} - y_{\max}(\mathbf{X}) = 1.905 - y_{\max}(\mathbf{X}), \tag{22}$$

where $y_{\max}(\mathbf{X})$ is the maximum lateral displacement at the top of the frame. For the strength limit state, the reliability of the weakest member (beam c–d in Fig. 6) is investigated.

Initially, all nine variables shown in Table 1 are considered to be random. From the sensitivity analysis, the plastic section modulus of the beams and the yield stress of the frame (Z_x^b and F_y) for the serviceability limit state, and the sectional area of the beams and columns (A^b and A^c) for the strength limit state were found to have very low sensitivity indexes. They are considered to be deterministic in the subsequent reliability analysis.

The frame is first analyzed assuming all the connections are FR type. The statistical characteristics of all the random variables for both limit states are given in Table 1. The term ξ in Table 1 represents the viscous damping coefficient expressed as a percentage of the critical damping. The probabilities of failure of the frame are estimated using Scheme 2. The results for both limit states are given in Table 2 in terms of failure probability, reliability index, error, CPU time, and the total number of experimental sampling points (TNSP) required for the evaluation. For the verification purpose the results are compared with Monte Carlo simulation (MCS) results using 100 000 simulations. The error estimation for the serviceability limit state is based on the failure probability. Since the probability of failure is very small for the strength limit state, the reliability index is used to compare the error. A supercomputer (SGI Origin 2000) was used for the numerical calculation using the proposed algorithm and MCS. Results in Table 2 clearly indicate that the probabilities of failure estimated using the proposed algorithm are very similar to those of MCS. It is reasonable to conclude that the proposed method is viable and efficient for the reliability of nonlinear frame structures subjected to dynamic loading including seismic loading applied in time domain.

Table 2.   Reliability analysis results of the frame with FR connections.

Limit state		Serviceability limit state	Strength limit state
Monte Carlo simulation	P_f	0.02887 ($\beta = 1.898$)	0.00002 ($\beta = 4.107$)
	CPU	98183 sec	72949 sec
Proposed algorithm	Scheme	Scheme 2	Scheme 2
	P_f	0.027792	0.000021
	β	1.914	4.098
	TNSP	173	188
	CPU (sec)	182.8	152.3
	Error	3.73%	0.23%

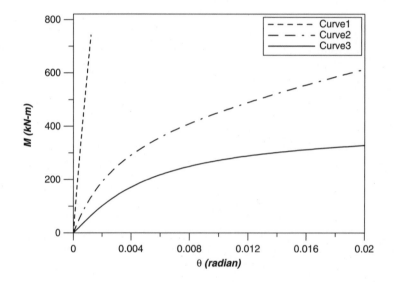

Fig. 8. M-θ curves for connections.

5.1.2. *Seismic risk of the frame with PR connections*

The same frame shown in Fig. 6 is studied again and the four beam-to-column connections at a, b, c, and d are considered to be PR type. Three M-θ curves, Curve 1, Curve 2, and Curve 3, shown in Fig. 8, are considered to represent the effects of different rigidities in the connections. Curve 1 represents high rigidity, Curve 3 represents very low rigidity, and Curve 2 represents intermediate rigidity.

The probabilistic descriptions of the four parameters of the Richard model representing the three curves are listed in Table 3. The statistical descriptions of all other random variables remain the same. As in the previous example, the frame is excited by the same acceleration time history shown in Fig. 7.

The probabilities of failure of the frame with PR connections for the serviceability and strength limit states are evaluated using the proposed method using Scheme 1 with 11 random variables. The reliability indexes of the frame with FR and three different PR connections are summarized in Table 4. From the results, it can be

Table 3.   Statistical description of the four parameters in the Richard model.

Random variables	Mean value			COV	Distribution
	Curve 1	Curve 2	Curve 3		
k (kN·m/rad)	1.13×10^6	1.47×10^5	5.65×10^4	0.15	Normal
k_p (kN·m/rad)	1.13×10^5	1.13×10^4	1.13×10^3	0.15	Normal
M_0 (kN·m)	508.64	452.12	339.09	0.15	Normal
N	0.50	1.00	1.5	0.05	Normal

Table 4.   Reliability analysis results of the frame with FR and PR connections.

Limit state	FR connections	PR connections		
		Curve 1	Curve 2	Curve 3
Serviceability limit state	$\beta = 1.914$	$\beta_1 = 1.274$	$\beta_2 = -0.008$	$\beta_3 = -0.899$
Strength limit state	$\beta = 4.098$	$\beta_1 = 2.351$	$\beta_2 = 2.558$	$\beta_3 = 3.156$

observed that the reliability indexes for the serviceability limit state decreased significantly with the decrease in the rigidity of the PR connections. The frame became very weak in serviceability, particularly when PR connections were represented by the very flexible Curve 3. Due to the redistribution of moment in beam *c–d*, the reliability indexes for the strength limit state also changed. However, the frame is found to be more prone to failure in serviceability than in strength. It can be concluded that connection rigidity should be taken into account appropriately in the reliability analysis of steel frame structures subjected to seismic loading.

5.2. *Example 2 — In-filled steel frame structures*

5.2.1. *Reliability analysis of the frame without shear walls*

A two-story two-bay steel frame without shear walls, shown in Fig. 9, is considered in this example. The frame is excited for 5.12 seconds with the El Centro Earthquake (N-S component) of 1940 as shown in Fig. 10.

For this example, the prescribed horizontal drift at the top floor is considered not to exceed $h/400$, where h is the height of the frame. Thus, $\delta_{\text{allow}}^{\text{drift}}$ is equal to 1.83 cm and the corresponding limit state is:

$$g(\mathbf{X}) = \delta_{\text{allow}}^{\text{drift}} - y_{\max}(\mathbf{X}) = 1.83 - y_{\max}(\mathbf{X}). \qquad (23)$$

The statistical characteristics of all the random variables present in the problem are shown in Table 5. Considering all the random variables identified in Table 5 and the serviceability limit states function represented by Eq. (23), the probability of failure of the frame due to horizontal deflection at Node *a* in Fig. 9 is calculated using the proposed algorithm with Scheme 1. The results are summarized in Table 6. The probability of failure for the overall horizontal deflection is compared with MCS results using 100 000 simulations. The probability of failure of the frame is very close to 1.0 indicating that it is unable to carry the seismic load. The probability of failure

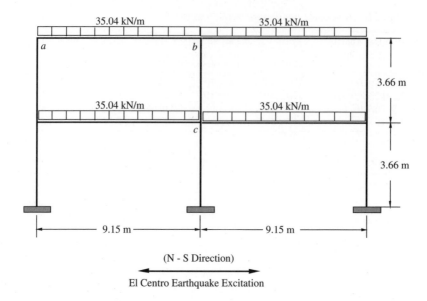

Fig. 9.   A frame structure without shear walls.

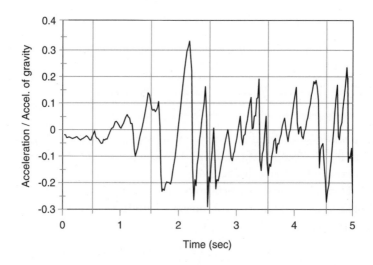

Fig. 10.   El Centro earthquake (N-S) time history.

for the overall horizontal deflection serviceability limit state using the proposed
algorithm is very similar to the MCS results, indicating that the algorithm is viable
and accurate. However, the CPU time required for the proposed algorithm (134 sec)
is about 0.136% of that of MCS (98459 sec), indicating that the proposed algorithm
is very efficient.

Table 5. Statistical description of random variables.

Item	Random variables	Mean value	COV	Distribution	Comment
Frame	E (kN/m^2)	1.999×10^8	0.06	Lognormal	
	A^b (m^2)	1.135×10^{-2}	0.05	Lognormal	Beam W18×60
	I_x^b (m^4)	4.096×10^{-4}	0.05	Lognormal	
	Z_x^b (m^3)	2.016×10^{-3}	0.05	Lognormal	
	A^c (m^2)	0.761×10^{-2}	0.05	Lognormal	Column W12×40
	I_x^c (m^4)	1.290×10^{-4}	0.05	Lognormal	
	Z_x^b (m^3)	0.942×10^{-3}	0.05	Lognormal	
	F_y (kN/m^2)	2.482×10^5	0.10	Lognormal	
Shear wall	E_C (kN/m^2)	2.140×10^7	0.18	Lognormal	$f'_C = 2.068 \times 10^4$ (kN/m^2)
	ν		0.17	0.10	Lognormal	
Dynamic	ξ	0.02	0.15	Lognormal	Without shear wall
property	g_e	1.00	0.20	Type I	
	ξ	0.05	0.15	Lognormal	With shear wall
	g_e	1.00	0.20	Type I	

Table 6. Reliability analysis result of the frame without and with shear walls.

Steel frame structure	Scheme	Proposed algorithm		Monte Carlo simulation	
		P_f	CPU (s)	P_f	CPU (s)
Without shear walls	1	0.9999	134	1.0	98459
With shear walls	1	0.0057	202	0.0049	117832
	2	0.0094	295		

5.2.2. *Reliability analysis of the frame with shear walls*

The frame without shear walls is unable to carry the seismic load applied to it. The frame is reinforced with RC shear walls as shown in Fig. 11. As discussed before, two additional random variables related to the shear walls, E_c and ν, need to be considered. Their statistical properties are given in Table 5.

The building is assumed to contain five similar frames connected by rigid diaphragms at the floor levels. Only the center frame of the building is assumed to have shear walls. Although the physical thickness of the shear wall is considered to be 12.7 cm, considering the presence of five similar frames and rigid behavior of diaphragms, the effective thickness per frame is assumed to be 2.54 cm for this example. After the tensile stress of each shear wall exceeds the prescribed tensile stress of concrete, the degradation of the shear wall stiffness is assumed to be reduced to 40% of the original stiffness.[21]

The frame is again excited for 5.12 seconds with the same El Centro Earthquake (N-S component) of 1940 as used in the previous case. The probability of failure of the combined system for the serviceability limit state is calculated using the proposed algorithm with both Scheme 1 and Scheme 2. The horizontal deflection

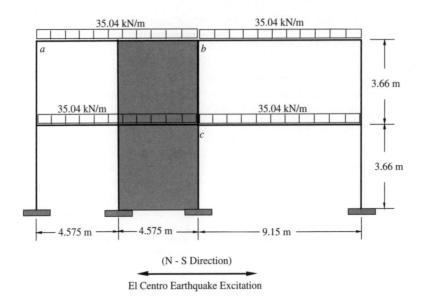

Fig. 11.   A frame structure with shear walls.

at the top of the combined system (point a in Fig. 11) is evaluated. The results
are summarized in Table 6. The results indicate that the presence of shear walls
significantly improves the serviceability behavior of the steel frame. This behavior
is expected. However, the amount of improvement in terms of probability of failure can be quantified using the proposed algorithm considering major sources of
uncertainty. Again, the MCS simulation technique is used to verify the algorithm
using 100 000 simulation cycles. The reliability indexes estimated by the proposed
algorithm and the MCS technique are almost identical. The result clearly indicates
that the proposed algorithm can accurately estimate the probability of failure of
a combined system consisting of frame and shear walls under dynamic loading.
However, considering CPU time required for the proposed algorithm (202 sec) is
significantly smaller than the MCS (117832 sec). Thus, the proposed algorithm is
an attractive alternative to MCS. This example clearly indicates that the proposed
algorithm is efficient without sacrificing any accuracy.

Two realistic examples clearly demonstrate the power of the proposed algorithm. The authors believe that the proposed algorithm will be very useful in the
performance-based design guidelines under development by the profession. It will
help to develop risk-consistent design requirements. Assuming the reliability of a
structure should be similar for both the serviceability and strength limit states, the
algorithm will help to develop the allowable deflection criterion. For example, the
current allowable lateral deflection criterion of $h/400$ may be too conservative.

6. Conclusions

A nonlinear finite element-based hybrid reliability analysis algorithm is proposed to evaluate the reliability of real structures subjected to seismic loading applied in time domain considering all major sources of uncertainty and nonlinearity. The unique feature of the technique is that the seismic loading is applied in the time domain. The procedure rationally and effectively combines the concepts of the response surface method (RSM), the finite element method (FEM), the first-order reliability method (FORM), and an iterative linear interpolation scheme.

The four-parameter Richard model is used to represent the flexibility of connections of real steel frames. The uncertainties in the loading and resistance-related parameters and the parameters in the Richard model are incorporated in the algorithm. The algorithm was elaborated by evaluating the probabilities of failure of steel frames with FR and PR connections in the first example. The applicability of the methods to estimate the reliability of real structures is extended in the second example by considering a steel frame reinforced with concrete shear panels. The algorithms found to be capable of estimating the reliability of any real structure that can be represented by finite elements.

This is a very advanced form of reliability analysis technique for real structures excited by earthquake loadings. The procedure appears to be very useful in the performance-based design guidelines under development by the profession. The member sizes and arrangements can be established based on the performance requirements. The flexibility of connections and the uncertainty in modeling them have considerable influence on the overall behavior of frames, particularly under seismic loading. In the analysis and design of structures, uncertainty in seismic loading should not be overlooked.

Acknowledgments

This work is partially supported by the National Science Foundation under Grant CMS-9526809. Any opinions, findings, conclusions, or recommendations expressed in this publication are those of the authors and do not necessarily reflect the views of the sponsors.

References

1. A. Haldar and S. Mahadevan, *Probability, Reliability and Statistical Methods in Engineering Design*, John Wiley & Sons, New York, NY (2000a).
2. K. Kondoh and S. N. Atluri, Large-deformation, elasto-plastic analysis of frames under nonconservative loading, using explicitly derived tangent stiffnesses based on assumed stresses, *Computational Mechanics* **2**, 1 (1987) 1–25.
3. A. Haldar and K. M. Nee, Elasto-plastic large deformation analysis of PR steel frames for LRFD, *Computers and Structures* **31**, 5 (1989) 811–823.

 4. J. Huh, *Dynamic Reliability Analysis for Nonlinear Structures Using Stochastic Finite Element Method*, PhD dissertation, Department of Civil Engineering and Engineering Mechanics, The University of Arizona, Tucson, Arizona, USA (1999).
 5. A. Haldar and L. Gao, Reliability evaluation of structures using nonlinear SFEM, *Uncertainty Modeling in Finite Element, Fatigue, and Stability of Systems*, A. Haldar, A. Guran and B. M. Ayyub, Editors, World Scientific Publishing Co., River Edge, NJ (1997) 23–50.
 6. P. Leger and S. Dussault, Seismic energy dissipation in MDOF structures, *Journal of Structural Engineering*, ASCE **118**, 5 (1992) 1251–1269.
 7. C. G. Bucher and U. Bourgund, A fast and efficient response surface approach for structural reliability problems, *Structural Safety* **7** (1990) 57–66.
 8. M. R. Rajashekhar and B. R. Ellingwood, A new look at the response surface approach for reliability analysis, *Structural Safety* **12** (1993) 205–220.
 9. J. Huh and A. Haldar, Stochastic finite-element-based seismic risk of nonlinear structures, *Journal of Structural Engineering*, ASCE **127**, 3 (2001) 323–329.
10. A. Haldar and S. Mahadevan, *Reliability Assessment using Stochastic Finite Element Analysis*, John Wiley & Sons, New York, NY (2000b).
11. American Institute of Steel Construction, *Manual of Steel Construction: Load and Resistance Factor Design*, Chicago, Illinois (1994).
12. J. Huh and A. Haldar, Seismic reliability of nonlinear frames with PR connections using systematic RSM, *Probabilistic Engineering Mechanics* **17** (2002) 177–190.
13. R. M. Richard and B. J. Abbott, Versatile elastic-plastic stress-strain formula, *Journal of Engineering Mechanics*, ASCE **101**, EM4 (1975) 511–515.
14. A. Colson, Theoretical modeling of semirigid connections behavior, *Journal of the Construction steel Research* **19** (1991) pp. 213–224.
15. M. K. El-Salti, *Design of Frames with Partially Restrained Connections*, PhD dissertation, Department of Civil Engineering and Engineering Mechanics, The University of Arizona, Tucson, Arizona, USA (1992).
16. S. Y. Lee and A. Haldar, Reliability of frame and shear wall structural systems. I: Static loading, *Journal of Structural Engineering*, ASCE **129**, 2 (2003) 224–232.
17. S. Y. Lee, *Static and Dynamic Reliability Analysis of Frame and Shear Wall Structural Systems*, PhD dissertation, Department of Civil Engineering and Engineering Mechanics, The University of Arizona, Tucson, Arizona, USA (2000).
18. A. K. Gupta and H. Akbar, Cracking in reinforced concrete analysis, *Journal of Structural Engineering*, ASCE **107**, ST1 (1983) 1735–1746.
19. T. C. Liauw and K. H. Kwan, Static and cyclic behaviors of multistory infilled frames with different interface conditions, *Journal of Sound and Vibration* **99**, 2 (1985) 275–283.
20. F. J. Vecchio, Nonlinear finite element analysis of reinforced concrete membranes, *ACI Structural Journal* (1989) 26–35.
21. D. Lefas, D. Kotsovos and N. Ambraseys, Behavior of reinforced concrete structural walls: Strength, deformation characteristics, and failure mechanism, *ACI Structural Journal* **87**, 1 (1990) 23–31.
22. N. Inoue, K. Yang and A. Shibata, Dynamic nonlinear analysis of reinforced concrete shear wall by finite element method with explicit analytical procedure, *Earthquake Engineering and Structural Dynamics* **26** (1997) 967–986.

CHAPTER 10

MESHFREE METHODS IN COMPUTATIONAL STOCHASTIC MECHANICS

SHARIF RAHMAN

Department of Mechanical and Industrial Engineering
The University of Iowa, Iowa City, IA 52245, USA
E-mail: rahman@engineering.uiowa.edu

This chapter provides an exposition of stochastic meshfree methods that involves deterministic meshfree formulation, spectral representation of random fields, multivariate function decomposition, statistical moment analysis, and reliability analysis. Numerical results indicate that stochastic meshfree methods, employed in conjunction with dimension-reduction and decomposition methods, yield accurate and computationally efficient estimates of statistical moments and reliability. Although significant strides have been made, breakthrough research on enhancing speed and robustness of meshfree methods is essential for their successful implementation into stochastic mechanics.

1. Introduction

During the last decade, much attention has been focused on collocation[1,2]- or Galerkin-based[3-8] meshfree or meshless methods to solve computational mechanics problems without using a structured grid. Among these methods, the element-free Galerkin method (EFGM)[4] is particularly appealing, due to its simplicity and use of a formulation that corresponds to the well-established finite element method (FEM). Similar to other meshfree methods, EFGM employs moving least-squares approximation[9] that permits the resultant shape functions to be constructed entirely in terms of arbitrarily placed nodes. Since no element connectivity data are needed, burdensome meshing or remeshing required by FEM is avoided. This issue is particularly important for crack propagation in solids for which FEM may be ineffective in addressing substantial remeshing.[10-15] Hence, EFGM and other meshfree methods provide an attractive alternative to FEM in solving computational-mechanics problems.

However, most meshfree development has focused on deterministic problems. Research in probabilistic modeling using EFGM or other meshfree methods has not been widespread and is only now gaining attention.[16-20] For example, using the perturbation and first-order reliability methods, Rahman and Rao[16,17] developed stochastic meshless formulations to predict both the second-moment and reliability of stochastic structures. An alternative approach involving spectral representation of random fields and Neumann series expansion has also appeared for second-moment meshless analysis.[18] Due to their inherent advantages, most stochastic meshless development has been focused on linear-elastic[19,20] and nonlinear[21] fracture-mechanics problems. More recently, new stochastic solutions integrating meshfree formulation and dimension-reduction techniques and decomposition methods have been reported.[22-24] Nevertheless, meshfree methods for probabilistic analysis present a rich and relatively unexplored area for future research in computational stochastic mechanics.

This chapter provides an exposition of meshfree methods for stochastic mechanics and reliability applications. Section 2 reviews deterministic formulation of EFGM. Section 3 discusses the Karhunen-Loève representation of a random field, meshfree solution of an integral equation, and modeling of Gaussian and translation fields. Section 4 informs the reader on function decomposition that facilitates lower-variate approximations of a general multivariate function. Using function decomposition, dimension-reduction methods for statistical moment analysis are presented in Sec. 5. A Monte Carlo simulation using response surface models of lower-variate approximations is examined in Sec. 6. Several numerical examples are presented to illustrate various methods developed. Finally, Sec. 7 concludes the chapter with the impact of current research and future research needs in stochastic meshfree analysis.

2. The Element-Free Galerkin Method

2.1. *Moving least squares and meshless shape function*

Consider a real-valued, continuous, differentiable function $u(\boldsymbol{x})$ over a domain $\mathcal{D} \subset \mathbb{R}^K$, where $K = 1, 2,$ or 3. Let $\mathcal{D}_x \subseteq \mathcal{D}$ denote a subdomain describing the neighborhood of a point $\boldsymbol{x} \in \mathcal{D} \subset \mathbb{R}^K$. A moving least-squares (MLS) approximation $u^h(\boldsymbol{x})$ of $u(\boldsymbol{x})$ is[9]

$$u^h(\boldsymbol{x}) = \sum_{i=1}^{m} p_i(\boldsymbol{x})a_i(\boldsymbol{x}) = \boldsymbol{p}^T(\boldsymbol{x})\boldsymbol{a}(\boldsymbol{x}), \tag{1}$$

where $\boldsymbol{p}^T(\boldsymbol{x}) = \{p_1(\boldsymbol{x}), \ldots, p_m(\boldsymbol{x})\}$ is a vector of complete basis functions of length m and $\boldsymbol{a}(\boldsymbol{x}) = \{a_1(\boldsymbol{x}), \ldots, a_m(\boldsymbol{x})\}^T$ is a vector of unknown parameters that depend on $\boldsymbol{x}$. For example, basis functions commonly used in two-dimensional $(K = 2)$ solid mechanics with $x_1 - x_2$ coordinates are $\boldsymbol{p}^T(\boldsymbol{x}) = \{1, x_1, x_2\}$; $m = 3$ and $\boldsymbol{p}^T(\boldsymbol{x}) = \{1, x_1, x_2, x_1^2, x_1x_2, x_2^2\}$; $m = 6$, which represent linear and quadratic basis functions, respectively. The basis functions need not be polynomial. When solving problems involving cracks, trigonometric basis functions consistent with singular crack-tip fields have been developed for both linear-elastic[11] and nonlinear[14] fracture-mechanics applications.

The coefficient vector $\boldsymbol{a}(\boldsymbol{x})$ in Eq. (1) is determined by minimizing a weighted error norm, defined as:

$$J(\boldsymbol{x}) \equiv \sum_{I=1}^{l} w_I(\boldsymbol{x})\left[\boldsymbol{p}^T(\boldsymbol{x}_I)\boldsymbol{a}(\boldsymbol{x}) - d_I\right]^2 = [\boldsymbol{P}\boldsymbol{a}(\boldsymbol{x}) - \boldsymbol{d}]^T\boldsymbol{W}[\boldsymbol{P}\boldsymbol{a}(\boldsymbol{x}) - \boldsymbol{d}], \tag{2}$$

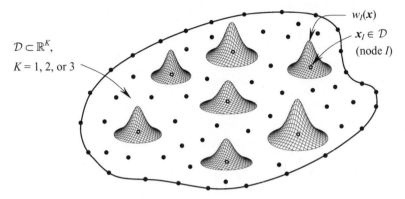

Fig. 1. A schematic illustration of meshfree discretization and weight functions with compact support.

where $\boldsymbol{x}_I$ denotes the coordinates of node I, $\boldsymbol{d}^T = \{d_1, \dots, d_l\}$ with d_I representing the nodal parameter for node I, $\boldsymbol{W} = \text{diag}[w_1(\boldsymbol{x}), \dots, w_l(\boldsymbol{x})]$ with $w_I(\boldsymbol{x})$ being the weight function associated with node I, such that $w_I(\boldsymbol{x}) > 0$ for all $\boldsymbol{x}$ in the support $\mathcal{D}_{\boldsymbol{x}}$ of $w_I(\boldsymbol{x})$ and *zero* otherwise, l is the number of nodes in $\mathcal{D}_{\boldsymbol{x}}$ for which $w_I(\boldsymbol{x}) > 0$, and $\boldsymbol{P} = [\boldsymbol{p}^T(\boldsymbol{x}_1), \dots, \boldsymbol{p}^T(\boldsymbol{x}_l)] \in \mathbb{L}(\mathbb{R}^l \times \mathbb{R}^m)$. The weight function has a compact support and is schematically depicted in Fig. 1. A number of weight functions are available in the current literature.[3–21] For example, a weight function proposed by Rao and Rahman is[12]

$$w_I(\boldsymbol{x}) = \begin{cases} \dfrac{\left(1+\beta^2 \frac{z_I^2}{z_{mI}^2}\right)^{-\left(\frac{1+\beta}{2}\right)} - \left(1+\beta^2\right)^{-\left(\frac{1+\beta}{2}\right)}}{1-(1+\beta^2)^{-\left(\frac{1+\beta}{2}\right)}}, & z_I \le z_{mI} \ , \\ 0, & z_I > z_{mI} \end{cases} \tag{3}$$

where β is a shape controlling parameter, $z_I = \|\boldsymbol{x} - \boldsymbol{x}_I\|$, and z_{mI} is the domain of influence of node I. The stationarity of $J(\boldsymbol{x})$ with respect to $\boldsymbol{a}(\boldsymbol{x})$ yields

$$\boldsymbol{A}(\boldsymbol{x})\boldsymbol{a}(\boldsymbol{x}) = \boldsymbol{C}(\boldsymbol{x})\boldsymbol{d}, \tag{4}$$

where

$$\boldsymbol{A}(\boldsymbol{x}) = \sum_{I=1}^{l} w_I(\boldsymbol{x})\boldsymbol{p}(\boldsymbol{x}_I)\boldsymbol{p}^T(\boldsymbol{x}_I) = \boldsymbol{P}^T \boldsymbol{W} \boldsymbol{P}, \tag{5}$$

and

$$\boldsymbol{C}(\boldsymbol{x}) = [w_1(\boldsymbol{x})\boldsymbol{p}(\boldsymbol{x}_1), \cdots, w_l(\boldsymbol{x})\boldsymbol{p}(\boldsymbol{x}_l)] = \boldsymbol{P}^T \boldsymbol{W}. \tag{6}$$

Solving for $\boldsymbol{a}(\boldsymbol{x})$ in Eq. (4) and then substituting into Eq. (1) yields

$$u^h(\boldsymbol{x}) = \sum_{I=1}^{l} \Phi_I(\boldsymbol{x})d_I = \boldsymbol{\Phi}^T(\boldsymbol{x})\boldsymbol{d}, \tag{7}$$

where

$$\boldsymbol{\Phi}^T(\boldsymbol{x}) = \{\Phi_1(\boldsymbol{x}), \dots, \Phi_l(\boldsymbol{x})\} = \boldsymbol{p}^T(\boldsymbol{x})\boldsymbol{A}^{-1}(\boldsymbol{x})\boldsymbol{C}(\boldsymbol{x}) \tag{8}$$

is a vector with its Ith component

$$\Phi_I(\boldsymbol{x}) = \sum_{j=1}^{m} p_j(\boldsymbol{x})\left[\boldsymbol{A}^{-1}(\boldsymbol{x})\boldsymbol{C}(\boldsymbol{x})\right]_{jI}, \tag{9}$$

representing the shape function of the MLS approximation corresponding to node I.
The partial derivatives of $\Phi_I(\boldsymbol{x})$ can also be obtained as

$$\Phi_{I,i}(\boldsymbol{x}) = \sum_{j=1}^{m} \left\{ p_{j,i}(\boldsymbol{A}^{-1}\boldsymbol{C})_{jI} + p_j(\boldsymbol{A}_{,i}^{-1}\boldsymbol{C} + \boldsymbol{A}^{-1}\boldsymbol{C}_{,i})_{jI} \right\}, \tag{10}$$

where $\boldsymbol{A}_{,i}^{-1} = -\boldsymbol{A}^{-1}\boldsymbol{A}_{,i}\boldsymbol{A}^{-1}$ and $()_{,i} = \partial()/\partial x_i$. From Eqs. (6) and (9), $\Phi_I(\boldsymbol{x}) = 0$
when $w_I(\boldsymbol{x}) = 0$. In other words, $\Phi_I(\boldsymbol{x})$ vanishes for $\boldsymbol{x}$ not in the support of nodal
point $\boldsymbol{x}_I$, thus preserving the local character of the MLS approximation.

2.2. *Variational formulation and discretization*

For small displacements in two-dimensional, isotropic, and linear-elastic solids, the
equilibrium equations and boundary conditions are

$$\boldsymbol{\nabla} \cdot \boldsymbol{\sigma} + \boldsymbol{b} = \boldsymbol{0} \quad \text{in } \mathcal{D} \text{ and} \tag{11}$$

$$\begin{aligned}
\boldsymbol{\sigma} \cdot \boldsymbol{n} &= \bar{\boldsymbol{t}} \quad \text{on } \Gamma_t \text{ (natural boundary conditions)} \\
\boldsymbol{u} &= \bar{\boldsymbol{u}} \quad \text{on } \Gamma_u \text{ (essential boundary conditions)}
\end{aligned}, \tag{12}$$

respectively, where $\boldsymbol{\sigma} = \boldsymbol{D}\boldsymbol{\varepsilon}$ is the stress vector, $\boldsymbol{D}$ is the material property matrix,
$\boldsymbol{\varepsilon} = \boldsymbol{\nabla}_s\boldsymbol{u}$ is the strain vector, $\boldsymbol{u}$ is the displacement vector, $\boldsymbol{b}$ is the body force
vector, $\bar{\boldsymbol{t}}$ and $\bar{\boldsymbol{u}}$ are the vectors of prescribed surface tractions and displacements,
respectively, $\boldsymbol{n}$ is a unit normal to the domain $\mathcal{D}$, Γ_t and Γ_u are the portions of
boundary Γ where tractions and displacements are respectively prescribed, $\boldsymbol{\nabla}^T = \{\partial/\partial x_1, \partial/\partial x_2\}$ is the vector of gradient operators, and $\boldsymbol{\nabla}_s\boldsymbol{u}$ is the symmetric part
of $\boldsymbol{\nabla}\boldsymbol{u}$. The variational or weak form of Eqs. (11) and (12) is

$$\begin{aligned}
&\int_{\mathcal{D}} \boldsymbol{\sigma}^T \delta\boldsymbol{\varepsilon}\, d\mathcal{D} - \int_{\mathcal{D}} \boldsymbol{b}^T \delta\boldsymbol{u}\, d\mathcal{D} - \int_{\Gamma} \bar{\boldsymbol{t}}^T \delta\boldsymbol{u}\, d\Gamma_t \\
&+ \sum_{x_k \in \Gamma_u} \boldsymbol{f}^T(\boldsymbol{x}_k)\delta\boldsymbol{u}(\boldsymbol{x}_k) + \sum_{x_k \in \Gamma_u} \delta\boldsymbol{f}^T(\boldsymbol{x}_k)\left[\boldsymbol{u}(\boldsymbol{x}_k) - \bar{\boldsymbol{u}}(\boldsymbol{x}_k)\right] = 0,
\end{aligned} \tag{13}$$

where $\boldsymbol{f}^T(\boldsymbol{x}_k)$ is the vector of reaction forces at the constrained node k on Γ_u
and δ denotes the variation operator. From Eq. (7), the MLS approximation of
$\boldsymbol{u}(\boldsymbol{x}) = \{u_1(\boldsymbol{x}), u_2(\boldsymbol{x})\}^T$ in two dimensions is

$$\boldsymbol{u}^h(\boldsymbol{x}) = \boldsymbol{\Phi}^T\boldsymbol{d}, \tag{14}$$

where

$$\boldsymbol{\Phi}^T(\boldsymbol{x}) = \begin{bmatrix} \Phi_1(\boldsymbol{x}) & 0 & \cdots & \Phi_M(\boldsymbol{x}) & 0 \\ 0 & \Phi_1(\boldsymbol{x}) & \cdots & 0 & \Phi_M(\boldsymbol{x}) \end{bmatrix}, \tag{15}$$

$d = \{d_1^1, d_1^2, \ldots, d_M^1, d_M^2\}^T \in \mathbb{R}^{2M}$ is the vector of nodal parameters or generalized displacements, and M is the total number of nodal points in $\mathcal{D}$. Applying the MLS approximation of Eq. 14 to discretization of Eq. 13 yields a linear system of equilibrium equations

$$KY = F, \tag{16}$$

or

$$\underbrace{\begin{bmatrix} k & G \\ G^T & 0 \end{bmatrix}}_{K} \underbrace{\begin{Bmatrix} d \\ f_R \end{Bmatrix}}_{Y} = \underbrace{\begin{Bmatrix} f^{\text{ext}} \\ g \end{Bmatrix}}_{F}, \tag{17}$$

where

$$k = \begin{bmatrix} k_{11} & k_{12} & \cdots & k_{1M} \\ k_{21} & k_{22} & \cdots & k_{2M} \\ \vdots & \vdots & \vdots & \vdots \\ k_{M1} & k_{M2} & \cdots & k_{MM} \end{bmatrix} \in \mathbb{L}(\mathbb{R}^{2M} \times \mathbb{R}^{2M}) \tag{18}$$

is the stiffness matrix with

$$k_{IJ} = \int_{\mathcal{D}} B_I^T D B_J d\mathcal{D} \in \mathbb{L}(\mathbb{R}^2 \times \mathbb{R}^2), \tag{19}$$

$$G^T = \begin{bmatrix} \Phi_1(x_1) & 0 & \cdots & \Phi_1(x_M) & 0 \\ 0 & \Phi_1(x_1) & \cdots & 0 & \Phi_1(x_M) \\ \vdots & \vdots & \ddots & \vdots & \vdots \\ \Phi_L(x_1) & 0 & \cdots & \Phi_L(x_M) & 0 \\ 0 & \Phi_L(x_1) & \cdots & 0 & \Phi_L(x_M) \end{bmatrix} \tag{20}$$

is a matrix comprising shape function values of nodes at which the displacement boundary conditions are prescribed, L is the total number of nodes on Γ_u, $f_R = \{f(x_{k_1}), \ldots, f(x_{k_L})\}^T \in \mathbb{R}^{2L}$ is the vector of reaction forces on Γ_u,

$$f^{\text{ext}} = \int_{\mathcal{D}} \Phi^T b d\mathcal{D} + \int_{\Gamma_t} \Phi^T \bar{t} d\Gamma_t \in \mathbb{R}^{2M} \tag{21}$$

is the force vector, $g = \{\bar{u}(x_{k_1}), \ldots, \bar{u}(x_{k_L})\}^T \in \mathbb{R}^{2L}$ is the vector of prescribed displacements on Γ_u, and

$$B_I = \begin{bmatrix} \Phi_{I,1} & 0 \\ 0 & \Phi_{I,2} \\ \Phi_{I,2} & \Phi_{I,1} \end{bmatrix}. \tag{22}$$

To perform numerical integrations in Eqs. (19) and (21), a background mesh is required, which can be independent of the arrangement of meshfree nodes. However, in forthcoming numerical examples, the nodes of the background mesh coincide with the meshless nodes. Standard 4×4 Gaussian quadratures are used to evaluate the integrals for assembling the stiffness matrix and the force vector.

2.3. *Essential boundary conditions*

In solving for d, the essential boundary conditions must be enforced. The lack of Kronecker delta properties in meshless shape functions presents some difficulty in imposing essential boundary conditions in EFGM. Nevertheless, several methods are currently available for enforcing essential boundary conditions. In this work, a full transformation method[12,25] is employed for stochastic applications.

It should be noted that the generalized displacement vector d represents the nodal parameters, and not the actual displacements at meshfree nodes. Let $\hat{d} = \{u^h(x_1), \ldots, u^h(x_M)\}^T \in \mathbb{R}^{2M}$ represent the vector of nodal displacements. From Eq. (14)

$$\hat{d} = \Lambda d, \tag{23}$$

where $\Lambda = [\Phi^T(x_1), \ldots, \Phi^T(x_M)]^T \in \mathbb{L}(\mathbb{R}^{2M} \times \mathbb{R}^{2M})$ is the transformation matrix. Hence, $\hat{d}$ can be easily calculated when d is known.

In summary, shape functions and resultant matrix equilibrium equations have been created without using any structured mesh, a key advantage of meshfree methods over FEM. However, the computational effort in generating these matrix equations is typically higher than that required by low-order FEM. Therefore, breakthrough research focused on enhancing speed and robustness of meshfree methods is required for their effective implementation.

3. Random Field and Parameterization

3.1. *Karhunen-Loève representation*

Let $(\Omega, \mathcal{F}, P)$ be a probability space, where Ω is the sample space, $\mathcal{F}$ is the σ-algebra of subsets of Ω and P is the probability measure. Defined on the probability space and indexed by a spatial coordinate $x \in \mathcal{D} \subset \mathbb{R}^K, K = 1, 2,$ or 3, consider a real-valued random field $\alpha(x)$ with mean *zero* and covariance function $\Gamma(x_1, x_2) \equiv \mathbb{E}[\alpha(x_1)\alpha(x_2)]$, which is continuous over $\mathcal{D}$. Denote by $\mathcal{L}_2(\Omega, \mathcal{F}, P)$ or simply $\mathcal{L}_2$ a collection of random variables α for each $x \in \mathcal{D}$ such that $\mathbb{E}[|\alpha|^2] < \infty$, where $\mathbb{E}$ represents the expectation operator. If α is in $\mathcal{L}_2$, then $\Gamma(x_1, x_2)$ is square integrable and hence a bounded function.

Let $\{\lambda_i, f_i(x)\}, i = 1, 2, \ldots, \infty$, be the eigenvalues and eigenfunctions of $\Gamma(x_1, x_2)$, which satisfy the integral equation[26]

$$\int_{\mathcal{D}} \Gamma(x_1, x_2) f_i(x_2) dx_2 = \lambda_i f_i(x_1), \quad \forall i = 1, 2, \ldots, \infty. \tag{24}$$

The eigenfunctions are orthogonal in the sense that

$$\int_{\mathcal{D}} f_i(x) f_j(x) dx = \delta_{ij}, \quad \forall i, j = 1, 2, \ldots, \infty, \tag{25}$$

with δ_{ij} representing the Kronecker delta. The Karhunen-Loève (K-L) representation of $\alpha(\boldsymbol{x})$ is[26]

$$\alpha(\boldsymbol{x}) = \sum_{i=1}^{\infty} V_i \sqrt{\lambda_i} f_i(\boldsymbol{x}), \tag{26}$$

where $V_i, i = 1, \ldots, \infty$ is an infinite sequence of uncorrelated random variables, each of which has *zero* mean and *unit* variance. In practice, the infinite series of Eq. (26) must be truncated, yielding a K-L approximation or expansion

$$\hat{\alpha}_M(\boldsymbol{x}) = \sum_{i=1}^{N} V_i \sqrt{\lambda_i} f_i(\boldsymbol{x}), \tag{27}$$

which approaches $\alpha(\boldsymbol{x})$ in the mean square sense for $\boldsymbol{x} \in \mathcal{D}$ as $N \to \infty$. According to Eq. (27), the K-L expansion provides a parametric representation of a random field with N random variables.

3.2. *Gaussian and translation random fields*

The K-L approximation captures only the second-moment properties of a random field. Hence, a random field that is completely described by its second-moment properties, such as a Gaussian random field, can be effectively approximated by K-L expansion. For example, if a random field is Gaussian, its K-L approximation in Eq. (27) forms a *zero*-mean, independent sequence of standard Gaussian random variables V_i, $i = 1, \ldots, N$. For a general non-Gaussian field, K-L representation cannot provide complete characterization and therefore may not be applicable. However, one class of non-Gaussian random fields for which the use of K-L expansion can be readily exploited is the class of translation random fields, where a non-Gaussian random field is defined as a nonlinear, memoryless transformation of a Gaussian random field.[27]

Let $Z(\boldsymbol{x})$ be a homogenous, non-Gaussian translation random field with mean μ_Z, standard deviation σ_Z, covariance function $\Gamma_Z(\boldsymbol{\xi}) \equiv \mathbb{E}[(Z(\boldsymbol{x}) - \mu_Z)(Z(\boldsymbol{x}+\boldsymbol{\xi}) - \mu_Z)]$, and marginal distribution F that has no atoms, and let $\alpha(\boldsymbol{x})$ be a homogeneous, *zero*-mean, Gaussian random field with *unit* variance and covariance function $\Gamma_\alpha(\boldsymbol{\xi}) \equiv \mathbb{E}[\alpha(\boldsymbol{x})\alpha(\boldsymbol{x}+\boldsymbol{\xi})]$. If G is a real-valued, monotonic, differentiable function, then,

$$Z(\boldsymbol{x}) = G[\alpha(\boldsymbol{x})] \tag{28}$$

can be viewed as a memoryless transformation of the Gaussian image field $\alpha(\boldsymbol{x})$. From the condition that the marginal distribution and the covariance function of $Z(\boldsymbol{x})$ coincide with specified target functions F and Γ_Z, respectively, it can be shown that[28,29]

$$G(\alpha) = F^{-1}[\Phi(\alpha)], \tag{29}$$

and

$$\mu_Z^2 + \Gamma_Z(\boldsymbol{\xi}) = \int_\infty^\infty \int_\infty^\infty (G(\alpha_1) - \mu_z)(G(\alpha_2) - \mu_z)\phi_2(\alpha_1, \alpha_2, \Gamma_\alpha(\boldsymbol{\xi}))d\alpha_1 d\alpha_2, \quad (30)$$

where Φ is the distribution function of a standard Gaussian variable and $\phi_2(\alpha_1, \alpha_2, \Gamma_\alpha(\boldsymbol{\xi}))$ is the bivariate standard Gaussian density function with correlation coefficient Γ_α. For given μ_Z, $\Gamma_Z(\boldsymbol{\xi})$, and F, G can be calculated from Eq. (29) and the required covariance function $\Gamma_\alpha(\boldsymbol{\xi})$ of $\alpha(\boldsymbol{x})$ can be solved from Eq. (30), if the target scaled covariance function $\Gamma_Z(\boldsymbol{\xi})/\Gamma_Z(\mathbf{0})$ is in the range $(\bar{\Gamma}^*, 1)$, where[29]

$$\bar{\Gamma}^* = \frac{\mathbb{E}\left[G(\alpha)G(-\alpha)\right] - \mathbb{E}\left[G(\alpha)\right]^2}{\mathbb{E}\left[G(\alpha)^2\right] - \mathbb{E}\left[G(\alpha)\right]^2}. \quad (31)$$

Once $\Gamma_\alpha(\boldsymbol{\xi})$ is determined, the parameterization of $Z(\boldsymbol{x})$ is achieved by K-L expansion of its Gaussian image, i.e.

$$Z(\boldsymbol{x}) \simeq G\left[\sum_{i=1}^N V_i \sqrt{\lambda_i} f_i(\boldsymbol{x})\right], \quad (32)$$

where V_i, $i = 1, \ldots, N$ are independent standard Gaussian random variables, and $\{\lambda_i, f_i(\boldsymbol{x})\}$, $i = 1, \ldots, N$ are eigenvalues and eigenfunctions of $\Gamma_\alpha(\boldsymbol{\xi})$.

3.3. Meshfree method for solving integral equation

The K-L expansion requires solution of an integral eigenvalue problem (Eq. (24)), which is not an easy task in general. Closed-form solutions are only available when the covariance kernel has simpler functional forms, such as exponential and linear functions, or domain $\mathcal{D}$ is rectangular. For arbitrary covariance functions or arbitrary domains, numerical methods are often needed to solve the eigenvalue problem. In this section, meshfree shape functions from EFGM are employed to solve the eigenvalue problem.[18]

For a random field $\alpha(\boldsymbol{x})$ indexed by $\boldsymbol{x} \in \mathcal{D} \subset \mathbb{R}^K$, $K = 1, 2$, or 3, consider an MLS approximation of the eigenfunction $f_i(\boldsymbol{x})$, given by

$$f_i(\boldsymbol{x}) = \sum_{I=1}^M \hat{f}_{iI} \Phi_I(\boldsymbol{x}), \quad (33)$$

where $\hat{f}_{iI}$ is the Ith nodal parameter for the ith eigenfunction, $\Phi_I(\boldsymbol{x})$ is the meshless shape function of the Ith node (see Sec. 2), and M is the total number of nodes. Hence, Eq. (24) becomes

$$\sum_{I=1}^M \hat{f}_{iI} \int_\mathcal{D} \Gamma(\boldsymbol{x}_1 \boldsymbol{x}_2)\Phi_I(\boldsymbol{x}_2)d\boldsymbol{x}_2 - \lambda_i \sum_{I=1}^M \hat{f}_{iI}\Phi_I(\boldsymbol{x}_1). \quad (34)$$

Define

$$\varepsilon_M = \sum_{I=1}^M \hat{f}_{iI}\left(\int_\mathcal{D} \Gamma(\boldsymbol{x}_1, \boldsymbol{x}_2)\Phi_I(\boldsymbol{x}_2)d\boldsymbol{x}_2 - \lambda_i \Phi_I(\boldsymbol{x}_1)\right) \quad (35)$$

as the residual error, which is associated with meshless discretization involving an M number of nodes. Following Galerkin approximation,

$$\int_{\mathcal{D}} \varepsilon_M \Phi_J(\boldsymbol{x}_1) d\boldsymbol{x}_1 = 0, \quad \forall J = 1, \ldots, M, \tag{36}$$

which, when combined with Eq. (35) can be expanded to yield the following matrix equation

$$\lambda_i \boldsymbol{R} \hat{\boldsymbol{f}}_i = \boldsymbol{S} \hat{\boldsymbol{f}}_i, \tag{37}$$

where $\hat{\boldsymbol{f}}_i = \{\hat{f}_{i1}, \ldots, \hat{f}_{iM}\}^T$ is the ith eigenvector, $\boldsymbol{R} = [R_{IJ}]$, $\boldsymbol{S} = [S_{IJ}]$,

$$R_{IJ} = \int_{\mathcal{D}} \int_{\mathcal{D}} \Gamma(\boldsymbol{x}_1, \boldsymbol{x}_2) \Phi_I(\boldsymbol{x}_2) \Phi_J(\boldsymbol{x}_1) d\boldsymbol{x}_1 d\boldsymbol{x}_2, \quad \forall I, J = 1 \ldots, M, \tag{38}$$

and

$$S_{IJ} = \int_{\mathcal{D}} \Phi_I(\boldsymbol{x}) \Phi_J(\boldsymbol{x}) d\boldsymbol{x}, \quad \forall I, J = 1, \ldots, M. \tag{39}$$

Equation (37) represents the matrix analog of the integral eigenvalue problem for a multi-dimensional random field with an arbitrary domain. Equation (37) can be formulated for any covariance function or domain and can be easily solved by standard methods. Hence, the meshfree method can solve problems involving a multi-dimensional random field with an arbitrary covariance function and an arbitrary domain. Once the eigenvector $\hat{\boldsymbol{f}}_i$ is calculated, Eq. (33) can be used to determine the eigenfunction $f_i(\boldsymbol{x})$.

Note that the meshless discretization proposed here is only intended for solving the integral eigenvalue problem, not for discretizing the random field. Matrices $\boldsymbol{R}$ and $\boldsymbol{S}$, which involve $2K$- and K-dimensional integration, respectively, can be computed using standard numerical quadrature. Integration involves meshless shape functions, which are already calculated and stored for meshless stress analysis. Hence, matrices $\boldsymbol{R}$ and $\boldsymbol{S}$ can be generated with little extra effort. However, for a large K, the computational effort in performing numerical integration can become intensive. Also note that for meshless stress analysis it is not necessary that the number and spatial distribution of nodes coincide with those for eigenfunction approximation. Different and selective discretizations can be employed, if necessary. However, in this study the same discretization is used for both meshless stress analysis and for solving the eigenvalue problem.

3.4. Example 1: Eigensolution for a two-dimensional domain

Consider a two-dimensional domain $\mathcal{D}$ that is constructed by subtracting a quarter of a circle of radius $a = 1$ unit from a square of size $L = 20$ units, as depicted by Fig. 2. A homogeneous Gaussian random field $\alpha(\boldsymbol{x})$ defined over $\mathcal{D}$ has mean *zero* and a bounded covariance function

$$\Gamma_\alpha(\boldsymbol{\xi}) = \sigma_\alpha^2 \exp\left(-\frac{\|\boldsymbol{\xi}\|}{bL}\right), \quad \forall \boldsymbol{x}, \boldsymbol{x} + \boldsymbol{\xi} \in \mathcal{D}, \tag{40}$$

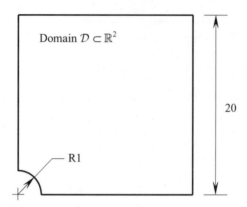

Fig. 2.　A two-dimensional domain $\mathcal{D}$ for random field $\alpha(\boldsymbol{x}), \boldsymbol{x} \in \mathcal{D} \subset \mathbb{R}^2$.

where $\sigma_\alpha = 0.1$ unit and $b = 0.5$. Since the domain is not rectangular, no analytical solution of eigenvalues and eigenfunctions exists for the above covariance function. Therefore, the meshfree method is needed to find a numerical solution. Figures 3(a) though 3(e) show five meshfree discretizations of $\mathcal{D}$ with total number of nodes $M = 9, 20, 30, 56$, and 90, respectively, which represent progressively increasing degrees of refinement.[18]

Figure 4 shows several eigenvalues calculated using the meshfree method (Eqs. 33–39) for $M = 9, 20, 30, 56$, and 90, for the given covariance kernel. Clearly, the eigenvalues converge with respect to M, as expected. Similar comparisons of the first four eigenfunctions $f_1(\boldsymbol{x})$, $f_2(\boldsymbol{x})$, $f_3(\boldsymbol{x})$, and $f_4(\boldsymbol{x})$, presented in Figs. 5(a), 5(b),

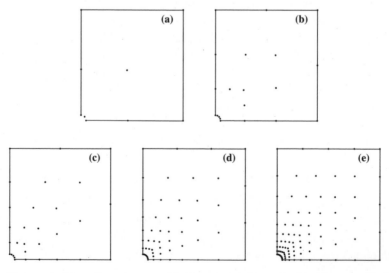

Fig. 3.　Various meshfree discretizatons; (a) $M = 9$; (b) $M = 20$; (c) $M = 30$; (d) $M = 56$; (e) $M = 90$.[18]

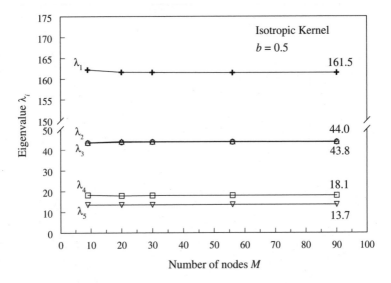

Fig. 4.   Eigenvalues for various meshfree discretizations.[18]

6(a), and 6(b), respectively, also demonstrate the convergence of eigenfunctions with respect to M. These convergent solutions of eigenvalues and eigenfunctions provide confidence in the following probabilistic results.

4. Multivariate Function Decomposition

Consider a continuous, differentiable, real-valued multivariate function $y(\boldsymbol{v})$ that depends on $\boldsymbol{v} = \{v_1, \cdots, v_N\}^T \in \mathbb{R}^N$. Suppose, $y(\boldsymbol{v})$ has a convergent Taylor expansion at an arbitrary reference point $\boldsymbol{c} = \{c_1, \cdots, c_N\}^T$. Applying the Taylor series expansion of $y(\boldsymbol{v})$ at $\boldsymbol{v} = \boldsymbol{c}$, $y(\boldsymbol{v})$ can be expressed by

$$y(\boldsymbol{v}) = y(\boldsymbol{c}) + \sum_{j=1}^{\infty} \frac{1}{j!} \sum_{i=1}^{N} \frac{\partial^j y}{\partial v_i^j}(\boldsymbol{c})(v_i - c_i)^j + R_2, \qquad (41)$$

or

$$y(\boldsymbol{v}) = y(c) + \sum_{j=1}^{\infty} \frac{1}{j!} \sum_{i=1}^{N} \frac{\partial^j y}{\partial v_i^j}(\boldsymbol{c})(v_i - c_i)^j$$

$$+ \sum_{j_1,j_2>0}^{\infty} \frac{1}{j_1!j_2!} \sum_{i_1<i_2} \frac{\partial^{j_1+j_2} y}{\partial v_{i_1}^{j_1} \partial v_{i_2}^{j_2}}(\boldsymbol{c})(v_{i_1} - c_{i_1})^{j_1}(v_{i_2} - c_{i_2})^{j_2} + R_3, \quad (42)$$

where the remainder R_2 denotes all terms with dimension two and higher and the remainder R_3 denotes all terms with dimension three and higher.

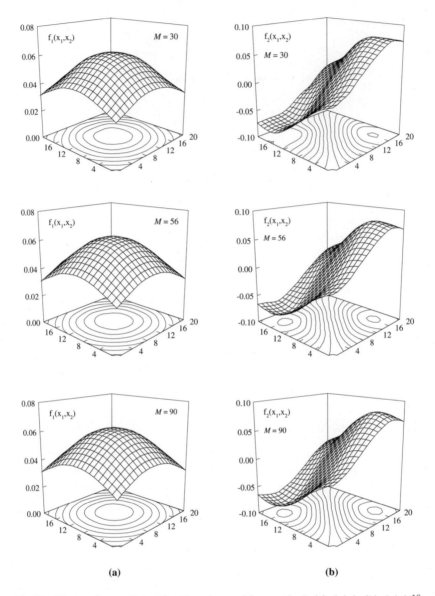

Fig. 5.   First and second eigenfunctions by meshfree method; (a) $f_1(\boldsymbol{x})$; (b) $f_2(\boldsymbol{x})$.[18]

4.1. Univariate approximation

Consider a univariate approximation of $y(\boldsymbol{v})$, denoted by

$$\hat{y}_1(\boldsymbol{v}) \equiv \hat{y}_1(v_1, \ldots, v_N) = \sum_{i=1}^{N} y(c_1, \ldots, c_{i-1}, v_i, c_{i+1}, \ldots, c_N) - (N-1)y(\boldsymbol{c}), \quad (43)$$

where each term in the summation is a function of only one variable and can be subsequently expanded in a Taylor series at $\boldsymbol{v} = \boldsymbol{c}$ yielding

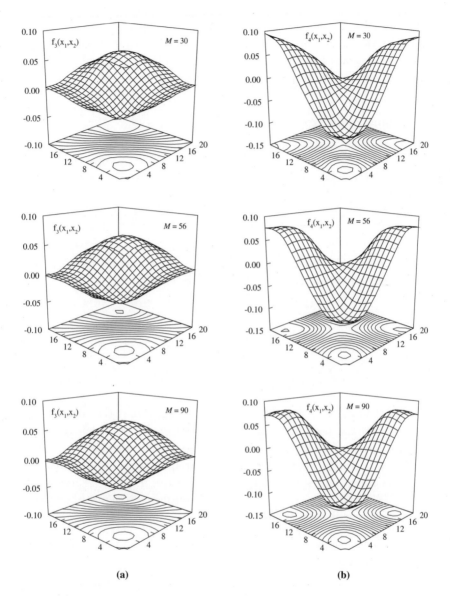

Fig. 6.   Third and fourth eigenfunctions by meshfree method; (a) $f_3(\boldsymbol{x})$; (b) $f_4(\boldsymbol{x})$.[18]

$$\hat{y}_1(\boldsymbol{v}) = y(\boldsymbol{c}) + \sum_{j=1}^{\infty} \frac{1}{j!} \sum_{i=1}^{N} \frac{\partial^j y}{\partial v_i^j}(\boldsymbol{c})(v_i - c_i)^j. \tag{44}$$

Comparison of Eqs. (41) and (44) indicates that the univariate approximation leads to the residual error $y(\boldsymbol{v}) - \hat{y}_1(\boldsymbol{v}) = R_2$, which includes contributions from terms of dimension two and higher. For a sufficiently smooth $y(\boldsymbol{v})$ having a convergent Taylor series, the coefficients associated with higher-dimensional terms are usually much smaller than those associated with one-dimensional terms. As such, higher-dimensional terms contribute less to the function, and therefore, can be neglected. Furthermore, Eq. (43) exactly represents $y(v) = \sum y_i(v_i)$, i.e. when $y(v)$ can be additively decomposed into functions $y_i(x_i)$ of single variables.

4.2. Bivariate approximation

In a similar manner, consider a bivariate approximation

$$\hat{y}_2(v) = \sum_{i_1 < i_2} y(c_1, \ldots, c_{i_1-1}, v_{i_1}, c_{i_1+1}, \ldots, c_{i_2-1}, v_{i_2}, c_{i_2+1}, \ldots, c_N)$$

$$- (N-2) \sum_{i=1}^{N} y(c_1, \ldots, c_{i-1}, v_i, c_{i+1}, \ldots, c_N) + \frac{(N-1)(N-2)}{2} y(c) \quad (45)$$

of $y(v)$, where each term on the right hand side is a function of at most two variables and can be expanded in a Taylor series at $v = c$, yielding

$$\hat{y}_2(v) = y(c) + \sum_{j=1}^{\infty} \frac{1}{j!} \sum_{i=1}^{N} \frac{\partial^j y}{\partial v_i^j}(c)(v_i - c_i)^j$$

$$+ \sum_{j_1, j_2 > 0}^{\infty} \frac{1}{j_1! j_2!} \sum_{i_1 < i_2} \frac{\partial^{j_1+j_2} y}{\partial v_{i_1}^{j_1} \partial v_{i_2}^{j_2}}(c)(v_{i_1} - c_{i_1})^{j_1}(v_{i_2} - c_{i_2})^{j_2}. \quad (46)$$

Again, the comparison of Eqs. (42) and (46) indicates that the bivariate approximation leads to the residual error $y(v) - \hat{y}_2(v) = R_3$, in which the remainder R_3 includes terms of dimension three and higher. The bivariate approximation includes all terms with no more than two variables, thus yielding higher accuracy than the univariate approximation. Furthermore, Eq. (45) exactly represents $y(v) = \sum \sum y_{ij}(v_i, v_j)$, i.e. when $y(v)$ can be additively decomposed into functions $y_{ij}(v_i, v_j)$ of at most two variables.

4.3. Generalized S-variate approximation

The procedure for univariate and bivariate approximations described in the preceding can be generalized to an S-variate approximation for any integer $1 \leq S \leq N$. The generalized S-variate approximation of $y(v)$ is[22-24]

$$\hat{y}_S(v) \equiv \sum_{i=0}^{S} (-1)^i \binom{N-S+i-1}{i} y_{S-i}(v), \quad (47)$$

where

$$y_R = \sum_{k=0}^{R} \binom{N-k}{R-k} t_k; \quad 0 \leq R \leq S, \quad (48)$$

with

$$t_0 = y(\mathbf{c})$$

$$t_1 = \sum_{j_1} \frac{1}{j_1!} \sum_{i_1=1}^N \frac{\partial^{j_1} y}{\partial v_{i_1}^{j_1}} (\mathbf{c}) \left(v_{i_1} - c_{i_1}\right)^{j_1}$$

$$t_2 = \sum_{j_1,j_2} \frac{1}{j_1! j_2!} \sum_{i_1 < i_2} \frac{\partial^{j_1+j_2} y}{\partial v_{i_1}^{j_1} \partial v_{i_2}^{j_2}} (\mathbf{c}) \left(v_{i_1} - c_{i_1}\right)^{j_1} \left(v_{i_2} - c_{i_2}\right)^{j_2} \tag{49}$$

$$\vdots \quad \vdots \quad \vdots$$

$$t_S = \sum_{j_1,\dots,j_S} \frac{1}{j_1! \cdots j_S!} \sum_{i_1 < \cdots < i_S} \frac{\partial^{j_1+\cdots+j_S} y}{\partial v_{i_1}^{j_1} \cdots \partial v_{i_S}^{j_S}} (\mathbf{c}) \left(v_{i_1} - c_{i_1}\right)^{j_1} \cdots \left(v_{i_S} - c_{i_S}\right)^{j_S}.$$

Using a multivariate function decomposition theorem, developed by the author's group, it can be shown that $\hat{y}_S(\mathbf{v})$ in Eq. (47) consists of all terms of the Taylor series of $y(\mathbf{v})$ that have less than or equal to S variables.[23] The expanded form of Eq. (47), when compared with the Taylor expansion of $y(\mathbf{v})$, indicates that the residual error in the S-variate approximation is $y(\mathbf{v}) - \hat{y}_S(\mathbf{v}) = R_{S+1}$, where the remainder R_{S+1} includes terms of dimension $S + 1$ and higher. When $S = 1$, Eq. (47) degenerates to the univariate approximation (Eq. (43)). When $S = 2$, Eq. (47) becomes the bivariate approximation (Eq. (45)). Similarly, trivariate, quadrivariate, and other higher-variate approximations can be derived by appropriately selecting the value of S. In the limit, when $S = N$, Eq. (47) converges to the exact function $y(\mathbf{v})$. In other words, the decomposition technique generates a convergent sequence of approximations of $y(\mathbf{v})$.

5. Statistical Moment Analysis

5.1. *General stochastic response*

Consider a mechanical system subject to a *zero*-mean independent random input vector $\mathbf{V} = \{V_1, \dots, V_N\}^T \in \mathbb{R}^N$, which characterizes uncertainty in loads, material properties, and geometry. Let $g(\mathbf{V})$ represent a general stochastic response of interest, for which the lth statistical moment

$$m_l \equiv \mathbb{E}\left[g^l(\mathbf{V})\right] = \int_{\mathbb{R}^N} g^l(\mathbf{v}) f_{\mathbf{V}}(\mathbf{v}) d\mathbf{v} \tag{50}$$

is sought, where $f_{\mathbf{V}}(\mathbf{v})$ is the joint probability density function of $\mathbf{V}$. If $y(\mathbf{V}) = g^l(\mathbf{V})$, the lth moment can also be evaluated from

$$m_l = \mathbb{E}[y(\mathbf{V})] = \int_{\mathbb{R}^N} y(\mathbf{v}) f_{\mathbf{V}}(\mathbf{v}) d\mathbf{v}. \tag{51}$$

Following the S-variate approximation procedure discussed in Eqs. (47)–(49) and using $\boldsymbol{c} = 0$ (mean input),

$$m_l \cong \mathbb{E}\left[\hat{y}_S(\boldsymbol{V})\right] = \sum_{i=0}^{S} (-1)^i \binom{N-S+i-1}{i} \times \sum_{k_1 < k_2 < \cdots < k_{S-i}}$$
$$\mathbb{E}\left[y\left(0, \ldots, 0, V_{k_1}, 0, \ldots, 0, V_{k_2}, 0, \ldots, 0, V_{k_{S-i}}, 0, \ldots, 0\right)\right]. \qquad (52)$$

If $f_{V_{k_j}}(v_{k_j})$ represents the marginal density of V_{k_j}, then by definition

$$\mathbb{E}\left[y\left(0, \ldots, 0, V_{k_1}, 0, \ldots, 0, V_{k_2}, 0, \ldots, 0, V_{k_{S-i}}, 0, \ldots, 0\right)\right]$$
$$\equiv \int_{-\infty}^{\infty} y\left(0, \ldots, 0, v_{k_1}, 0, \ldots, 0, v_{k_2}, 0, \ldots, 0, v_{k_{S-i}}, 0, \ldots, 0\right) \prod_{j=1}^{S-i} f_{V_{k_j}}(v_{k_j}) dv_{k_j},$$
$$(53)$$

which is valid for any independent random vector $\boldsymbol{V}$. If $\boldsymbol{V}$ comprises dependent variables, an appropriate transformation, such as the Rosenblatt transformation,[30] should be applied to map the dependent random vector $\boldsymbol{V}$ to an independent standard Gaussian random vector $\boldsymbol{U}$. Note that Eq. (53) only requires at most S-dimensional deterministic integration, which can be more easily evaluated using standard quadrature rules if $S \ll N$. For example, Gauss-Legendre and Gauss-Hermite quadratures are frequently used when V_j follows uniform and Gaussian distributions, respectively.[31] For an arbitrary distribution of V_j, a moment-based quadrature rule developed by the author can be used to evaluate the integral.[23]

The moment equation entails evaluating at most S-dimensional integrals, which is substantially simpler and more efficient than performing one N-dimensional integration when $S \ll N$. For practical problems involving a large number of input random variables (e.g. $N > 30$), the moment equation presents a promising method. The method does not require calculation of any partial derivatives of response and inversion of random matrices as compared with, respectively, the commonly used Taylor/perturbation and Neumann expansion methods. Hence, the computation effort in conducting statistical moment analysis is significantly reduced using the function decomposition technique. The method is coined "*S-variate* or *multivariate dimension-reduction method*," since calculation of an N-dimensional integral is essentially reduced to that of an at most S-dimensional integral.[23] When $S = 1$, the method degenerates to the *univariate dimension-reduction method* involving only one-dimensional integrations.[22] When $S = 2$, the method becomes the *bivariate dimension-reduction method* entailing at most two-dimensional integrations. Similarly, trivariate, quadrivariate, and other higher-variate methods can be derived by appropriately selecting the value of S. In the limit, when $S = N$, there is no dimension reduction and the method yields the exact solution.

5.2. Discrete equilibrium equations

Consider a linear mechanical system subject to a vector of input random parameters $\boldsymbol{V} \in \mathbb{R}^N \mapsto (\boldsymbol{\mu}, \boldsymbol{\gamma})$ characterizing uncertainty in the system and loads. Following discretization, let $Y \in \mathbb{R}^L \mapsto (m_Y, \gamma_Y)$ represent a displacement (response) vector associated with L degrees of freedom of the system, satisfying the linear equilibrium equation

$$K(V)Y(V) = F(V), \tag{54}$$

in which the stiffness matrix K and force vector F depend on V and were defined in Sec. 2 in the review of meshfree formulation for linear elasticity. Equation (54) is common in mesh-free methods when the system, loads, or both, are uncertain. The solution

$$Y(V) = K(V)^{-1}F(V) \tag{55}$$

is random and depends on V. Using the S-variate dimension-reduction method, the mean vector m_Y and covariance matrix γ_Y of Y can be derived as

$$m_Y \cong \mathbb{E}[\hat{Y}] = \sum_{i=0}^{S} (-1)^i \binom{N-S+i-1}{i} \sum_{k_1 < k_2 < \cdots < k_{S-i}} \mathbb{E}\left[K(\tilde{V}_i)^{-1}F(\tilde{V}_i)\right], \tag{56}$$

$$\gamma_Y = \mathbb{E}\left[YY^T\right] - m_Y m_Y^T, \tag{57}$$

where $\tilde{V}_i = \{0, \ldots, 0, V_{k_1}, 0, \ldots, 0, V_{k_{S-i}}, 0, \ldots, 0\}^T$ and

$$\mathbb{E}[YY^T] \cong \sum_{i=0}^{S} (-1)^i \binom{N-S+i-1}{i}$$
$$\times \sum_{k_1 < k_2 < \ldots < k_{S-i}} \mathbb{E}\left[K(\tilde{V}_i)^{-1}F(\tilde{V}_i)F(\tilde{V}_i)^T K(\tilde{V}_i)^{-T}\right]. \tag{58}$$

Note that the calculation of expected values on the right hand side of Eqs. (56)–(58) involves at most S-dimensional integrations.

5.3. *Example 2: Response statistics of a plate with a hole*

Consider a square plate with a circular hole, as shown in Fig. 7. The plate has dimension $2L = 40$ units, a hole with diameter $2a = 2$ units, and is subjected to a far-field, uniformly distributed stress of magnitude $\sigma^\infty = 1$ unit. The Poisson's ratio $\nu = 0.3$. The elastic modulus is a homogeneous Gaussian random field $E(x) = \mu_E[1 + \alpha(x)]$; $x \in \mathcal{D} \subset \mathbb{R}^2$, where $\mu_E = 1$ unit is the constant mean over domain $\mathcal{D}$ and $\alpha(x)$ is a homogeneous Gaussian random field with mean *zero* and covariance function defined by Eq. (40) in Example 1.[a] Furthermore, the modulus of elasticity is assumed to be symmetrically distributed with respect to x_1- and x_2-axes (see Fig. 7). Therefore, only a quarter of the plate needs to be analyzed.

---

[a]The Gaussian random field is adopted in Example 2 to allow direct comparison between dimension-reduction methods and Neumann expansion method,[32] of which the latter method entails the Gaussian assumption. The dimension-reduction methods do not require any specific distribution of random fields or variables.[23]

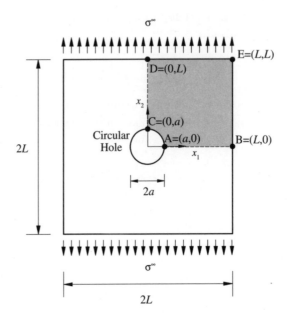

Fig. 7.   A square plate with a hole subjected to uniform tension.

Figures 3(a)–3(e) show several meshfree discretizations of the quarter plate with various degrees of refinement. A plane stress condition is assumed.

Based on the correlation parameter $b = 0.5$, a value of $N = 12$ was selected for the K-L approximation of $\alpha(\boldsymbol{x})$. The meshfree method using the finest discretization (i.e. $M = 90$; see Fig. 3(e) in Example 1) was employed to obtain both the stress field and eigensolutions $\{\lambda_j, \phi_j(\boldsymbol{x})\}$, $j = 1, \ldots, 12$. Hence, the input random vector $\boldsymbol{V} = \{V_1, \ldots, V_{12}\}^T$ becomes a twelve-dimensional standard Gaussian random vector.

Table 1 presents standard deviations of displacements and strains at points A, B, C, D, and E (see Fig. 7), predicted by the univariate and bivariate dimension-reduction methods, as well as results of a fourth-order Neumann expansion method and a Monte Carlo simulation (5000 samples).[23] The Neumann expansion solutions are obtained by following the development of Spanos and Ghanem.[32] As can be seen in Table 1, the Neumann expansion and dimension-reduction methods provide satisfactory results for prediction of standard deviations in comparison with simulation results. The accuracy of the response statistics from the bivariate dimension-reduction method is similar to the Neumann expansion method, and is slightly higher than the univariate dimension-reduction method. More importantly, however, a comparison of CPU times, shown in Fig. 8, indicates that the univariate dimension-reduction method is far more efficient than the Neumann expansion method. From Fig. 8, it can be seen that the bivariate dimension-reduction method also surpasses the computational efficiency of the fourth-order Neumann expansion method.

Table 1. Standard deviations of displacements and strains by various methods.[23]

Location	Response variable	Standard deviation of response			
		4th-order Neumann expansion method	Univariate dimension-reduction method	Bivariate dimension-reduction method	Monte Carlo simulation (5000 samples)
A	u_1	1.17×10^{-1}	1.15×10^{-1}	1.17×10^{-1}	1.19×10^{-1}
	ε_{11}	2.78×10^{-2}	2.72×10^{-2}	2.78×10^{-2}	2.79×10^{-2}
	ε_{22}	2.57×10^{-1}	2.51×10^{-1}	2.57×10^{-1}	2.58×10^{-1}
	ε_{12}	3.52×10^{-2}	3.45×10^{-2}	3.52×10^{-2}	3.54×10^{-2}
B	u_1	4.92×10^{-1}	4.83×10^{-1}	4.93×10^{-1}	4.95×10^{-1}
	ε_{22}	8.58×10^{-2}	8.41×10^{-2}	8.59×10^{-2}	8.49×10^{-2}
C	u_2	2.64×10^{-1}	2.58×10^{-1}	2.64×10^{-1}	2.66×10^{-1}
	ε_{11}	9.12×10^{-2}	8.92×10^{-2}	9.13×10^{-2}	9.28×10^{-2}
	ε_{22}	1.38×10^{-2}	1.35×10^{-2}	1.38×10^{-2}	1.41×10^{-2}
	ε_{12}	4.06×10^{-2}	3.97×10^{-2}	4.07×10^{-2}	4.13×10^{-2}
D	u_2	1.44	1.41	1.44	1.44
	ε_{22}	8.76×10^{-2}	8.53×10^{-2}	8.77×10^{-2}	8.52×10^{-2}
E	u_1	6.03×10^{-1}	5.91×10^{-1}	6.04×10^{-1}	5.98×10^{-1}
	u_2	1.46	1.44	1.47	1.46
	ε_{22}	8.74×10^{-2}	8.53×10^{-2}	8.76×10^{-2}	8.59×10^{-2}

Note: u_1 and u_2 are horizontal and vertical displacements, respectively; ε_{11} and ε_{22} represent normal tensorial strains; and ε_{12} represents tensorial shear strain.

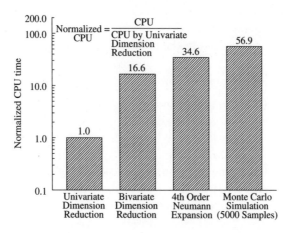

Fig. 8. Comparison of CPU times for moment analysis by various methods.[23]

6. Reliability Analysis

A fundamental problem in time-invariant reliability analysis entails calculation of a multi-fold integral[33–35]

$$P_F \equiv P(\boldsymbol{V} \in \Omega_F) = \int_{\Omega_f} f_V(\boldsymbol{v})d\boldsymbol{v}, \tag{59}$$

where $\Omega_F \subset \Omega$ defines the failure domain and P_F is the probability of failure. For component reliability analysis, $\Omega_F = \{\boldsymbol{v} : y(\boldsymbol{v}) < 0\}$, where $y(\boldsymbol{v})$ represents a single performance function. For system reliability analyses involving r performance functions, $\Omega_F = \{\boldsymbol{v} : \bigcup_{k=1}^{r} y^{(k)}(\boldsymbol{v}) < 0\}$ and $\Omega_F = \{\boldsymbol{v} : \bigcap_{k=1}^{r} y^{(k)}(\boldsymbol{v}) < 0\}$ for series and parallel systems, respectively, where $y^{(k)}(\boldsymbol{v})$ represents the kth performance function. Nevertheless, for most practical problems, the exact evaluation of this integral, either analytically or numerically, is not possible since N is large and $y(\boldsymbol{v})$ or $y^{(k)}(\boldsymbol{v})$ are highly nonlinear functions of $\boldsymbol{v}$.

The most common approach to compute the failure probability in Eq. (59) involves the first- and second-order reliability methods (FORM/SORM),[33-35] which are based on linear (FORM) or quadratic approximation (SORM) of the limit-state surface at a most probable point (MPP). Experience has shown that FORM/SORM are sufficiently accurate for engineering purposes, provided that the limit-state surface at the MPP is close to being linear or quadratic, and no multiple MPPs exist.[35] Otherwise, the results of FORM/SORM should be interpreted with caution. Simulation methods involving sampling and estimation are well known in the statistics and reliability literature.[36-38] While simulation methods do not exhibit the limitations of approximate reliability methods, such as FORM/SORM, they generally require considerably more extensive calculations than the latter methods. Consequently, simulation methods are useful when alternative methods are inapplicable or inaccurate, and have been traditionally employed as a yardstick for evaluating approximate methods.

In this work, innovative response-surface approximations using multivariate function decomposition, which provides accurate and computationally efficient reliability estimates, are presented in the following subsection.

6.1. *Response surface generation*

Consider the univariate terms $y_i(v_i) \equiv y(c_1, \ldots, c_{i-1}, v_i, c_{i+1}, \ldots, c_N)$ in Eqs. (43) and (45). If for $v_i = v_i^{(j)}$, n function values

$$y_i(v_i^{(j)}) = y(c_1, \ldots, c_{i-1}, v_i^{(j)}, c_{i+1}, \ldots, c_N); \quad j = 1, 2, \ldots, n \tag{60}$$

are given, the function value for arbitrary v_i can be obtained using the Lagrange interpolation as

$$y_i(v_i) = \sum_{j=1}^{n} \phi_j(v_i) y_i(v_i^{(j)}), \tag{61}$$

where the Lagrange shape function $\phi_j(v_i)$ is defined as

$$\phi_j(v_i) = \frac{\prod_{k=1, k \neq j}^{n} \left(v_i - v_i^{(k)}\right)}{\prod_{k=1, k \neq j}^{n} \left(v_i^{(j)} - v_i^{(k)}\right)}. \tag{62}$$

Using Eq. (61), arbitrarily numerous function values of $y_i(v_i)$ can be generated if n function values are given. This is defined as the *univariate method*.[24]

The same concept can be applied to the bivariate terms $y_{i_1 i_2}(v_{i_1}, v_{i_2}) \equiv y(c_1, \ldots, c_{i_1-1}, v_{i_1}, c_{i_1+1}, \ldots, c_{i_2-1}, v_{i_2}, c_{i_2+1}, \ldots, c_N)$ in Eq. (45). If for $v_{i_1} = v_{i_1}^{(j_1)}$ and $v_{i_2} = v_{i_2}^{(j_2)}$, n^2 function values

$$y_{i_1 i_2}(v_{i_1}^{(j_1)}, v_{i_2}^{(j_2)}) \equiv y(c_1, \ldots, c_{i_1-1}, v_{i_1}^{(j_1)}, c_{i_1+1}, \ldots, c_{i_2-1}, v_{i_2}^{(j_2)}, c_{i_2+1}, \ldots, c_N);$$
$$j_1 = 1, 2, \ldots, n; \quad j_2 = 1, 2, \ldots, n \tag{63}$$

are given, the function value $y_{i_1 i_2}(v_{i_1}, v_{i_2})$ for arbitrary point (v_{i_1}, v_{i_2}) can be obtained using the Lagrange interpolation as

$$y_{i_1 i_2}(v_{i_1}, v_{i_2}) = \sum_{j_2=1}^{n} \sum_{j_1=1}^{n} \phi_{j_1}(v_{i_1}) \phi_{j_2}(v_{i_2}) y_{i_1 i_2}(v_{i_1}^{(j_1)}, v_{i_2}^{(j_2)}), \tag{64}$$

where shape functions $\phi_{j_1}(v_{i_1})$ and $\phi_{j_2}(v_{i_2})$ are already defined in Eq. (62). The resulting approximation is defined as the *bivariate method*.[24] Note that there are n and n^2 performance function evaluations (e.g. meshfree analyses) involved in Eqs. (61) and (64), respectively. Therefore, the total maximum cost for univariate method entails $nN + 1$ function evaluations, and for bivariate method, $N(N-1)n^2/2 + nN + 1$ maximum function evaluations are required. More accurate multivariate methods, such as an S-variate $(S > 2)$ decomposition method, can be developed in a similar manner.

6.2. *Monte Carlo simulation*

For component reliability analysis, the Monte Carlo estimate of the failure probability employing S-variate response surface method is[24]

$$P_F \cong \frac{1}{N_S} \sum_{j=1}^{N_S} \mathbb{I}\left[\hat{y}_S(v^{(j)}) < 0\right], \tag{65}$$

where $v^{(j)}$ is the jth realization of V, N_S is the sample size, and $\mathbb{I}[\cdot]$ is an indicator function such that $\mathbb{I} = 1$ if $v^{(j)}$ is in the failure set (i.e. when $\hat{y}_S(v^{(j)}) < 0$) and *zero* otherwise.

For system reliability analysis involving the union and intersection of r failure sets, similar response surface approximations can be developed for the kth performance function $y^{(k)}(v)$. Hence, the Monte Carlo estimate of the failure probability employing S-variate response surface method for series and parallel systems is

$$h_i \cong \begin{cases} \dfrac{1}{N_S} \displaystyle\sum_{j=1}^{N_S} \mathbb{I}\left[\bigcup_{k=1}^{r} \hat{y}_S^{(k)}(v^{(j)}) < 0\right], & \text{series system} \\[3mm] \dfrac{1}{N_S} \displaystyle\sum_{j=1}^{N_S} \mathbb{I}\left[\bigcap_{k=1}^{r} \hat{y}_S^{(k)}(v^{(j)}) < 0\right], & \text{parallel system} \end{cases}, \tag{66}$$

where $\mathbb{I}[\cdot]$ is another indicator function such that $\mathbb{I} = 1$ if $v^{(j)}$ is in the system failure domain and *zero* otherwise. By setting $S = 1$ or 2, univariate or bivariate decomposition methods can be generated.

6.3. *Example 3: Reliability analysis of a plate with a hole*

Consider again the problem of a square plate with a hole under tension, as described in Example 2. In Example 3, the modulus of elasticity $E(x)$ is defined as a homogeneous, lognormal translation field $E(x) = c_\alpha \exp[\alpha(x)]$, which has mean $\mu_E = 1$ unit and standard deviation $\sigma_E = 0.2$ or 0.5 unit. The image field $\alpha(x)$ is a *zero*-mean, homogeneous, Gaussian random field with standard deviation $\sigma_\alpha = \sqrt{\ln(1 + \sigma_E^2/\mu_E^2)}$, $c_\alpha = \mu_E^2\sqrt{\mu_E^2 + \sigma_E^2}$, and covariance function $\Gamma_\alpha(\xi) = \sigma_\alpha^2 \exp[-|\xi_1|/(bL) - |\xi_2|/(bL)]$ with $b = 0.5$ unit. All other input parameters are the same as in Example 2.

The random field $\alpha(x)$ is parameterized using $N = 8$ in the K-L approximation. Hence, the input random vector $V = \{V_1, \ldots, V_8\}^T$ becomes an eight-dimensional standard Gaussian random vector. The failure condition is defined when the von Mises equivalent stress $\sigma_A(V_1, \ldots, V_8)$ at point A exceeds the uniaxial yield strength S_y of the material.[24]

Figures 9(a) and 9(b) present failure probabilities for various yield strengths, predicted by the mean-point-based univariate and bivariate methods ($c = 0$), as well as by the direct Monte Carlo simulation (10^5 samples). As can be seen in Fig. 9(a), when the uncertainty of elastic modulus is lower ($\sigma_E = 0.2$ unit), both univariate and bivariate methods provide satisfactory results in comparison with the simulation results. However, when a higher uncertainty is considered ($\sigma_E = 0.5$ unit), Fig. 9(b) indicates that the accuracy of the failure probability from the bivariate method is slightly higher than that from the univariate method. The number of function evaluations (i.e. meshfree analyses) for the proposed method with univariate and bivariate methods are only 33 and 481, respectively, when $n = 5$ and $N = 8$.

A comparison of total CPU times, shown in Fig. 10, indicates that both response decomposition methods are far more efficient than the Monte Carlo simulation. In calculating the CPU times, the overhead cost due to random field discretization, random number generation, and response surface approximations are all included. The overhead cost is comparable to the cost of conducting meshfree stress analysis in this particular problem. For this reason, the ratios of CPU times by bivariate and univariate methods and by Monte Carlo and univariate methods are respectively only 8 and 1080, as compared with 15 ($= 481/33$) and 3030 ($= 1\,000\,000/33$), when function evaluations alone are compared. For complex problems requiring more expensive response evaluations, the overhead cost is negligible. In that case, the CPU ratio should approach the ratio of function evaluations. Hence, the response decomposition methods are effective when a response evaluation entails costly meshfree or other numerical analysis.

The numerical results indicate that stochastic meshfree methods can generate accurate estimates of response moments and reliability. Although the same results can be produced using the well-established stochastic FEM, meshfree methods presented here do not require a structured mesh — a key advantage over FEM. It is generally recognized that successful meshing of complex geometric configurations

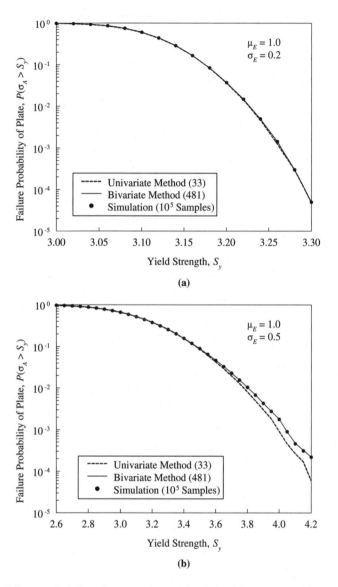

Fig. 9. Failure probability of square plate with a hole; (a) $\sigma_E = 0.2$; (b) $\sigma_E = 0.5$.[24]

can be difficult, time consuming, and expensive. This issue is further exacerbated when solving solid-mechanics problems characterized by a continuous change of the domain geometry, such as crack propagation in solids and metal forming, where a large number of automated remeshings are required due to moving cracks or mesh distortion. Stochastic meshfree methods are ideal candidates for solving these special classes of problems. Nevertheless, the computational cost of deterministic meshfree method is still much higher than that required by low-order FEM. Therefore, breakthrough research focused on enhancing speed and robustness of

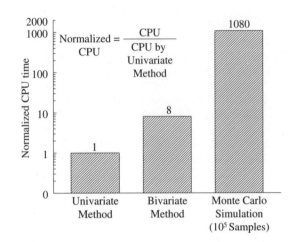

Fig. 10.   Comparison of CPU times for reliability analysis by various methods.[24]

meshfree methods is required for their effective implementation into stochastic mechanics.

7. Conclusions and Outlook

This article has focused on meshfree methods for potential applications in stochastic mechanics and reliability. An exposition involving a brief summary of meshfree formulation, spectral representation of random field, multivariate function decomposition, statistical moment analysis, and reliability analysis has been presented. By avoiding burdensome meshing or remeshing required by the commonly-used finite element method, meshfree methods provide an attractive alternative to the finite element method for solving computational mechanics problems. For the same reason, the meshfree methods are effective in solving integral equations for spectral representation of a random field over a complex arbitrary domain. Numerical results indicate that stochastic meshfree methods, employed in conjunction with dimension-reduction and response-decomposition methods, yield accurate and computationally efficient estimates of statistical moments and reliability.

Although significant strides have been made, stochastic meshfree methods still require considerable improvement before they equal the prominence of the stochastic finite element method. Breakthrough research on enhancing speed and robustness of meshfree methods is essential for their successful implementation into stochastic mechanics.

Acknowledgments

This work was supported by the U.S. National Science Foundation under Grant Nos. CMS-9900196 and DMI-0355487.

References

1. J. J. Monaghan, *Comput. Phys. Commun.* **48**, 89 (1988).
2. P. W. Randles and L. D. Libersky, *Comput. Method. Appl. M.* **139**, 375 (1996).
3. B. Nayroles, G. Touzot and P. Villon, *Comput. Mech.* **10**, 307 (1992).
4. T. Belytschko, Y. Y. Lu and L. Gu, *Int. J. Numer. Meth. Eng.* **37**, 229 (1994).
5. C. A. M. Duarte and J. T. Oden, *Numer. Meth. Part. D. E.* **12**, 673 (1996).
6. J. M. Melenk and I. Babuska, *Comput. Method. Appl. M.* **139**, 280 (1996).
7. W. K. Liu, S. Jun and Y. F. Zhang, *Int. J. Numer. Meth. Fl.* **20**, 1081 (1995).
8. S. N. Atluri and T. Zhu, *Comput. Method. Appl. M.* **22**, 117 (1998).
9. P. Lancaster and K. Salkauskas, *Math. Comput.* **37**, 141 (1981).
10. T. Belytschko, Y. Y. Lu and L. Gu, *Eng. Fract. Mech.* **51**, 295 (1995).
11. M. Fleming, Y. A. Chu, B. Moran and T. Belytschko, *Int. J. Numer. Meth. Eng.* **40**, 1483 (1997).
12. B. N. Rao and S. Rahman, *Comput. Mech.* **26**, 398 (2000).
13. B. N. Rao and S. Rahman, *Eng. Fract. Mech.* **70**, 1 (2003).
14. B. N. Rao and S. Rahman, *Int. J. Numer. Meth. Eng.* **59**, 197 (2004).
15. B. N. Rao and S. Rahman, *Int. J. Pres. Ves. Pip.* **78**, 647 (2001).
16. S. Rahman and B. N. Rao, *Int. J. Numer. Meth. Eng.* **50**, 1969 (2001).
17. S. Rahman and B. N. Rao, *Int. J. Solids. Struct.* **38**, 9313 (2001).
18. S. Rahman and H. Xu, *Int. J. Comput. Meth. Eng. Sci. Mech.* **6**, 41 (2005).
19. B. N. Rao and S. Rahman, *Comput. Mech.* **28**, 351 (2002).
20. S. Rahman and B. N. Rao, *Comput. Mech.* **28**, 365 (2002).
21. B. N. Rao and S. Rahman, *Comput. Mech.* **32**, 199 (2003).
22. S. Rahman and H. Xu, *Probabilist. Eng. Mech.* **19**, 393 (2004).
23. H. Xu and S. Rahman, *Int. J. Numer. Meth. Eng.* **61**, 1992 (2004).
24. H. Xu and S. Rahman, *Probabilist. Eng. Mech.* **20**, 239 (2005).
25. J. S. Chen and H. P. Wang, *Comput. Method. Appl. M.* **187**, 441 (2000).
26. W. B. Davenport and W. L. Root, *An Introduction to the Theory of Random Signals and Noise*, McGraw-Hill, New York, NY (1958).
27. M. Grigoriu, *J. Eng. Mech. ASCE.* **110**, 610 (1984).
28. M. Grigoriu, *Applied Non-Gaussian Processes: Examples, Theory, Simulation, Linear Random Vibration, and MATLAB Solutions*, PTR Prentice Hall, Englewood Cliffs, NJ (1995).
29. M. Grigoriu, *J. Eng. Mech. ASCE.* **124**, 121 (1998).
30. M. Rosenblatt, *Ann. Math. Stat.* **23**, 470 (1952).
31. M. Abramowitz and I. A. Stegun, *Handbook of Mathematical Functions*, 9th Edition, Dover Publications, Inc., New York, NY (1972).
32. R. G. Ghanem and P. D. Spanos, *Stochastic Finite Elements: A Spectral Approach*, Springer-Verlag New York, NY (1991).
33. H. O. Madsen, S. Krenk and N. C. Lind, *Methods of Structural Safety*, Prentice-Hall, Inc., Englewood Cliffs, NJ (1986).
34. R. Rackwitz, *Struct. Saf.* **23**, 365 (2001).
35. O. Ditlevsen and H. O. Madsen, *Structural Reliability Methods*, John Wiley & Sons Ltd., Chichester, United Kingdom (1996).
36. R. Y. Rubinstein, *Simulation and the Monte Carlo Method*, John Wiley & Sons, New York (1981).
37. H. Niederreiter and J. Spanier, *Monte Carlo and Quasi-Monte Carlo Methods*, Springer-Verlag, Berlin, Germany (2000).
38. W. R. Gilks, S. Richardson and D. J. Spiegelhalter, *Markov Chain Monte Carlo in Practice*, Chapman-Hall, London, United Kingdom (1996).

CHAPTER 11

RELIABILITY ANALYSIS USING INFORMATION FROM EXPERTS

JAMSHID MOHAMMADI* and EDUARDO DESANTIAGO[†]

Department of Civil and Architectural Engineering, Illinois Institute of Technology
Chicago, Illinois 60616, USA
*E-mails: *mohammadi@iit.edu*
[†]*desantiago@iit.edu*

The assessment of the reliability of an engineering system to a great extent depends on performance data. Such data is obtained in a variety of ways. Most conventional methods utilize field or laboratory data. In cases where such data is not readily available or cannot be obtained through conventional means at a reasonable cost, data from experts can be utilized. In addition to the data from the experts, there may be cases where the correct approach to a reliability analysis may not be clear. In such situations, experts may have to be called to provide a direction as to what methods would be most suitable for conducting a rational reliability analysis. This chapter focuses on reliability analysis using this approach. Specifically, the significance of using experts in providing information on the reliability methods, data collection process and performance data from opinions expressed by experts are described. Different statistical methods in conducting tests on such data, identification of biases and methods to improve the outcome of reliability analysis using information from experts are also discussed.

1. Introduction

The reliability assessment of most engineering systems to a great extent depends on performance data. The conventional methods of gathering such data are primarily through field and/or laboratory investigations. In cases where reliable data through conventional sources is not available, a group of experts can be used to provide information needed to complete the necessary system reliability analysis. In addition to providing performance data, experts can also be helpful in identifying the method of analysis that may be most suitable for assessing the reliability of a given system. In essence, the expert opinion approach involves: (1) Data gathering using experts' knowledge; and (2) developing a methodology that can suitably be used in implementing the reliability analysis. The domain of information gathered from experts for the method of analysis can be very focused and narrow or wide. If a problem is well-defined and all modes of failure are known, the information sought from the experts can only be on failure probabilities of individual modes. However, in certain cases, the data gathering process may be extensive and include estimates for probability models; types of distribution; treatment of variables as random; correlation among variables; development of event and fault trees

213

and results verifications. Yet in other cases where the parameters that control a system's performance are not well-defined or known, the information sought from the experts may be directed towards identification of parameters that control the system. In this case, the information may be gathered in two or more rounds. A preliminary round can be devoted in identifying the attributes of a system's performance and modes of failure that are believed to be more prevalent. Subsequent rounds can then be used to quantifying the failure probabilities associated with individual modes of failure.

The specific process via which the data from the experts is gathered depends on several factors [Mohammadi[1]]. Among these are: (1) Availability of information on modes of failure; (2) availability of experts to participate in the data gathering process; (3) type of data gathering method to be used; (4) time and cost restraints; and, (5) level of details expected in the data. Depending on the type of data, the information from the experts can be used as the final outcome in the process. However, there may be cases where the information from the experts can only be used as a basis to further explore the problem and conduct a second or even third round of data gathering. When an elaborate process is planned, the experts can be solicited for basic information on the probability of occurrence of failure events as well as on the type of probability distribution functions, inter-variables correlations, estimated values for means, coefficient of variation and other statistical parameters of individual random variables. The data gathering process can in fact continue for several rounds as long as the time and budget would allow for the continuation of the process.

The process of information gathering from experts has been widely used in various disciplines. Most notably, the process is often employed in association with new products, where opinions of certain consumers (as experts) are solicited in an effort to determine how the new products will be accepted by consumers. In such applications, the data gathering methods are usually through face to face interviews and group seminars. In engineering applications, the data gathering method in principal is the same; except that the data is often gathered through questionnaires. Several examples in civil engineering projects, in which the expert opinion data has been used, are discussed below. In some of these applications, the data gathering is conducted for structural condition assessment while in others the method was employed exclusively to gather information on failure probabilities in an effort to estimate the reliability of a system.

A typical example of the expert opinion data gathering process in civil engineering is in the area of structural inspections. Highway and railway bridges are often inspected every other year for the purpose of detecting the extent of damage to their components and identifying the type of remedial work needed to keep them in service. Municipalities often require the owners of tall buildings to conduct periodic inspections of the building exteriors for any signs of deterioration, cracks, and other problems that may impose a hazard to the public. In both these applications, experts (structural engineers) are hired to conduct the inspections. In bridge systems, often the areas prone to damage are known; and as such, standard forms are used to gather specific information on the condition of individual components. As it is expected, the data entered into these forms is subjective and reflects the opinion of the engineer conducting the inspection of a given bridge. Different agencies use their own rating systems for structural condition of a bridge component. For example, most highway agencies use a rating of 0–9 (where a rating of 9 indicates a component is in an excellent condition; whereas, a rating of zero is an indication that a component is in poor condition and needs replacement). In building inspections, the expert data is more in the form of an inspection report and may not specifically follow a standard format.

In several other applications, the use of expert opinion data has been considered exclusively for the purpose of gathering information for a specific parameter for which conventional methods could not be used due to cost or lack of time. One such application deals with the significance of human error and its impact upon system reliability. Along this type of application, Melhem, *et al.*[2] report on a problem where data on fabrication errors in steel bridge components were sought through experts' knowledge. In another application, Mohammadi, *et al.*[3] used data from experts to identify modes of failure and probabilities associated with these modes for gas distribution systems used in residential and commercial buildings. In this application, the experts consisted of technical personnel involved with gas distribution industry and were familiar with potential modes of failure with interior gas pipes in building. The results gathered from these experts were used in estimating the probability of gas leaks and potential cost associated with any follow-up events. In several other applications, the data from experts has been used in such problems as seismic performance evaluation of buildings [e.g. Miyasato, *et al.*[4] and Savy[5]]; in construction management problems and risk evaluation [Kangari[6]]; and in several structural reliability problems [Yamamoto and Ang[7]; Lind and Novak[8]; Kulkarani and Cornell[9] and Yao[10]].

As in any type of data, expert opinion data is subject to uncertainties and is thus qualified for various methods of statistical analyses. One major source of uncertainties that stands out in expert opinion data is due to the expert's bias. Various modes of an expert's bias are covered by Spetzler and Van Holstein.[11] By knowing the specific prevalence of one or more of such modes in a given problem, one may be able to design the data acquisition process in such a way to reduce the effect of the bias in the experts' opinions (Mohammadi, *et al.*[3]). Statistical analyses and test methods can also be used to identify any trends and skews in the data, which may indicate bias. Several other tests can also be applied to identify whether there are differences between two or more sets of data gathered from different teams of experts. Any statistical differences identified can be indicative of bias in the data. The uncertainties and biases, once identified, can be controlled by repeating the data acquisition process in multiple rounds (as explained later).

In system reliability problems, a rather simple method involves the use of the fault-tree analysis method [Ang and Tang[12]], where an event of concern (top event) is disintegrated into causative events and further down to a series of basic events.

Basic events are primarily events of occurrences of specific modes of failure. The expert opinion is then sought for probability of occurrence of individual basic events. Once the necessary data is compiled, analyzed and tested, the basic event probabilities are used to arrive at the probability of the top event, which will then provide the information on the system reliability (i.e. probability of survival). In cases where inter-correlation between causative or basic events exist, the procedure may be more involved and the data from the expert may include identification of random variables describing various events, correlation information and/or even distribution models for these random variables.

This chapter provides an overview of the expert opinion data gathering process and their use in structural reliability evaluation. A discussion of biases and statistical methods applied to such data are also presented.

2. Procedure

The procedure in conducting the expert opinion data depends on the type of problem and the specific information known about the problem. In system reliability applications, the procedure depends on whether the modes of failure and their significance on the overall system reliability are known. For example, in dealing with a simple structural truss system, one must know what modes of failure are prevalent and whether the occurrence of any one mode will cause the failure of the system. In non-redundant trusses, the occurrence of any one mode of failure is tantamount to the system failure; whereas, in redundant systems, more than one mode will be needed to occur before the system fails. Thus the procedure for these two systems will be different in the sense that the type of data that is required from the expert may be different. In problems where the failure modes are not known or well-defined, the expert opinion process must devote a separate effort to gather information from the expert on the prevalent modes of failure. For example, in an investigation of the reliability of a retaining wall system, although generic modes of failure may be known, expert opinion data will be helpful in identifying several other less known modes (e.g. soil washout, foundation settlement, concrete deterioration, rebar corrosion, unexpected heavy surcharge), the significance of these modes on the overall wall stability and any correlation between these modes. Furthermore, when formulating the system reliability in terms of individual modes of failure, expert opinion data may be needed for estimating the uncertainties associated with the individual modes. In any case, the fault tree analysis will be helpful to disintegrate events into basic events; since it is often easier to obtain data from the expert when events are simple and do not depend on too many other parameters. Figure 1 shows an example of a fault tree describing the failure of a simply-supported wood truss system which is supporting a highway advertising board. The reliability of this truss against strong wind loads is of concern. Our assumption is that there is no information on the type of wood used in making members. Furthermore, individual members are severely deteriorated and have knots in them, which are randomly distributed. There are n members in the truss. Thus the top event which is the

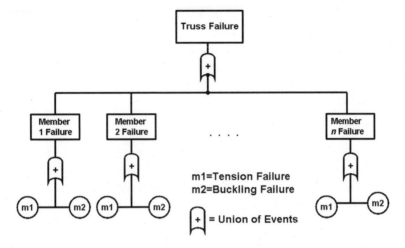

Fig. 1. Fault tree for failure of a simple truss.

failure of the truss is disintegrated into n events. The union of these events will result in the top event (i.e. failure of the truss). Two modes of failure (i.e. tension or buckling) are identified for each member. Experts are invited to express their opinions on the probability of occurrence of each mode. Note that for each member, the event of failure is considered to be the union of the events of occurrence of the two modes of failure. Also note that the failure mode events are shown inside circles. This means that these events cannot be disintegrated further; and as such they are considered as top event. Denoting the event of the failure of the truss as F_T and that of member i as f_i and those of the failure modes as f_{ci} and f_{ti}, the probability of truss failure can be mathematically written as

$$P(F_T) = P(f_1 \cup f_2 \cup \cdots \cup f_n), \tag{1}$$

where

$$P(f_i) = P(f_{bi} \cup f_{ti}). \tag{2}$$

The reliability of the truss is computed as one minus the probability of the truss failure.

3. Data Collection Process

The data can be obtained from the experts using various formats. A popular method is data collection through a questionnaire. However, the use of a questionnaire will be effective if the expert can be provided with enough information and there is no ambiguity on the nature of the problem. For example, in the wood truss example, a question for a mode of failure can be expressed as follows:

> "Given the occurrence of an 80 mph wind on the advertising board, in your opinion what will be the probability that a main truss member will fail due to buckling?"

The expert can be given more specific information on how to provide his/her numerical probability values. For example, the questionnaire may include a statement directing an expert to express his/her opinion in number of failures expected in 100 occurrences. In this case, the questionnaire may include the following expression:

"In providing your opinion on failure due to buckling, assume there will be 100 identical elements. Identify how many of these members may fail due to the 80 mph wind."

Of course, if the probability of failure is expected to be very small, the expert's response in terms of 1 in 1000 or 1 in 10000 occurrences can be requested. Furthermore, the questionnaire may include a series of values and the expert asked to mark the value that is closest to his or her opinion on the occurrence of the event.

An alternative method for compiling data from experts is through group or face-to-face interviews. The advantage of this method is that it allows for a direct interaction between the interviewer and the expert [Spetzler and Van Holstein[11]]. This also allows for an on-site analysis of the expert data, a quick verification of the data (for example, comparing an individual's response compared with others) and additional explanation on the problem for further clarification. A face-to-face interview or a group seminar is generally more difficult to arrange and involves a higher cost. It is also possible to combine a questionnaire for compiling a preliminary set of data and then conduct the second round of data collection in a group seminar.

4. Probability Encoding

Various methods can be used for obtaining data on probability values for basic events or statistical information (distribution functions, mean values, coefficients of variation, etc.) for random variables describing basic events. Spetzler and Van Holstein[11] discuss several methods through which the probability encoding can be achieved. These methods are divided into two types; namely, direct and indirect response techniques. In the indirect response technique, one can select an analog device or simply a numerical and/or measuring scheme to collect and encode probability values from experts. In a device referred to as the 'probability wheel,' measurements representing probability values are marked. The interviewer uses this wheel to mark different values and ask whether the expert agrees that a specific number represents his/her opinion on a desired probability. In a different scheme, the expert can be provided with a paper lined to form, say, a 10 × 10 table (i.e. containing 100 squares). The expert is asked to darken as many squares needed to represent percentage probability for the event of concern. Most such devices are only applicable when the data gathering process is done in a face-to-face interview or in a group seminar. It is also emphasized that the use of these devices may not become effective if the probability values are expected to be very small. When probabilities are expected to be very small, a direct response technique (as described below) may work better.

In the direct response technique, the experts are asked explicitly to provide their opinions on a value for the event of concern. They can even be asked to provide their opinions on a type of probability distribution model suitable for a random variable or estimates for such statistical quantities as the mean or coefficient of variation. This method can work well with experts who are involved in probabilistic modeling applications; and as such can comment on the modeling aspects of the problem as well as providing data. Spetzler and Van Holstein[11] mention that the choice of probability functions is generally a modeling consideration than being part of the encoding process. Thus if the experts are only identified with their experience with the nature of the problem but not with the modeling process, the data gathering should perhaps focus only on encoding probability values.

When applied to the evaluation of system reliability, some form of direct response technique seems to be appropriate. This is provided that the problem modeling is primarily based on a fault tree analysis procedure and that the probability values for the basic events can be obtained directly from experts. In cases where each basic event is modeled through the introduction of one or more random variables, the direct response technique can still be used, if the experts understand the nature of modeling through random variables and estimation of statistical parameters (mean and coefficient of variation) needed for each random variable. As an example, we refer to the problem depicted in Fig. 1. Suppose the intention is to decode a probability value for failure of member 1 due to the buckling mode. An expert may be asked to directly express his opinion on this probability. We can give the expert a scenario in which some 100 identical members are involved and ask him/her to express how many of those may fail given a specific loading condition.

In an alternate method, we may formulate the probability of failure of any member i for the buckling mode of failure using the following equation (see, for example, Ang and Tang[12] for the derivation of this equation)

$$P(f_i) = \Phi \left[\frac{-\ln(\bar{R}i/\bar{S}i)}{\sqrt{\Omega_{Ri}^2 + \Omega_{Si}^2}} \right], \tag{3}$$

in which, Ri and Si are random variables describing the resistance and the applied load on member I, respectively. The bar indicates the corresponding mean values for these random variables. The parameters Ω_{Ri} and Ω_{Si} are the coefficients of variation for the resistance and applied load; respectively; and $\Phi(\cdot)$ is the standard normal probability distribution function. In computing the probability of failure for individual members using Eq. (3), experts may be requested to provide, for example, their opinions on the mean value and coefficient of variation for the resistance of members. Alternatively, the experts may be asked to provide their opinions directly for the ratio of resistance to the load, $\theta_i = Ri/Si$ (which is a measure for the margin of safety) and the corresponding uncertainty associated with this ratio $\Omega_{\theta i}$ (which will be the denominator in Eq. (3)). It is emphasized that since usually more than one expert is approached for data, there will be a variation for the mean value of θ_i and also a variation involved in opinions expressed for $\Omega_{\theta i}$. We can take an average of all responses to arrive at a single value for an estimate of the mean of θ_i and one for $\Omega_{\theta i}$. However, the effect of variation in these responses must be included in the process. Denoting these variances as Δ_1 and Δ_2, the overall estimate of uncertainty for θ_i can be written as follows:

$$\Omega_{\theta i} = \sqrt{\bar{\Omega}_{\theta i}^2 + \Delta_1^2 + \Delta_2^2}, \tag{4}$$

in which $\bar{\Omega}_{\theta i}^2$ indicates the average value from all responses obtained for $\Omega_{\theta i}$, Δ_1 is the coefficient of variation associated with the data gathered for the mean value of θ_i and Δ_2 is the coefficient of variation associated with responses obtained for $\Omega_{\theta i}$.

The computation of the system reliability through Eq. (1) depends on the inter-correlation conditions between events f_i. If f_i are statistically independent, the system failure probability is[12]

$$P(F_T) = 1 - \prod_{i=1}^{n} [1 - P(f_i)]. \tag{5}$$

However, if the events f_i are perfectly correlated, the system failure probability will be equal to the maximum value[12] among all $P(f_i)$, i.e.

$$P(F_T) = \text{Max}[P(f_i)]. \tag{6}$$

In reality, the conditions between the failure events of members are only partially correlated and the true system probability of failure will be bounded by the two values from Eqs. (5) and (6). To ultimately arrive at a single value for the system failure probability, experts can be approached again. The data gathering in this case will be direct. We can provide the experts with the significance of and potential sources that contribute to correlation between events f_i. Then, they can be asked to provide their estimate for the most likely value of system failure probability between the two bounds established through Eqs. (5) and (6).

A similar approach can also be used in obtaining data on the correlation between modes of failure for each structural member and the corresponding failure probability through Eq. (2). The basic formulations used in the derivation of Eqs. (1)–(6) can be found, for example, in Ang and Tang.[12]

5. Biases in Data

As described earlier, expert opinion data is subject to bias. Spetzler and Van Holstein[11] express sources of bias to be either 'motivational' or 'cognitive.' The motivational bias is initiated from the expert's desire to change the outcome of the survey to his/her favor. In system reliability applications, a motivational bias can arise when questions used to gather experts' data may appear to be in direct conflicts with an expert's interests. For example, an expert may belong to a certain manufacturing group which specializes in fabricating structural truss components; and such components have been mentioned in the survey questionnaire as a poten-tial source of failure. The concern over the perception of public in regard to the risk might cause the expert to think that his/her company's product will be associated with failure.[1]

Cognitive biases are reported to depend on the expert's modes of judgment.[11] This type of bias is systematically introduced by the way the expert responds to the survey questions. The expert may be influenced by most recent occurrences of the event of interest and thus would respond accordingly. For example, when asked to express an opinion for the failure probability of a truss component due to material defects, the expert may provide an unusually low value for the failure probability only to pretend that there is no particular problem with the material. In this way, the expert tries to make the point that the material is sound, even though there has been several recent failure cases alluded to material defects on this specific type of trusses. Contrary to this example, in another scenario, recent failure cases on the material defect may influence the thought process of the expert; and as such, he/she may provide a large value for the failure probability to emphasize how critical the material defect can be in playing a crucial role in a potential failure of this particular type of trusses. Cognitive biases are reported to have five different modes (Spetzler and Van Holstein,[11] Kahneman and Tversky,[13] and Tversky and Kahneman.[14]) These are: (1) Availability; (2) adjustment and anchoring; (3) representativeness; (4) unstated assumptions; and (5) coherence. The availability and the adjustment and anchoring modes concern occurrences the expert recalls. Availability deals with events the expert recalls in general; whereas, the adjustment and anchoring deals with most recent occurrences. The unstated assumptions mode of judgment deals with the expert's assumptions on which the response is based. In a system reliability problem, for example, an expert may simply assume only one factor (among several other prevalent factors) as the factor affecting the occurrence of a base event. For example, in providing an estimate for the probability of settlement of a foundation, the expert may assume that the applied load and its duration as the factor causing settlement not considering other factors such as water contents, soil type and the past history of loading and unloading. In the representativeness mode of judgment, the expert's opinion is based on the similarity of the raised issues to the general population these issues belong to. This mode may be prevalent when a specific factor is often mentioned to have caused failure in cases similar to the one being discussed with the experts. The coherence mode of judgment concerns the reasonableness of a series of events that lead to the development of the event for which expert's opinion is sought.

Biases can be measured by detecting any skews or shift in the distribution of responses from the experts. For example, in Fig. 2, when the distribution of responses for a probability of occurrence of an event is plotted, a nearly symmetric distribution is observed for values between x_1 to x_5. Several responses indicate values such as x_6 and x_7 that are considered to be "outliers" and may indicate biases.

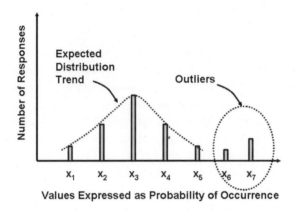

Fig. 2. Distribution of responses for detection of biases.

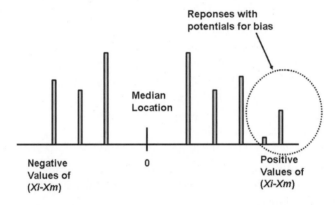

Fig. 3. Detection of bias by examining symmetry of responses about median.

When there is no distinct distribution in the responses; and the response population does not show any trend, the median can be used to determine whether the responses are symmetric about the median. In Fig. 3, the median is estimated to be X_m. This is estimated either based on a separate survey or from past knowledge. The expert Reponses (X_i) are compared with the median. Using the median as the center of the location for all the responses, symmetric values for the quantities $X_i - X_m$ are expected when all the responses are plotted. The outliers in Fig. 3 are shown to the right and are subject to bias and need to be further investigated. A statistical test (e.g. the Wilcoxon Signed-Rank Test) can be used to determine whether there is enough statistical evidence to support the symmetry of the data about the expected median.[15] If the test indicates that there is not enough evidence to support the symmetry of the data, the outliers may indicate elements of bias in the responses.

Biases can also be detected by dividing the experts into several groups and conducting statistical tests to identify whether significant differences across various group responses exist. This is especially the case when experts within each group share similar levels of responsibilities or job definitions. For example, in compiling the data for the probability of failure of the timber truss member cited earlier, one group of experts is made up of engineers from the county highway department, while another group is made up of engineers who do designs as outside consultants or contractors to the highway department.

Differences between group responses can be tested using statistical methods used for comparing means of two or more populations. Considering two populations with unequal variances, the test statistic for comparing the means is written as

$$D = \frac{m_1 - m_2 - d_0}{\sqrt{S_1^2/n_1 + S_2^2/n_2}}, \tag{7}$$

in which m_1 and m_2 are, respectively, the mean values of the two populations, d_0 is the hypothesized difference (taken as zero, for example), S_1, and S_2 are the standard deviations of the two populations; n_1, and n_2 are the respective sample sizes. The null hypothesis in testing the difference between the mean is $D = 0$; whereas the alternate hypothesis is $D \neq 0$. The test statistic is compared with the critical values from the t distribution with α significance (i.e. $\pm t_{\alpha/2}$) and ν degrees of freedom. For a t-distribution function, usually, the degree of freedom is $n - 1$ (in which n is the sample size). However, in this case, because two populations are involved, the degree of freedom is computed as[15]

$$\nu = \frac{(S_1^2/n_1 + S_2^2/n_2)^2}{\frac{(S_1^2/n_1)^2}{n_1-1} + \frac{(S_1^2/n_1)^2}{n_2-1}}. \tag{8}$$

The degree of freedom ν is rounded down to the nearest whole number. As shown in Fig. 4, the area bounded by the critical values is $(1 - \alpha)$.

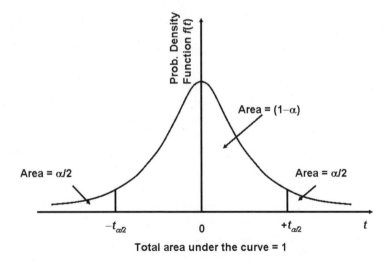

Fig. 4. A typical t-distribution.

This indicates that there is $(1 - \alpha)$ probability that D will be between the two critical values $\pm t_{\alpha/2}$. If the test statistic D is bounded by the critical values from the t distribution, there is enough evidence to assume that the difference between the means is zero. Statistically-speaking, we conclude that the null hypothesis cannot be rejected based an α significance. We further conclude that the responses from two groups do not indicate a major element of bias. If, however, D is either less than $-t_{\alpha/2}$ or larger than $+t_{\alpha/2}$, there is enough evidence to conclude that D is too large. This indicates that the difference between the means from the two populations is significant. Statistically, we conclude that the null hypothesis *is* rejected. Thus a bias may be present in one of the two group responses. For example, at $\alpha = 0.10$, and a degree of freedom equal to 22, the critical values of the t distribution are $t_{0.05} = \pm 1.717$ (Tables for the values of the t-distribution function can be found in Khisty and Mohammadi[15]). With a test statistic equal to -3.12, we conclude that the difference is statistically significant because -3.12 is outside of the critical range ± 1.717. Thus responses need to be examined further to determine whether this difference indicates a bias. It is emphasized that comparing the means from the two populations using Eqs. (7) and (8) is done when each population is tested for and passed the normal distribution function. Furthermore, one must conduct a separate test to determine whether the variances of the two populations are statistically different.

If a statistical test indicates that the variances are identical, then the test statistic is obtained from the following equation:

$$D = \frac{m_1 - m_2 - d_0}{\sqrt{S_P^2(1/n_1 + 1/n_2)}}, \tag{9}$$

in which S_P is an estimate of the common standard deviation of the two populations and is obtained from the following equation:

$$S_P^2 = \frac{(n_1 - 1)S_1^2 + (n_2 - 1)S_2^2}{n_1 + n_2 - 2}, \tag{10}$$

in which S_1, and S_2 are the standard deviations of the two populations. The degree of freedom in this case is computed as $\nu = n_1 + n_2 - 2$.

For cases where more than two groups of experts are involved, additional statistical tests may have to be conducted to determine if there are differences between groups and between what two groups major differences exist. An analysis of variance (ANOVA) needs to be conducted to find if populations are statistically different. A multiple range test will further identify where the difference is more prevalent (i.e. between what two populations the difference is significant). The details of these procedures can be found in Khisty and Mohammadi.[15]

When populations do not follow a distribution trend or they do not pass the test of normality, one can use distribution free methods to determine whether responses from two groups of experts are significantly different. If there are significant differences, then an element of bias may be present. The Wilcoxon Rank-Sum Test for Unmatched Data can be used for this purpose. The two populations are tested for their locations. If their locations are different (based on comparing the difference in the location with a critical value), then there is enough statistical evidence that the two populations are different. And as such, an element of bias may be detected. The procedure for this method requires pooling the data, then ranking them and computing the sum of the ranks for each population for comparison. The details of this method are also explained in Khisty and Mohammadi.[15]

Once biases are detected, additional information can be provided to the experts for further refinement of their responses. This can be done in a second round of survey. For example, the distribution models similar to that in Fig. 2 can be made based on the first round of survey. Upon conducting statistical tests and detecting biases, each expert's response, suspecting of having a bias, can be marked on the graph. The expert is then provided with this information and is requested to evaluate the results before providing his/her response in the second round of survey. This process can be repeated as many times as necessary, if the time and budget would of course allow conducting a multi-round of surveys. Upon conducting each round, the form of data gathering (i.e. the format of a questionnaire or types of questions asked in a face-to-face interview) can be modified to better explain the problem to the expert and thus gather more meaningful and refined data.

If the bias is identified to be motivational, the expert must be assured of the confidentiality of the results. This may motivate the expert to provide an unbiased response. If this tactic is not helpful, we may decide to exclude the expert from the study. When biases are determined to be cognitive, several measures can be taken to control them.[1] For example, for a mode of judgment known as available or adjustment and anchor, the survey questionnaire or the interview can explicitly asked the experts to provide answers irrespective of any recent occurrences. Furthermore, the experts may be asked to specify confidence levels for their responses. These confidence levels may then be used to adjust the responses from the individual experts. If the mode of judgment is one called representativeness, the bias may be controlled by re-framing the problem to remove the element of representativeness.[11] For example, in gathering data for the failure of a foundation, specific parameters can be outlined to direct the expert to provide probability values for, say, the mode of failure because of the load effect, time effect or saturation effect. This approach can also be helpful in controlling biases that are believed to be from the unstated assumptions and the coherence modes of judgment.

6. Summary and Conclusions

A procedure for gathering data from experts is explained in this chapter. Specifically, the data is used in conjunction with the reliability analysis of an engineering system. The reliability is measured as one minus the probability of the system failure. The event of failure of the system is modeled with a series of basic events using a fault tree analysis method. The probability values from experts are gathered for the basic events. Experts can be asked to directly provide a value for the probability of occurrence of any one basic event. Alternatively, they may be asked to provide their estimates for the mean, coefficient of variation or even a distribution model of individual random variables that are used to formulate the probability of failure for a given basic event. The chapter further explains methods used for treating uncertainty in the expert opinion data and a procedure to detect and control biases. Expert opinion data is treated like any other types. And as such, various methods of data analysis and test, such as the ANOVA, can be used for the data.

References

1. J. Mohammadi, Expert opinion data and their uncertainties, *Uncertainty Modeling and Analysis in Civil* Engineering, B. M. Ayyub, Editors, CRC Press, Boca Raton, FL (1998).
2. H. G. Melhem, W. M. K. Roddis, S. Nagarja and M. R. Hess, *Computing in Civil Engineering* **10** (1996).
3. J. Mohammadi, A. Longinow and T. A. Williams, *Structural Safety* **9** (1991).
4. G. H. Miyasato, W. H. Dong, R. E. Levitt, A. C. Boissonnade and H. C. Shah, *Expert Systems in Civil* Engineering, C. N. Kostem and M. L. Maher, Editors, American Society of Civil Engineers, Reston, VA (1986).
5. J. B. Savy, An application of the use of expert opinion to seismic hazard assessment, *Abstract Proceedings, Engineering Mechanics 6th Conference*, American Society of Civil Engineers, Reston, VA (1987).
6. R. Kangari, *Expert Systems in Civil* Engineering, C. N. Kostem and M. L. Maher, Editors, American Society of Civil Engineers, Reston, VA (1986).
7. M. Yamamoto and A. H.-S. Ang, Significance of gross errors on reliability of structures, *Proceedings 4th International Conference on Structural Safety and Reliability*, I. Konishi, A. H.-S. Ang and M. Shinozuka, Editors, Vol. III, Columbia University, New York, NY (1985).
8. N. C. Lind and A. S. Novak, Pooling expert opinions on probability distributions, *Abstract Proceedings, Structures Congress*, American Society of Civil Engineers, Reston, VA (1986).
9. R. B. Kulkarani and C. A. Cornell, A statistical procedure for combining expert opinions, *Abstract Proceedings, Engineering Mechanics 6th Conference*, American Society of Civil Engineers, Reston, VA (1987).
10. J. T. P. Yao, *Safety and Reliability of Existing Structures*, Pitman Publishing, Inc., Marshfield, MA (1985).
11. C. S. Spetzler and C. A. S. Van Holstein, *Management Science* **22** (1975).
12. A. H.-S. Ang and W. H. Tang, *Probability Concepts in Engineering Planning and Design,* Vol. II, John Wiley and Sons, New York, NY (1984).
13. D. Kahneman and A. Tversky, *Cognitive Psychology* **3** (1972).
14. A. Tversky and D. Kahneman, *Cognitive Psychology* **5** (1973).
15. C. J. Khisty and J. Mohammadi, *Fundamentals of Systems Engineering with Economics, Probability and Statistics*, Prentice Hall, Upper Saddle River, NJ (2001).

CHAPTER 12

RISK-BASED OPTIMIZATION OF LIFE-CYCLE COST FOR DETERIORATING CIVIL ENGINEERING INFRASTRUCTURES

RÜDIGER RACKWITZ

Institut fuer Baustoffe und Konstruktion
Technische Universitaet Muenchen, Arcisstr. 21
D-80290 Muenchen, Germany
E-mail: rackwitz@mb.bv.tum.de

Risk-based optimization of life-cycle cost for civil engineering infrastructures requires a sustainable replacement strategy. Since all cost must be discounted down to the decision point a sustainable, intergenerationally acceptable discount rate must be used. A rate decaying from some market rate down to the real economic growth rate per capita in a country based on intra- and inter-generational equity is proposed. The renewal model established elsewhere for setting up suitable objective functions is adjusted to time-dependent interest rates. Optimal replacement strategies including inspection and repair are proposed and the theory is extended to series systems of dependent or independent deteriorating structural components.

1. Introduction

The civil engineering profession has clearly recognized that the infrastructure in a country or region has to be maintained and finally renewed because failures occur not only due to extreme external events but primarily due to wear out, deterioration and, in some cases, obsolescence. The strategies adopted by engineers to manage the life cycle of a structure were widely technical, i.e. by improving reliability and durability and by proposing design solutions which should enable a longer time span of full use, possibly with other uses than initially foreseen. This is, no doubt, an important aspect of life-cycle engineering.

In addition, there is growing awareness of the fact that our world is a limited one in the sense that it has only limited non-renewable natural resources. This lead to the co-called Brundland Commission[1] to conclude in their famous report "Our Common Future" in 1987 that a sustainable development is one "that meets the needs of the present without compromising the ability of future generations to meet their own needs." This statement has widely become a new ethical standard. The immediate implications for the planning, design and operation of civil engineering infrastructures are clear: Save energy, save non-renewable resources and find out about recycling of building materials, do not pollute the air, water or soil with toxic substances, save or even regain arable land and much more. Furthermore, our generation must not leave the burden of maintenance or replacement of short-lived structures to future generations and it must not use more financial resources than there are available and affordable in a sustainable manner. Finally, it is assumed that civil engineering infrastructures are financed by the public via taxes, public charges, tolls or other. It is in any case the citizen who pays and, of course, who also enjoys the benefits derived from their existence.

In this paper a renewal model for sustainable life-cycle costing is reviewed and extended to time-dependent discount rates. A scheme for the derivation of intra- and inter-generationally acceptable discount rates is proposed. The renewal model is then extended to cover corrective and preventive replacements for series systems of deteriorating components.

2. Cost-Benefit Optimal Technical Facilities

In 1971 Rosenblueth and Mendoza[2] proposed optimization with respect to benefits and cost as the final goal of setting up structural codes and for the direct design and operation of structures. A technical facility is financially optimal if the following objective is maximized:

$$Z(\mathbf{p}) = B(\mathbf{p}) - C(\mathbf{p}) - D(\mathbf{p}). \tag{1}$$

It is assumed that all quantities in Eq. (1) can be measured in monetary units. $\mathbf{p}$ is the vector of all safety relevant parameters or actions. $B(\mathbf{p})$ is the benefit derived from the existence of the facility, $C(\mathbf{p})$ is the cost of design and construction and $D(\mathbf{p})$ is the cost in case of failure. The quantities $B(\mathbf{p})$, $C(\mathbf{p})$ and especially $D(\mathbf{p})$ involve uncertainties. Statistical decision theory then dictates that expected values are to be taken. It is assumed that $B(\mathbf{p})$, $C(\mathbf{p})$ and $D(\mathbf{p})$ are differentiable in each component of $\mathbf{p}$. The cost as well as the benefits may differ for the different parties involved. The different parties, e.g. the owner, the builder, the user and society, may also have different economic objectives. A facility makes sense only if $Z(\mathbf{p})$ is positive within certain parameter ranges for all parties involved. In this paper we will primarily focus on an optimization for and in the name of the public.

In view of sustainability one has to distinguish between at least four replacement strategies:

- The facility is given up after service or failure.
- The facility is systematically replaced after failure.
- The facility is renewed (repaired) after deterioration.
- The facility is renewed due to obsolescence.

Further, we distinguish between facilities which can fail upon completion and facilities which can fail at a random point in time later due to service loads, extreme external disturbances or deterioration. The option "failure upon completion or never" implies that loads and resistances on the facility are time-invariant which is not considered further. The option "facility given up after service or failure" is also not considered further because infrastructure facilities must remain functioning for all foreseeable future in accordance with the sustainability requirement.

3. The Renewal Model for Systematic Replacement

3.1. *Discounting*

The facility has to be optimized during design and construction at the decision point, i.e. at time $t = 0$. Therefore, all cost need to be discounted down to the decision point. For analytical convenience continuous discounting is assumed which is accurate enough for all practical purposes. Let the discount rate $\gamma(t)$ be a function of time. If damage $D(t)$ occurs at time t its present value $D(0)$ can be determined from

$$D(0) = D(t) \exp\left[-\int_0^t \gamma(\tau)d\tau\right].\tag{2}$$

For a constant (time-averaged) discount rate one has $D(0) = D(t) \exp[\gamma t]$. Discount rates are understood as real rates net of any taxes. If a interest rate γ' for discrete discounting is given one converts by $\gamma = \ln(1 + \gamma')$.

Because we are mainly interested in public works and long financing times the discount rate $\gamma(t)$ should be decomposed into a subjective time preference rate ρ and the rate δ related to economical growth. In particular, we will decompose the discount rate as

$$\gamma(t) = \rho(t) + \varepsilon\delta > 0\tag{3}$$

as proposed in the economic growth theory. Assume $\varepsilon\delta$ as characteristic and constant for each country and we take the subjective rate $\rho(t)$ as time-dependent in sufficient generality. The parameter ε is close to one. The long-term average economic growth rate (per capita) δ is close to 0.02 for most developed countries.[3] Following Bayer[4] we assume constant preference rates ρ for living generations in an overlapping generation model but omit ρ for all future generations as being ethically indefensible. The demand for intergenerational equity lets us then define an equivalent time-variant rate for easy understanding and analytical convenience.[5] This can be approximated by

$$\gamma(t) = \rho_0 \exp[-0.013t] + \varepsilon\delta.\tag{4}$$

3.2. *Basic renewal model*

We assume random events in time forming a renewal process. The times between failure (renewal) events have identical distribution function $F(t, \mathbf{p})$, $t \geq 0$, with probability density $f(t, \mathbf{p})$ and are independent. Renewal theory allows us to distinguish between the densities of the first renewal and all subsequent renewals. This extension will not considered herein. The independence assumption needs to be verified carefully. In particular, one has to assume that loads and resistances in the system are independent for consecutive renewal periods and there is no change in the design rules after the first and all consecutive failures (renewals). Even if designs change failure time distributions must remain the same. Neglecting finite (re)construction times the objective function for systematic reconstruction is in full generality:

$$Z(\mathbf{p}) = B^* - C(\mathbf{p}) - \sum_{n=1}^{\infty}(C(\mathbf{p}) + H)\int_0^{\infty} e^{-\int_0^t \gamma(\tau)d\tau} f_n(t, \mathbf{P})dt, \qquad (5)$$

where $f_n(t, \mathbf{p})$ is the density of the time to the nth renewal. It is assumed that construction cost $C(\mathbf{p})$ exclude the cost of financing. They can easily be included. H is the monetary loss in case of failure, including direct failure cost, demolition cost, cost of removal of debris, loss of business and other indirect cost and, of course, the cost to reduce the risk to human life and limb. Therefore, it is useful to decompose H into physical losses H_M and losses H_F associated with losses of human life and limb. Also, it is assumed that $C(\mathbf{p})$ and H are independent of time.

3.3. *Constant benefit and discount rates*

For constant benefit per time unit $b(t) = b$ and a constant discount rate $\gamma(t) = \gamma$ the objective function simplifies greatly, especially because one can make use of the convolution theorem for Laplace transforms in the damage term. Laplace transforms are defined by $f^*(\gamma) = \int_0^{\infty} e^{-\gamma t} f(t)dt$ and there is $0 \le f^*(\gamma) \le 1$ if $f(t), t \ge 0$, is a probability density for which $f^*(0) = 1$ and $f^*(\infty) = 0$. Then, the Laplace transform can also be written as $f^*(\gamma) = E[e^{-\gamma T}]$. In the transformed space there is $h^*(\gamma) = f(\gamma)^* g^*(\gamma)$ for $h(t) = \int_0^t f(t-\tau)g(\tau)d\tau$, an operation necessary to determine $f_n(t) = \int_0^t f_{n-1}(t-\tau)f(\tau)d\tau$. In the following formula $r^*(\gamma, \mathbf{p}) = \frac{f^*(\gamma, \mathbf{p})}{1-f^*(\gamma, \mathbf{p})}$ is the Laplace transform of the renewal density (renewal intensity) $r(t, \mathbf{p}) = \sum_{k=1}^{\infty} f_k(t, \mathbf{p})$.

$$\begin{aligned}
Z(\mathbf{p}) &= \int_0^{\infty} be^{-\gamma t}\, dt - C(\mathbf{p}) - (C(\mathbf{p}) + H)\sum_{n=1}^{\infty}\int_0^{\infty} e^{\gamma t} f_n(t, \mathbf{p})dt \\
&= \int_0^{\infty} be^{-\gamma t}dt - C(\mathbf{p}) - (C(\mathbf{p}) + H)\sum_{n=1}^{\infty} f^*(\gamma, \mathbf{p})^{n-1} f^*(\gamma, \mathbf{p}) \\
&= \frac{b}{\gamma} - C(\mathbf{p}) - (C(\mathbf{p}) + H)\frac{f^*(\gamma, \mathbf{p})}{1 - f^*(\gamma, \mathbf{p})} \\
&= \frac{b}{\gamma} - C(p) - (C(p) + H)r^*(\gamma, \mathbf{p})
\end{aligned} \qquad (6)$$

If failures occur according to a Poisson process with occurrence rate $\lambda(\mathbf{p})$ Eq. (6) simplifies to:

$$Z(\mathbf{p}) = \frac{b}{\gamma} - C(\mathbf{p}) - (C(\mathbf{p}) + H)\frac{\lambda(\mathbf{p})}{\gamma}, \qquad (7)$$

because $f^*(\gamma, \mathbf{p}) = \frac{\lambda(\mathbf{p})}{\gamma+\lambda(\mathbf{p})}$ for $f(t, \mathbf{p}) = \lambda(\mathbf{p})\exp[-\lambda(\mathbf{p})t]$. This result is especially relevant because the parameter $\lambda(\mathbf{p})$ may be replaced asymptotically by the stationary outcrossing rate $\nu^+(\mathbf{p})$ frequently used in time-variant structural reliability analysis.

Without derivation another useful result exists for random Poissonian disturbances. If the disturbances occur with rate λ and can cause failure with probability $P_f(\mathbf{p})$ we have[6]

$$Z(\mathbf{p}) = \frac{b}{\gamma} - C(\mathbf{p}) - (C(\mathbf{p}) + H)\frac{\lambda P_f(\mathbf{p})}{\gamma}. \tag{8}$$

This model will be used later in an example.

An important asymptotic result for arbitrary failure models is[7]

$$\lim_{t \to \infty} r(t, \mathbf{p}) = \lim_{\gamma \to 0} \gamma r^*(\gamma, \mathbf{p}) = \frac{1}{E[T(\mathbf{p})]}, \tag{9}$$

where $E[T(\mathbf{p})]$ is the mean time between renewals.

The precise details of this and more general renewal models can be found in Ref. 8. Many other objective functions can be formulated. For example, serviceability failure, obsolescence, aging, deterioration and inspection and maintenance, finite renewal times, repeated renewal during construction and finite service times can be dealt with. Benefit and damage term can be functions of time.[6,9,10] Also, multiple failure modes can be considered. Some more important cases especially relevant for life cycle costing and sustainability are summarized in the following sections.

3.4. *Non-constant discounting*

For non-constant discount rates Eq. (5) is still valid but the elegant methodology with the Laplace transforms do not apply because the convolution theorem does not hold any more. As will be shown the consideration of time-variant discount rates is necessary in the context of sustainable life cycle cost-benefit analysis. Therefore, we first study the behavior of the term $\sum_{n=1}^{\infty} \exp[-\int_0^t \gamma(\tau)d\tau] f_n(t, \mathbf{p})dt$ in Eq. (5). A general analysis appears not feasible but we can study it at an important example. Assume that the times between renewals are exponentially distributed with parameter λ. Then, the density function to the nth renewal is the density of the Γ-distribution, i.e. $f_n(t) = \frac{\lambda(\lambda t)^{n-1}}{\Gamma(n)} \exp[-\lambda t]$. It is, therefore, easy to determine the sum-terms of $\sum_{n=1}^{\infty} \int_0^{\infty} \exp[-\int_0^t \gamma(\tau)d\tau] f_n(t, \mathbf{p})dt$. Clearly, we have to ensure that all integrals in Eq. (5) converge. This is the case if the integral $\int_0^t \gamma(\tau)d\tau$ asymptotically grows linearly or $\gamma(t) \to \gamma_0 > 0$ for $t \to \infty$ as is the case for Eq. (4). Alternatively, one could apply the convolution theorem for Laplace transforms to the new "transformation" $\int_0^{\infty} \exp[-\int_0^t \gamma(\tau)d\tau] f_n(t, \mathbf{p})dt$, i.e. $f^\#(\gamma)^n$. Figure 1 shows the ratio of the exact result $\sum_{j=1}^n f_j^\#(\gamma)$ and the application of the convolution theorem $\sum_{j=1}^n f^\#(\gamma)^j$ for increasing n upto $n = 20$ for the discount rate $\gamma(\tau) = \rho_0 \exp[-at] + \delta$.

It is seen that: (i) Only a few terms in the sum need to be considered and (ii) the application of the convolution theorem for Laplace transforms to the modified transformation yields an excellent approximation, even for mean failure times and decay times for the interest rate in the same order of magnitude ($1/a \approx 1/\lambda$)

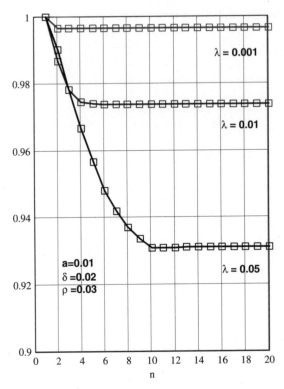

Fig. 1.   Ratio of exact terms in Eq. (5) and by Laplace convolution theorem using Eq. (10) for $\gamma(t) = \rho \exp[-at] + \delta$ and exponential times between failures with rate λ.

although slightly on the unconservative side. At most an error of a few percent has been found under realistic conditions. Only if the decay time, i.e. $1/a$ of the discount rate is much smaller than the mean time between renewals one must expect larger errors and must include a larger number of sum-terms in Eq. (5). This suggests that all results of the last section should be valid and the additional models studied in the next section are used as very good approximations provided that the new transformation

$$f^{\#}(\gamma, \mathbf{p}) = \int_0^\infty e^{-\int_0^t \gamma(\tau)d\tau} f(t, \mathbf{P})dt \tag{10}$$

is used instead of $f^*(\gamma, \mathbf{p})$.

3.5. Non-constant benefit

The benefit term requires special treatment depending on how the benefit per time unit evolves in time. A realistic model is to assume that at each renewal the benefit function starts from its initial value $b(0)$. The time to the nth renewal is $T_n = \sum_{i=1}^n U_i$. Therefore, the total discounted benefit is the sum of the integrated benefits in the various renewal intervals. For $\gamma(t) = \gamma$ we have

$$
B = E\left[\sum_{i=1}^{\infty} e^{-\gamma T_{i-1}} \int_0^{U_i} e^{-\gamma t} b(t) dt\right] = \sum_{i=1}^{\infty} E\left[\left(\prod_{k=1}^{i-1} e^{-\gamma U_k}\right) \int_0^{U_i} e^{-\gamma t} b(t) dt\right]
$$

$$
= \sum_{i=1}^{\infty} E\left[e^{-\gamma U}\right]^{i-1} E\left[\int_0^U e^{-\gamma t} b(t) dt\right] = \frac{E\left[\int_0^U e^{-\gamma t} b(t) dt\right]}{1 - E[e^{-\gamma U}]} \tag{11}
$$

$$
= \frac{\int_0^{\infty} \int_0^t e^{-\gamma \tau} b(\tau) d\tau f(t) dt}{1 - f^*(\gamma)}.
$$

If it is assumed that the benefit function $b(t)$ is unaffected by any renewal in the future one obtains

$$
B^* = \int_0^{\infty} b(t) e^{-\int_0^t \gamma(\tau) d\tau} dt. \tag{12}
$$

4. Deteriorating Structures and Numerical Laplace Transforms

Deteriorating structures are characterized by an increasing hazard function $h(t) = \frac{f(t)}{1-F(t)}$. Let the failure probability at a given time t be computed by FORM or SORM[11] so that

$$
P_f(t) = P(T \leq t) = F_T(\mathbf{p}, t) = \Phi(-\beta(\mathbf{p}, t)) C_{SORM}, \tag{13}
$$

where

$$
\beta(\mathbf{p}, t) = \|\mathbf{u}^*\| = \min\{\|u\|\} \quad \text{for } g(\mathbf{u}, \mathbf{p}, t) \leq 0, \tag{14}
$$

with $g(\mathbf{u}, \mathbf{p}, t)$ as a monotonically decreasing state function, $\mathbf{u} = \mathbf{T}(\mathbf{x})$ a vector of independent standard normal variables and $\mathbf{x}$ the vector of uncertain variable in the original space. $\mathbf{u} = \mathbf{T}(\mathbf{x})$ is a unique probability distribution transformation. C_{SORM} is the second-order correction which is usually neglected. The mean time to failure is

$$
E[T(\mathbf{p})] = \int_0^{\infty} (1 - F_T)(\mathbf{p}, t) dt = \int_0^{\infty} \Phi(\beta(\mathbf{P}, t)) dt. \tag{15}
$$

It can be shown that the density of the time to failure is to first-order[12]

$$
f_T(\mathbf{p}, t) = -\varphi(\beta(\mathbf{p}, t)) \frac{d\beta(\mathbf{p}, t)}{dt} = -\varphi(\beta(\mathbf{p}, t)) \frac{\frac{\partial}{\partial t} g(\mathbf{u}^*, \mathbf{p}, t)}{\|\nabla_u g(\mathbf{u}^*, \mathbf{p}, t)\|}, \tag{16}
$$

so that the Laplace transform can be determined numerically from

$$
f^*(\gamma, \mathbf{p}) \approx \Delta \sum_{j=0}^{m} w_j \exp[-\gamma t_j] f_T(\mathbf{p}, t_j), \tag{17}
$$

with w_j the weights of a suitable integration formula (trapezoid, Simpson, Newton,...) and Δ is an appropriate spacing in time. For the variant with time-dependent interest rate one has:

$$f^{\#}(\gamma, \mathbf{p}) \approx \Delta \sum_{j=0}^{m} w_j \exp\left[-\sum_{i=0}^{j} \gamma(t_i)\Delta\right] f_T(\mathbf{p}, t_j). \tag{18}$$

Of course, more refined numerical integration formulae can be used.

5. Renewal Models for Series Systems of Deteriorating Components

5.1. Independent failure modes and different failure causes including obsolescence

Assume for the moment two independent failure modes, denoted by "V1" and "V2", respectively, each requiring renewal after failure. The times between renewals are then distributed as $F(t) = 1 - (1 - F_{V_1}(t))(1 - F_{V_2}(t)) = 1 - \bar{F}_{V_1}(t)\bar{F}_{V_2}(t)$. The corresponding density is $f(t) = f_{V_1}(t)\bar{F}_{V_2}(t) + f_{V_2}(t)\bar{F}_V(t)$ and its Laplace transform is $f^{**}(\gamma, \mathbf{p}) = f^{**}_{V_1 \mid \bar{V}_2}(\gamma) + f^{**}_{V_2 \mid \bar{V}_1}(\gamma)$. It follows that

$$D(\mathbf{p}) = \frac{(C_1(\mathbf{p}) + H_1)f^{**}_{V_1\mid\bar{V}_2}(\gamma) + (C_2(\mathbf{p}) + H_2)f^{**}_{V_2\mid\bar{V}_1}(\gamma)}{1 - (f^{**}_{V_1\mid\bar{V}_2}(\gamma) + f^{**}_{V_2\mid\bar{V}_1}(\gamma))}. \tag{19}$$

One can easily generalize to more (independently) caused renewals. Here, we distinguish between ordinary Laplace transforms $f^*(\gamma)$ for densities and modified Laplace transforms $f^{**}(\gamma)$ for which $f^*(\gamma) \leq f^{**}(\gamma)$.

Obsolescence occurs if the technical facility no longer fulfills its function. For example, a bridge may become too narrow for the increasing traffic, a fabrication hall is replaced because the machinery inside this hall has to be modernized and restructured, certain vehicles are put out of service because they become too uncomfortable, too uneconomical or unserviceable because of outdated equipment. Usually, this happens despite full system integrity. In fact, most structural facilities will be replaced not because they fail or deteriorate but because they become obsolete. Unfortunately, very few data are available about this well-known fact. Obsolescence is almost always completely independent of the system state. But this is just the case dealt with in Eq. (19) where one of the failure modes, i.e. cause for renewal, is treated as obsolescence. With $A = H_2$ denoting all cost for demolishment and removal of debris Eq. (19) holds.

5.2. Dependent failure modes

Multiple mode failures (series systems) with stationary failure models or even non-stationary failure models with dependent modes can also be considered.[9] Here, only the case of deteriorating components is presented. Assume that there are s

time-dependent failure modes and whose state functions are given by $g_k(\mathbf{u}, t) \approx \boldsymbol{\alpha}_k^T(t)\mathbf{u} + \beta_k(t)$ so that

$$V_k(t) = P(T_k \le t) = P(g_k(\mathbf{U}, t) \le 0) = P(Z_k \le -\beta_k(t)). \tag{20}$$

The failure probability at time t is then

$$
\begin{aligned}
F(t) &= P\left(\bigcup_{k=1}^{s}(t)\{Z_k \le -\beta_k(t)\}\right) \\
&= 1 - P\left(\bigcap_{k=1}^{s}\{Z_k \le \beta_k(t)\}\right) \approx 1 - \Phi_s(\boldsymbol{\beta}(t); \mathbf{R}),
\end{aligned}
\tag{21}
$$

where $\boldsymbol{\beta}(t) = \{\boldsymbol{\alpha}_k^T \mathbf{u}_k^*(t); k = 1, 2, \ldots, s\}$, $\|\boldsymbol{\alpha}_k\| = 1, k = 1, 2, \ldots, s; \mathbf{u}_k^*(t) = \min\{\|\mathbf{u}\|\}$ for $\{\mathbf{u} : g_k(\mathbf{u}, t) \le 0\}$ and $\mathbf{R} = E[\mathbf{ZZ}^T] = \{\rho_{ij}\} = \{\boldsymbol{\alpha}_i^T \boldsymbol{\alpha}_j; i, j = 1, 2, \ldots, s\}$. In good approximation it is assumed that the matrix of correlation coefficients $\mathbf{R}$ varies little with time so that $\frac{\partial}{\partial t}\boldsymbol{\alpha}_k(t) \approx 0$ and, hence, $\frac{\partial}{\partial t}\rho_{ij}(t) \approx 0$ and there is $g_k(\mathbf{0}, t) > 0$ for all k. The general case of $\frac{\partial}{\partial t}\rho_{ij}(t) \ne 0$ is given in Ref. 9. The failure density is

$$
\begin{aligned}
f_s(t) &= \frac{d}{dt}(1 - \Phi_s(\boldsymbol{\beta}(t); \mathbf{R})) = -\sum_{k=1}^{s} \frac{\partial}{\partial \beta_k(t)} \Phi_s(\boldsymbol{\beta}(t); \mathbf{R}) \frac{\partial \beta_k(t)}{\partial t} \\
&= -\sum_{k=1}^{s} \frac{\partial}{\partial \beta_k(t)} \int_{-\infty}^{\beta_k(t)} \Phi_{s-1}(\boldsymbol{\beta}(t); \mathbf{R}\,|Z_k = \beta_k(t))\varphi_1(z_k)\,dz_k \frac{\partial \beta_k(t)}{\partial t} \\
&= -\sum_{k=1}^{s} \varphi_1(\beta_k(t))\Phi_{s-1}(\hat{\mathbf{c}}_k; \hat{\mathbf{R}}_k)\frac{\partial \beta_k(t)}{\partial t} \\
&= \sum_{k=1}^{s} \varphi_1(\beta_k(t))\Phi_{s-1}(\hat{\mathbf{c}}_k; \hat{\mathbf{R}}_k)\left(\frac{-\frac{\partial}{\partial t}g_k(\mathbf{u}^*, t)}{\|\nabla_{\mathbf{u}}g_k(\mathbf{u}^*, t)\|}\right) \\
&\le \sum_{k=1}^{s} \varphi_1(\beta_k(t))\left(\frac{-\frac{\partial}{\partial t}g_k(\mathbf{u}^*, t)}{\|\nabla_{\mathbf{u}}g_k(\mathbf{u}^*, t)\|}\right),
\end{aligned}
\tag{22}
$$

with $\hat{\mathbf{c}}_k = \boldsymbol{\beta}^k(t) - \beta_k(t)\boldsymbol{\rho}_k^k$; and $\hat{\mathbf{R}}_k = \mathbf{R} - \boldsymbol{\rho}_k^k(\boldsymbol{\rho}_k^k)^T$, where $\boldsymbol{\rho}_k$ is the kth column vector of $\mathbf{R}$ and the superscript means that the kth row and column, respectively, are deleted from the original vector and matrix, respectively. This result is obtained from regression analysis. Note that $\hat{\mathbf{R}}_k$ needs to be re-normalized and therefore also $\hat{\mathbf{c}}_k$. The result $\frac{\partial}{\partial \beta_k(t)}\Phi_s(\boldsymbol{\beta}(t); \mathbf{R}) = \varphi_1(\beta_k(t))\Phi_{s-1}(\hat{\mathbf{c}}_k; \hat{\mathbf{R}}_k)$ is due to Ref. 13. Here, the $(s-1)$-dimensional normal integrals have to be evaluated for each t. Suitable computation schemes for $\Phi_r(\mathbf{b}; \mathbf{B})$ have been given in Ref. 14 and elsewhere. Dropping the terms $\Phi_{s-1}(\hat{\mathbf{c}}_k; \hat{\mathbf{R}}_k)$, i.e. the survival probabilities in the other failure modes, corresponds to the upper bound solution:

$$D = \sum_{i=1}^{s}(C_i + H_i)\frac{f_{1,i}^{**}(\gamma)}{1 - \sum_{i=1}^{s} f_{1,i}^{**}(\gamma)} \le \sum_{i=1}^{s}(C_i + H_i)\frac{f_{1,i}^{*}(\gamma)}{1 - \sum_{i=1}^{s} f_{1,i}^{*}(\gamma)}, \tag{23}$$

where

$$f_{1,i}^{**}(\gamma) = \int_0^\infty \exp[-\gamma t]\, \varphi_1(\beta_i(t))\Phi_{s-1}(\hat{\mathbf{c}}_i;\,\hat{\mathbf{R}}_i)\left(\frac{-\frac{\partial}{\partial t}g_i(u^*,t)}{\|\nabla_u g_i(u^*,t)\|}\right)dt$$

$$\leq f_{1,i}^*(\gamma) = \int_0^\infty \exp[-\gamma t]\,\varphi_1(\beta_i(t))dt. \tag{24}$$

5.3. *r-of-s-system of deteriorating components*

The model of a series system of deteriorating components is realistic if structural collapse occurs once a component fails. In many cases, however, deterioration leads more or less only to some sort of loss of serviceability and subsequent repair. It is understood that the components of a system under deterioration are identified as a finite number of "hot spots." In such cases one probably does not define system failure if the first failure occurs. Instead, a number of hot spots must show severe signs of deterioration before repairs are undertaken. Then, the theory for ideal series systems is easily extended to r-of-s-systems which are said to be intact if at least any combination of $r \leq s$ components in a system of s components are intact. Consequently, system survival has probability

$$1 - F(t)$$

$$= \sum_{m=r}^s \sum_{\{N_m^s\}} P\left(\bigcap_{k \in I_m(C_m^s)} \{Z_k > -\beta_k\} \cap \left\{\bigcap_{k \in I_s(C_m^s)\backslash I_m(C_m^s)} \{Z_k \leq -\beta_k\}\right\}\right), \tag{25}$$

where $N_m^s = \frac{s!}{m!(s-m)!}$ is the total number of combinations, $I_m(C_m^s)$ the index sets of all combinations of r elements in a set of s components and $I_s(C_m^s)\backslash I_m(C_m^s)$ the index sets of all remaining components. Specializing now to equicorrelated and equireliable components we have

$$F(t) = 1 - \sum_{m=r}^s \frac{s!}{m!(s-m)!}\Phi_{s,m}\left((\beta_{\{m\}},-\beta_{\{s-m\}})^T;\mathbf{R}\right), \tag{26}$$

with $\beta_{\{m\}} = (\beta_1,\ldots,\beta_m)$, $\beta_{\{s-m\}} = (\beta_{m+1},\ldots,\beta_s)$, $\beta_k = \beta$ and $\mathbf{R} = \{\rho\}$. The failure density is

$$f(t) = \sum_{m=r}^s \frac{s!}{m!(s-m)!}\frac{d}{dt}\Phi_{s,m}((\beta_{\{m\}},-\beta_{\{s-m\}})^T;R)$$

$$= \sum_{m=r}^s \frac{s!}{m!(s-m)!}\sum_{k=1}^s \varphi_1(\beta_k(t))\Phi_{s-1,m}(\hat{\mathbf{c}}_k;\,\hat{\mathbf{R}}_k)\left(\frac{\frac{\partial}{\partial t}g_k(u^*,t)}{\|\nabla_u g_k(u^*,t)\|}\right). \tag{27}$$

In this case one can use a result in Ref. 15 for the multinormal integral

$$\Phi_{s-1,m}(\hat{\mathbf{c}};\,\hat{\mathbf{R}}) = \int_{-\infty}^\infty \varphi_1(t)\Phi_1\left(\frac{\hat{c}-\sqrt{\hat{\rho}}t}{\sqrt{1-\hat{\rho}}}\right)^m \Phi_1\left(\frac{-(\hat{c}-\sqrt{\hat{\rho}}t)}{\sqrt{1-\hat{\rho}}}\right)^{s-1-m}dt, \tag{28}$$

where $\hat{c} = \frac{\beta-\beta\rho}{\sqrt{1-\rho^2}}$ and $\hat{\rho} = \frac{\rho}{1+\rho}$. This theory is used later in an example.

6. Inspection and Repair of Aging Components

Assume repairs at regular intervals $a, 2a, 3a, \ldots$ Repairs occur only if renewals have not occurred before due to obsolescence or failure. Assume further that repairs, if undertaken, restore the properties of a component to its original (stochastic) state, i.e. repairs are equivalent to renewals. Inspection and repair times are assumed negligibly short. Of course, it makes only sense to consider aging components with increasing risk function $h(t)$.

A renewal (repair) occurs either after failure or at times $a, 2a, 3a, \ldots$ Renewal (repair) times are assumed negligibly short. In Ref. 16 this is denoted by age replacement. Then, one obtains

$$Z(p, a) = B - C(\mathbf{p}) - \frac{(C(\mathbf{p}) + H) f_V^{***}(\gamma, \mathbf{p}, a) + I_1(\mathbf{p}) \exp[-\gamma a] \bar{F}_V(\mathbf{p}, a)}{1 - (f_V^{***}(\gamma, \mathbf{p}, a) + \exp[-\gamma a] \bar{F}_V(\mathbf{p}, a))}, \quad (29)$$

with $I_1(\mathbf{p}) < (C(\mathbf{p}) + H)$ the cost of repair, $\bar{F}_V(\mathbf{p}, a)$ the probability of survival up to a and $f_X^{***}(\gamma, \mathbf{p}, a) = \int_0^a \exp[-\gamma t], f_X(t, \mathbf{p}) dt$ the incomplete Laplace transform of $f_X(t)$. It is seen that repair is treated as a second failure (and renewal) mode. This important result was already obtained by Fox[17] and later by Van Noortwijk.[18] Under suitable conditions the quantity $f_X^{***}(\gamma, \mathbf{p}, a)$ can be replaced by $f_X^{\#\#\#}(\gamma, \mathbf{p}, a)$.

If there are regular inspections there is not necessarily a repair because inspections are uncertain (or the signs of deterioration are vague). Then, inspection and repair cost must also be included in the damage term:

$$Z(\mathbf{p}, a) = B(\mathbf{p}, a) - C(\mathbf{p}) - D(\mathbf{p}, a). \quad (30)$$

Including one failure mode "V" with subsequent renewal and assuming independence between failure and repair events results in Ref. 19:

$$D(\mathbf{p}, a) = \frac{ND}{D}$$

$$ND = (C(\mathbf{p}) + H)(f_V^{***}(\gamma, \mathbf{p}, a) + A1)$$
$$+ I_0((1 - P_R(a)) \exp[-\gamma a] \bar{F}_V(\mathbf{p}, a) + A21)$$
$$+ (I_0 + I_1(\mathbf{p}))(P_R(a) \exp[-\gamma a] \bar{F}_V(\mathbf{p}, a) + A22)$$

$$D = 1 - \begin{pmatrix} f_V^{***}(\gamma, p, a) + A1 + \\ + P_R(a) \exp[-\gamma a] \bar{F}_V(p, a) + A22 \end{pmatrix}, \quad (31)$$

$$A1 = \sum_{n=2}^{\infty} \prod_{j=1}^{n-1} (1 - P_R(ja)) f_V^{****}(\gamma, \mathbf{p}, (n-1)a \leq t \leq na)$$

$$A21 = \sum_{n=2}^{\infty} (1 - P_R(na)) \prod_{j=1}^{n-1} (1 - P_R(ja)) \exp[-\gamma(na)] \bar{F}_V(\mathbf{p}, na)$$

$$A22 = \sum_{n=2}^{\infty} P_R(na) \prod_{j=1}^{n-1} (1 - P_R(ja)) \exp[-\gamma na] \bar{F}_V(\mathbf{p}, na)$$

where

$P_R(a)$ = probability of repair after inspection increasing in a

$\bar{P}_R(a) = 1 - P_R(a)$ = probability of no repair after inspection

a = deterministic inspection interval

I_0 = cost per inspection

$I_1(\mathbf{p})$ = repair cost including inspection cost

$f_X^{***}(\gamma, \mathbf{p}, a) = \int_0^a \exp[-\gamma t] f_X(t, \mathbf{p}) dt$ = incomplete Laplace transform of $f_X(t)$

$$f_X^{****}(\gamma, \mathbf{p}, (n-1)a \leq t \leq na) = \int_{(n-1)a}^{na} \exp[-\gamma t] f_X(t, \mathbf{p}) dt.$$

Here, one has to extend the renewal interval to $2a, 3a, \ldots$ if an inspection is not followed by repair. The terms $A1$, $A22$ and $A22$ vanish for $P_R(a) \to 1$ and are significant only for relatively small a.

If the benefit is constant in time we simply have $B(\mathbf{p}, a) = \frac{b}{\gamma}$. For non-constant benefit $b(t)$ as in Eq. (11) it is in analogy to Eq. (31):

$$B(\mathbf{p}, a) = \frac{NB}{D}$$

$$NB = \int_0^\infty B_D(t) f_V(t, \mathbf{p}) dt + B11 + B_D(a)(1 - F_V(a, \mathbf{p})) + B2$$

$$B11 = \sum_{n=2}^\infty \int_{(n-1)a}^{na} B_D^*(t) \left(\prod_{j=1}^{n-1} (1 - P_R(ja)) \right) f_V(t, p) \, dt$$

$$B2 = \sum_{n=2}^\infty B_D^*(t) \left(\prod_{j=1}^{n-1} (1 - P_R(ja)) \right) (1 - F_V(t, \mathbf{p}))$$

$$B_D(t) = \int_0^t e^{-\gamma t} b(t) dt$$

$$B_D^*(t) = \int_{(n-1)a}^t e^{-\gamma t} b(t) dt.$$

(32)

The denominator D is the same as in Eq. (31). The terms $B11$ and $B2$ also vanish for $P_R(a) \to 1$ and are significant only for relatively small a.

The repair probability depends on the magnitude of a suitable damage indicator. For cumulative damage phenomena $P_R(a, \mathbf{p})$ increases with a. For example, $P_R(a, \mathbf{p}) = P(S(a, X, \mathbf{p}) > s_c)$ with $S(a, X, \mathbf{p})$ a monotonically increasing damage indicator, X a random variable taking into account of all uncertainties during inspection and s_c a given threshold level. Frequently, the length of inspection intervals is taken as an optimization parameter. Repair after inspection is interpreted as preventive renewal (replacement of an aging component after a finite time of use a). Renewal after failure is called corrective renewal. It must be mentioned that optimal inspection/repair intervals do not always exist. Preventive renewals must, in fact, be substantially cheaper than corrective renewals. Also, the repair probability must be sufficiently high at the optimum a.

7. Numerical Techniques of Optimization

7.1. *Principles of a one-level approach*

Let $\mathbf{p}$ be a parameter vector which enters the cost function and the limit state function $g(\mathbf{u}, \mathbf{p}) = 0$. Benefit, construction and damage function as well as the limit state function(s) are differentiable in $\mathbf{p}$ and $\mathbf{u}$. The conditions for the application of FORM/SORM hold.[11] In the so-called β-point $\mathbf{u}^*$ the optimality conditions (Kuhn-Tucker conditions) are[20]:

$$g(\mathbf{u}, \mathbf{p}) = 0$$
$$\frac{\mathbf{u}}{\|u\|} = -\frac{\nabla_u g(\mathbf{u}, \mathbf{p})}{\|\nabla_u g(\mathbf{u}, \mathbf{p})\|}. \tag{33}$$

The geometrical meaning of (33) is that the gradient of $g(\mathbf{u}, \mathbf{p}) = 0$ is perpendicular to the vector of direction cosines of $\mathbf{u}^*$. The basic idea mentioned first in Ref. 21 and elaborated in Ref. 20 now is to use these conditions as constraints in the cost optimization problem thus avoiding a bi-level optimization.

It is important to reduce the set of the gradient conditions in the Kuhn-Tucker conditions by one. Otherwise the system of Kuhn-Tucker conditions is overdetermined. It is also important that the remaining Kuhn-Tucker conditions are retained under all circumstances, for example, if one or more gradient Kuhn-Tucker conditions become co-linear with one or more of the other constraints possibly included in the cost-benefit optimization task. Otherwise the so-called β-point conditions are not fulfilled.

7.2. *Formulations for time-variant problems*

For the general case as in Eq. (6) we have:

$$Z(\mathbf{p}) \approx B - C(\mathbf{p}) - (C(\mathbf{p}) + H) \cdot \frac{f^*(\gamma, \mathbf{p})}{1 - f^*(\gamma, \mathbf{p})}$$
$$g(\mathbf{u}_j, \mathbf{p}, t_j) = 0 \quad \text{for } j = 0, 1, \ldots, m$$
$$u_{i,j} \|\nabla_u g(\mathbf{u}_j, \mathbf{p}, t_j)\| + \nabla_u g(\mathbf{u}_j, \mathbf{p}, t_j)_i \|\mathbf{u}_j\| = 0$$
$$i = 1, \ldots, n-1; \quad j = 0, \ldots, m$$
$$h_\ell(\mathbf{p}) \leq 0, \quad \ell = 1, \ldots, q \tag{34}$$
$$\frac{1}{E[T(\mathbf{p})]} \leq r_{\text{admissible}}$$

or

$$\nabla_p C(\mathbf{p}) + G_{x\bar{E}}(g, w, \delta, \rho) k N_{PE} \nabla_p \left(\frac{1}{E[T(\mathbf{p})]} \right) \geq 0.$$

Here, the failure rate acceptability criterion must use the asymptotic failure rate Eq. (9). Alternatively, one can add a criterion derived from socio-economic considerations on life quality and necessary and affordable investments into risk reduction.[22] In this criterion $G_{x\bar{E}}(g, w, \delta, \rho)$ is a constant in the order of 2 to 5 Mill.

PPP US\$ for industrialized countries depending on the part g of the GDP available for risk reduction, the fraction w of life expectancy necessary for (paid) work, the economic growth rate per capita δ and the time-preference rate ρ. k is the probability of being killed in a failure event and N_{PE} the number of potentially endangered people by the event. More discussion about this criterion can be found in Refs. 5, 22–24 The damage cost H can also include so-called life saving cost SLSC discussed in detail in Refs. 22. In countries with a fully developed social system SLSC is approximately the amount to support the (not working) surviving dependents of an adverse event by the social system, mostly by redistribution. If no social system is present, it is useful to think of the amount an insurance should cover after an event. It should be close to the mean of the lost earnings. For example, if GDP $\approx 25\,000$ PPP US\$ and thus, $g \approx 15\,000$ PPP US\$, life expectancy $e \approx 77$ years and $w \approx 0.15$ one calculates SLSC $\approx 600\,000$ PPP US\$. The mean value of interarrival times of renewal in Eq. (34) is computed by Eq. (15) and similar equations for dependent series systems. $f^*(\gamma, \mathbf{p})$ must be replaced by $f^{\#}(\gamma, \mathbf{p})$ whenever a time-dependent discount rate must be used.

8. Examples

8.1. *Random capacity and random demand*

The example has already been given[5,23] in somewhat different form and with different parameters. A single-mode system is considered where failure is defined if a random resistance or capacity is exceeded by a random demand. The demand is modeled as a one-dimensional, stationary marked Poissonian rectangular wave renewal process of disturbances (earthquakes, wind storms, explosions, etc.) with stationary renewal rate λ and random, independent sizes of the disturbances $S_i, i = 1, 2, \ldots$ Disturbances are assumed to be short as compared to their mean interarrival times. The capacity is lognormally distributed with mean p and varies with coefficient of variation V_R. The disturbances are also independent and lognormally distributed with mean equal to unity and coefficient of variation V_S so that p can be interpreted as central safety factor. A disturbance can cause failure with probability:

$$P_f(p) = \Phi\left(\frac{\ln\left\{p\sqrt{\frac{1+V_S^2}{1+V_R^2}}\right\}}{\sqrt{\ln((1+V_S^2)(1+V_R^2))}}\right). \tag{35}$$

An appropriate objective function based on the model in Eq. (8) is

$$Z(p) = \frac{b}{C_0}\int_0^\infty e^{-\int_0^t \gamma(\tau)d\tau}dt - \left(1 + \frac{C_1}{C_0}p^a\right)$$

$$- \left(1 + \frac{C_1}{C_0}p^a + \frac{H}{C_0}\right)\int_0^\infty e^{-\int_0^t \gamma(\tau)d\tau}g(t,p)\,dt, \tag{36}$$

with $g(t,p) = \sum_{n=1}^{\infty} P_f(p)f_n(t)(1 - P_f(p))^{n-1}$, $f_n(t) = \frac{\lambda(\lambda t)^{n-1}}{\Gamma(n)} e^{-\lambda t}$ and $\gamma(t) = \varepsilon\delta + \rho_0 e^{-\alpha t}$; $\varepsilon\delta = 0.03$; $\alpha = 0.013$. In Eq. (36) the first term is the benefit derived from the existence of the facility. A constant benefit rate b is assumed and a time-depending discount rate $\gamma(t)$. The second term is the erection cost where C_0 is the constant part and $C_1 p^a$ the part depending on the design parameter p. The third term are the damage cost. Because systematic reconstruction is chosen the cost term includes reconstruction cost and cost due to physical damage. Reconstruction times are assumed to be negligibly small. The details of this model are discussed in Ref. 5. The acceptability criterion in Eq. (34) has the form:

$$\frac{d}{dp}(C_0 + C_1 p^a) \geq -G_{x\bar{E}}(g, w, \delta, \rho)kN_{PE}\frac{d}{dp}(\lambda P_f(p)). \tag{37}$$

Some more or less realistic, typical parameter assumptions are: $C_0 = 10^6$, $C_1 = 10^4$, $a = 1.25$, $H = 3C_0$, $V_R = 0.2$, $V_S = 0.3$, $b = 0.05\,C_0$ and $\lambda = 1$. The LQI-data is $e = 77$, GDP $= 25\,000$, $g = 15\,000$, $w = 0.13$. Also there is $N_{PE} = 100$, $k = 0.1$. so that $H_F = \text{SLSC}kN_{PE} = 5.4 \times 10^6$ and $G_{x\bar{E}}(g, w, \delta, \rho)kN_{PE} = 5 \times 10^7$.

The value of N_{PE} is chosen relatively large for demonstration purposes. Monetary values are in US\$.

From Eq. (36) one concludes that there is a maximum discount rate in order to render it positive. Projects whose objective is negative even at the optimum do not make sense. Keeping $\varepsilon\delta = 0.02$ constant one can determine the maximum time preference rate from

$$\rho_{\max} = \text{solution of } Z(p_{\text{opt}}, \rho), \tag{38}$$

which is $\rho_{\max} = 0.035^6$.

The stochastic model and the variability of capacity and demand also play an important role for the magnitude and location of the optimum as well as on the position of the acceptability limit. The specific marginal cost (rate of change) of a safety measure and its effect on a reduction of the failure rate are equally important as pointed out already in Ref. 23. This is shown in Fig. 3 where the acceptable failure rates are plotted over the parameters C_1 and V_R for $G_{x\bar{E}}(g, w, \delta, \rho)kN_{PE} = 5 \times 10^6$, i.e. $kN_{PE} = 1$.

Performing the optimization with the (maximum) discount rate gives $p_{\text{opt}} = 4.04$, $(r(p_{\text{opt}}) = 3.0 \times 10^{-5})$ leaving the objective function positive but close to zero. Criterion (37) requires $p_{\lim} = 3.61$, $(r(p_{\lim}) = 1.110^{-4})$. It is interesting to see that in this case one can do better in adopting the optimal solution rather than just realizing the facility at its acceptability limit (see Fig. 2). If, however, life saving cost are not considered in Eq. (36) then the limiting design parameter according to criterion (38) is situated to the right of the optimum.

This example also allows to derive risk-consequence curves by varying the number of fatalities in an event. For the same data as before we first vary the cost effectiveness of the safety measure. Only the ratio C_1 is changed. The upper bounds (solid lines) are derived from Eq. (37). The lower bounds (dashed lines) correspond

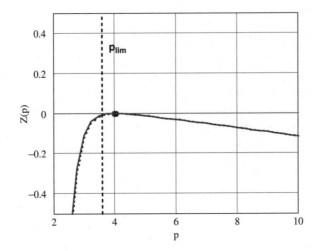

Fig. 2.   Objective function of capacity-demand example for $\rho_{max} = 0.035$.

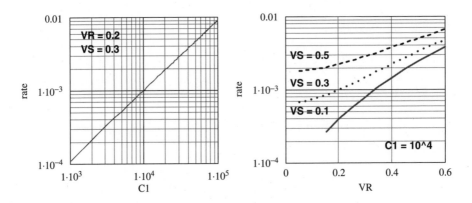

Fig. 3.   Acceptable failure rates versus C_1 and V_R.

to the optimum according to Eq. (36) (see left hand side of Fig. 4). In general, the use of a time-dependent discount rate moves the lower bounds upwards as can be compared with earlier studies.

ρ_{max} also must vary (decrease) with the failure consequences (number of fatalities N_F for a given ratio C_1/C_0. This is taken into account on the right hand side of Fig. 4 indicating that the area between solid and dashed lines broadens for very high failure consequences.

Most realistic is probably a ratio of around $C_1/C_0 = 0.001$. A ratio of $C_1/C_0 = 0.01$ or higher may apply for earthquake resistant structures. Note that in these figures the failure rate is given by $\lambda P_f(p)$ and the number of fatalities is given by $N_F = kN_{PE}$. Therefore, these figures cover the full range of λ and $P_f(p)$ as well as k and N_{PE}. Figure 4 is not a theoretically based F-N-curve because the variable at the ordinate is a failure rate and not an exceedance probability.

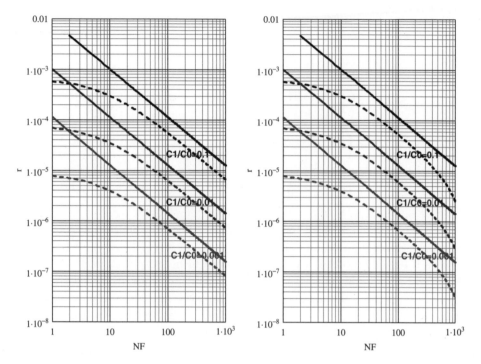

Fig. 4. Risk-consequence curves (solid lines = acceptability limit, dashed lines = optimal solutions).

8.2. *Optimal replacement of a series system of corroding dependent expansion joints*[10]

A long multi-span bridge has s expansion joints which are exposed to corrosion due to heavy winter salting. For illustration purposes the state function is taken as $g(\mathbf{X}, t) = R(1 - C\sqrt{t}) - (S_1 + S_2)$ where $R \sim LN(m_R, 2)$, $C \sim UN(0.085, 0.115)$, $S_1 \sim N(1, 0.3)$ and $S_2 \sim GU(0, 0.2)$. If any of the expansion joints fails the bridge must be closed off. The optimization variables are taken as the mean of resistance m_R and the repair interval a and it is assumed that there are $s = 10$ joints. For simplicity, no uncertainty when inspecting the bridge is assumed, i.e. the detection probability $P_R(a)$ is set to one. The objective function for a time-variant discount rate

$$\gamma(t) = \begin{cases} 0.05 & \text{for } t \leq 40 \\ 0.02 + 0.03 \exp\left[-0.01t\right] & \text{for } t > 40 \end{cases}, \tag{39}$$

and (constant) benefit $b/C_0 = 0.1$ can be written as

$$Z(m_R, a) = \frac{b}{C_0} \int_0^\infty \delta(t)\, dt - \frac{C(m_R)}{C_0}$$
$$- \frac{1}{C_0} \frac{C(m_R)\delta(a)P_s(a) + (C(m_R) + H)f_S^{\#\#\#}(m_R, \gamma)}{1 - \left(\delta(a)P_S(a) + f_S^{\#\#\#}(m_R, \gamma)\right)}, \tag{40}$$

in which:

$$\delta(t) = \exp\left[-\int_0^t \gamma(\tau)d\tau\right],$$

$$P_S(a) = P\left(\bigcap_{k=1}^s \{R(1 - C\sqrt{a}) - (S_{1,k} + S_{2,k}) > 0\}\right),$$

$$f_S^{\#\#\#}(m_R, \gamma) = \int_0^a \delta(t)f_s(t)dt$$

$$f_s(t) = \frac{d}{dt}F_S(t) = \frac{d}{dt}P\left(\bigcup_{k=1}^s \{R(1 - C\sqrt{t}) - (S_{1,k} + S_{2,k}) \le 0\}\right).$$

R and C are common to all joints while the other variables are assumed to be independent from joint to joint and, therefore, $\rho_{ij}(t) = \alpha_R^2(t) + \alpha_C^2(t) \ge 0$. For b/C_0 between 0.10 and 0.09 the objective function will be close to zero. For $C(m_R) = C_0 + C_1 m_R^2$, $C_0 = 10^6$, $C_1 = 10^4$, $H = 10^7$ one determines $m_R^* \approx 7.8$, $a^* \approx 40$. Figure 5 shows the total damage cost, the corrective cost and the preventive cost.

In this case a constant discount rate of $\gamma = 0.03$ leads to roughly the same results for m_R^* and a^* but the benefit differs. The objective function is positive at the optimum. Systematic replacement after failure or at the optimal repair interval can thus save more than 50% of the cost.

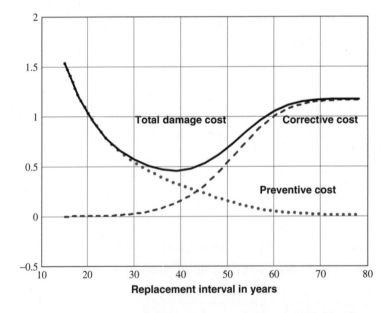

Fig. 5.   Optimization of replacement interval, all cost divided by C_0.

8.3. *Optimal replacement of a reinforced concrete structure (r-of-s-system) subject to chloride corrosion in warm sea water*

Following Ref. 19 a simplified failure criterion for chloride corrosion in the splash zone of some reinforced concrete harbor installation in warm sea water is:

$$C_{cr} - C_s \left(1 - \text{erf}\left(\frac{c}{2\sqrt{Dt}} \right) \right) \leq 0, \tag{41}$$

where C_{cr} = critical chloride content, C_s = surface chloride content, c = concrete cover in cm and D = diffusion parameter. The stochastic model is

Variable	Distr. function	Parameters
C_{cr}	Uniform $[a, b]$	0.125, 0.175
C_s	Uniform $[a, b]$	0.2, 0.4
c	Lognormal	$m_c, \sigma_c = 1$
D	Uniform $[a, b]$	0.085, 0.115

The uniform distributions reflect the large uncertainty in the variables C_{cr}, C_s and D. The standard deviation of concrete cover is 1 cm. The units are chosen such that t is in years. Inspections are performed at regular intervals a. They are followed by renewals (repairs) with probability $P_R(a) = 1 - \exp[-a_R a^2]$. The optimization variables are the mean concrete cover m_c and the length a of the inspection (or repair) interval. Erection cost are $C(m_c) = C_0 + C_1 m_c^2$, inspection cost are $I_0 = 0.1C_0$, repair cost are $I_1 = 0.5C_0$ and there is $C_0 = 10^6, C_1 = 10^4, H = 10C_0$, a benefit function decreasing in time and a time-variant interest rate, i.e. $B = C_0 \int_0^\infty b(t) \exp[\int_0^t \gamma(\tau)d\tau]dt, b(t) = b_0 \Phi(-\frac{t-50}{25}), b_0 = 0.1, \gamma(t) = 0.02 + 0.03 \exp(-0.012t)$, and $a_R = 0.005$. The relatively large damage cost essentially reflect loss of use. The solution is $a^* \approx 65$ and $m_c^* \approx 5.7$. It turns out that preventive repairs should be performed every 65 years which saves considerable cost. The same example with constant discount and constant benefit rate $\gamma = 0.03$ gave $a^* \approx 66$ and $m_c^* \approx 6.5$. These results comply well with practical experience with such structures. The contributions to the total damage cost are shown in Fig. 6.

Relatively small variations in the repair model or in the cost factors will, however, result in designs where it is better not to inspect and repair but just wait for failure. It is noted that for the given failure model no mean time to failure exists. Therefore, no acceptability criterion can be defined theoretically. But this is not necessary because there are clear signs of deterioration and, thus, pre-warnings. The parameter a_R has been chosen such that $P_R(a^*) \approx 1$.

This example is also used to illustrate the procedure for a r-of-s-system. 10 critical components are assumed (for example, piles). C_{cr} and C_s are common to all

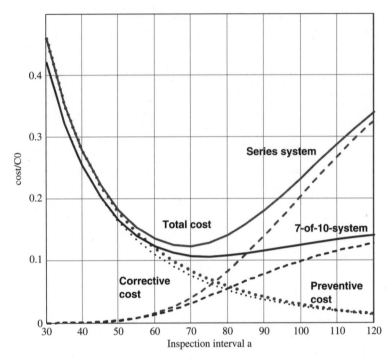

Fig. 6. Total cost for regular inspections and renewals for a series system (upper curves) and a 7-out-of-10-system (lower curves).

components whereas c and D vary independently from component to component. If serious deterioration is found in more than any three of the components with probability $P_R(a) = 1 - \exp[-a_R a^2]$, the inspections are followed by renewals (repairs) for all components. One expects a probability smaller than the probability for a simple component and also smaller than the probability of an ideal series system. This should make m_c^* smaller and a^* larger as compared to the single component case. One finds $m_c^* \approx 4.4$ and $a^* \approx 75$. But the total cost curve (and thus the benefit is larger) is lower for the r-of-s-system.

The optimal values for a* do not exactly correspond to the minimum values in Fig. 6 because the benefit has also been made time-dependent.

9. Conclusions

Technical facilities should be optimal with respect to benefits and cost. The only replacement strategy fulfilling the requirement of sustainability is systematic reconstruction after failure or obsolescence or, alternatively, preventive repair. If the structure is deteriorating a suitable inspection and maintenance strategy has to be adopted. Repairs must be already planned in the design phase and must be such that they correspond to a renewal, i.e. re-establish the (stochastically) initial state. The cost of preventive and/or corrective repairs should be included in life-cycle costing. The repair intervals together with other design parameters can also be optimized. In some cases, however, no optimal interval for preventive repairs exists.

If time is involved all monetary quantities need to be discounted down to the decision point. Discount rates γ must be long term averages in view of the time horizon of some 20 to more than 100 years and net of inflation and taxes. While the operator may use long term averages from the financial market for his cost-benefit analysis the assessment of interest rates for investments of the public is more difficult. The classical Ramsey model decomposes the output growth rate into the rate of time preference of consumption and the rate of economical growth multiplied by the elasticity of marginal utility of consumption. It is found that the rate of time preference of consumption should be as small as possible. The public interest rate should be smaller than the sum of the population growth rate and the long term growth rate of a national economy which is around 2 to 3% for most industrial countries. Alternatively, one can use so-called generation-adjusted discounting models. These models assume that all generations should be treated according to their own preferences but discounting is only by the economic growth rate if future generations are concerned which is also intergenerationally acceptable. An equivalent time-dependent discount rate can be defined decaying from some market value down to the economic growth rate. It is also shown that given a certain output growth rate there is a corresponding maximum real interest rate in order to maintain non-negativity of the objective function. Some numerical tools for cost-benefit optimization are also presented.

References

1. UN World Commission on Environment and Development (known as the Brundland Commission) *Our Common Future* (1987).
2. E. Rosenblueth and E. Mendoza, Reliability optimization in isostatic structures, *J. Eng. Mech. Div., ASCE, 97*, EM6 (1971) 1625–1642.
3. A. Maddison, *Monitoring the World Economy 1820–1992*, OECD, Paris (1995).
4. S. Bayer, Generation-adjusted discounting in long-term decision-making, *Int. J. Sustainable Development* **6**, 1 (2002) 133–149.
5. R. Rackwitz, A. Lentz and M. Faber, Sustainable civil engineering infrastructures by optimization, *Structural Safety* **27**, (2005) 187–229.
6. A. M. Hasofer and R. Rackwitz, Time-dependent models for code optimization, *Proc. ICASP'99*, R. E. Melchers & M. G. Stewart, Editors, Balkema, Rotterdam, **1** (2000) 151–158.
7. D. R. Cox, *Renewal Theory*, Methuen, London (1962).
8. R. Rackwitz, Optimization — The basis of code making and reliability verification, *Structural Safety* **22**, 1 (2000) 27–60.
9. H. Streicher and R. Rackwitz, Time-variant reliability-oriented structural optimization and a renewal model for life-cycle costing, *J. Probabilistic Engineering Mechanics* **19**, 1–2 (2004) 171–183.
10. H. Streicher and R. Rackwitz, Objective functions for reliability-oriented optimization, *Proc. Workshop on Reliability-Based Optimization*, Warsaw (2003), S. Jendo and

K. Dolinski (eds.), Institute of Fundamental Technological Research, Warsaw (2004), 345–405.

11. M. Hohenbichler, S. Gollwitzer, W. Kruse and R. Rackwitz, New light on first- and second-order reliability methods, *Structural Safety* **4**, 4 (1987) 267–284.

12. M. Hohenbichler and R. Rackwitz, Sensitivity and importance measures in structural reliability, *Civil Engineering Systems* **3** (1986) 203–209.

13. G. Schall, S. Gollwitzer and R. Rackwitz, Integration of multinormal densities on surfaces, *Proc. 2nd IFIP WG-7.5 Conf.*, Springer, London, Heidelberg (1988) 235–248.

14. S. Gollwitzer and R. Rackwitz, An efficient numerical solution to the multinormal integral, *Probabilistic Engineering Mechanics* **3**, 2 (1988) 98–101.

15. C. W. Dunnet and M. Sobel, Approximation to the probability integral and certain percentage points of multivariate analogue of student's distribution, *Biometrika* **42** (1955) 258–260.

16. R. E. Barlow and F. Proschan, *Mathematical Theory of Reliability*, Wiley, New York (1965).

17. B. Fox, Age replacement with discounting, *Operations Research* **14** (1966) 533–537.

18. J. M. Van Noortwijk, Cost-based criteria for obtaining optimal design decisions, *Proc. ICOSSAR 01*, Corotis *et al.* Editors, Sweets & Zeitlinger, Lisse (2001).

19. H. Streicher and R. Rackwitz, Renewal models for optimal life-cycle cost of aging civil infra-structures, *3rd Int. IABMAS Workshop*, Lausanne (2003).

20. N. Kuschel and R. Rackwitz, Two basic problems in reliability-based structural optimization, *Mathematical Methods of Operations Research* **46** (1997) 309–333.

21. P. Friis Hansen and H. O. Madsen, A comparison of some algorithms for reliability-based structural optimization and sensitivity analysis, *Proc. 4th IFIP WG 7.5 Conf. Munich*, R. Rackwitz and P. Thoft-Christensen, Editors, Springer Verlag, Berlin (1992) 443–451.

22. R. Rackwitz, *Socio-economic Risk Acceptance Criteria*, in this volume.

23. R. Rackwitz, Optimization and risk acceptability based on the life quality index, *Structural Safety* **24** (2002) 297–331.

24. R. Rackwitz, Optimal and acceptable technical facilities involving risks, *J. Risk Analysis* **24**, 3 (2004) 675–695.

CHAPTER 13

STRUCTURAL HEALTH ASSESSMENT
UNDER UNCERTAINTY

HASAN KATKHUDA* and ACHINTYA HALDAR[†]

*Department of Civil Engineering and Engineering Mechanics,
The University of Arizona, Tucson, AZ 85719, USA
E-mails: *hasan@email.arizona.edu
[†]haldar@u.arizona.edu*

A generalized system identification procedure is presented to identify structural stiffness parameters at the element level using only limited noise-contaminated response information and completely ignoring the excitation information. The authors called it a GILS-EKF-UI method. The structures are represented by finite elements. The procedure detects defects by tracking the changes in the stiffness property of each element. The method can identify defect-free and defective structures even in the presence of relatively large amount of noise in the responses. Defects could be minor in nature. The method is very robust and can identify defects caused by different types of loadings including seismic loading. The research team at the University of Arizona is in the process of developing a nondestructive defect assessment procedure for existing structures and the GILS-EKF-UI procedure will be an essential component of that effort.

1. Introduction

Structural health assessment of existing structures as they age is important to maintain our way of life. Health assessment is also essential just after a natural disaster, such as strong earthquakes or high winds, or after man made events such as blasts or explosions. Visual inspections are commonly used for this purpose. However, for large structural systems, there are no professional guidelines on what to inspect. This strategy is not expected to be effective if defects are not visible to the naked eyes or are hidden behind obstacles such as false ceilings. The quality of inspections or the experience of inspectors can be called in to question. A simple and economical yet objective health assessment technique is necessary for this purpose. For proper and accurate assessment and removing the subjectivity in the evaluation process, the information on structural responses and the uncertainty in collecting response information need to be integrated. Defects that change structural responses can then be identified. To locate the defect spots, a finite element representation of structures will be ideal. By tracking the dynamic properties (mass, stiffness, and damping) of all the elements, the defect spots and the health of structures can be assessed. The discussions lead to the development of a system identification (SI) — based algorithm. The basic three components of any SI technique are the system to be identified, the response information, and the excitation force that caused the system to vibrate. However, in most cases the information on excitation force is not available or it contains so much uncertainty that the SI concept cannot be used. The collection of input excitation forces could be very expensive in some cases. The desirability of the health assessment technique will be significantly enhanced if it can identify a structure with response information only, completely ignoring the excitation information. For large dynamic systems, collection of response information at all dynamic degrees of freedom (DDOFs) may not be practical. Available SI-based approaches cannot satisfy all these requirements. A new finite element-based approach which can identify structures using only noise-laden limited response information is proposed for this purpose. The method is discussed in this chapter.

The SI is a multidisciplinary research area and the existing literature is very extensive [Doebling et al.[1] and Housner et al.[2]]. Most of the available SI techniques are in frequency domain [Alampalli et al.,[3] Law et al.,[4] Lam et al.[5] and Barroso and Rodriguez[6]]. There are many advantages of this approach including: (1) The modal information can be expressed in countable form in terms of frequencies and mode shape vectors instead of using enormous amount of data, (2) it provides a simple basis for comparing modal measurements, and (3) there may be an averaging effect on the presence of noise in the response measurements. However, they suffer from many disadvantages including: (1) These methods identify the defects in the global sense without identifying the location of the defects, (2) a large fraction of the structural members could be broken without the fundamental period changing by more than 2%, and (3) the higher order calculated modes are unreliable for large complicated structural systems. Since the objective of the structural health assessment is to detect the defects in the structures at the local element level, the time domain approach is more appropriate and is used in developing a new method presented in this chapter.

Several time domain approaches have been proposed in the last two decades based on this assumption [Yun and Shinozuka,[7] Hoshiya and Saito,[8] Hoshiya and Maruyama,[9] Toki et al.,[10] Oreta and Tanabe,[11,12] Koh et al.,[13,14] Hoshiya and Sutoh[15] and Loh et al.[16]]. They require excitation information for the identification purpose and will not satisfy the objective of the study. Wang and Haldar.[17] Ling and Haldar,[18] Vo and Haldar,[19] and Katkhuda et al.[20] proposed a linear time domain finite element-based SI approach to identify structures using only response information. They called it the iterative least square with unknown input (ILS-UI). All types of structures including shear-type buildings, trusses, beams and plane frames can be identified using this approach. It was verified using both computer generated theoretical response information and measured responses collected in laboratory experiments. The method was observed to be very accurate compared with other currently available methods even when they use excitation information. From the practical implementation point of view, the only drawback of this method is that the response measurements at all DDOFs must be available to identify a structure. The method needs to be extended so that it can identify structures with noise-laden limited responses information.

Kalman Filter-based approach is commonly used when response information is limited. However, it requires that the excitation information and the state vector must be known. Thus, the ILS-UI and the Kalman Filter-based methods need to be combined to meet the objective of the study. Wang and Haldar[21] proposed such a method. They called it the Iterative Least Square Extended Kalman Filter with Unknown Input (ILS-EKF-UI) method. If the extension is done properly, the ILS-EKF-UI method can provide an ideal platform to identify any structural systems using only limited response information and without suing any excitation information. The concept behind the ILS-EKF-UI procedure is discussed next.

2. ILS-EKF-UI Method

Kalman filter is an optimal recursive data processing algorithm which processes the available response measurements, regardless of their precision [Maybeck[22]]. To implement the algorithm, the information on the state vector and the excitation force, $\mathbf{f}(t)$ must be available. The algorithm uses the prior knowledge about the system and limited response information to produce an estimation of the desired variables by statistically minimizing the error. To increase the efficiency of the optimization algorithm, Hoshiya and Saito[8] proposed the Extended Kalman Filter with Weighted Global Iteration (EKF-WGI) method. The ILS-UI method can be intelligently used to satisfy all the requirements of the EKF-WGI method. A two-stage approach is proposed as discussed below.

Stage 1: Based on the response information, develop a sub-structure that will satisfy all the requirements for the ILS-UI method and then extract information on the unknown excitation force $\mathbf{f}(t)$ and the initial values of the state vectors.

Stage 2: Identify the whole structure using the EKF-WGI method using the information generated in Stage 1.

This endeavor will result the desired ILS-EKF-UI method. A system now can be identified with noise-contaminated limited response information and without using any information on the excitation. Wang and Haldar[21] used the method to identify shear-type buildings. The representation of a multi story building as a shear-type structure is the simplest mathematical representation and consists of many assumptions. This type of building deflects under shear forces only. The total mass of the structure is lumped at the floor levels, assigning 1 DDOF for each floor; the girders/floors are assumed infinitely rigid compared to columns; all the columns in a floor are represented by one column, and the deformation of the structure is considered to be independent of the axial force present in the columns.

The governing equation of motion for this type of structures using viscous damping can be written as:

$$\mathbf{M}\ddot{\mathbf{x}}(t) + \mathbf{C}\dot{\mathbf{x}}(t) + \mathbf{K}\mathbf{x}(t) = \mathbf{f}(t), \tag{1}$$

where $\mathbf{M}$ is the mass matrix; $\mathbf{C}$ is the viscous damping matrix; $\mathbf{K}$ is the stiffness matrix; $\ddot{\mathbf{x}}(t)$, $\dot{\mathbf{x}}(t)$, and $\mathbf{x}(t)$ are vectors containing the dynamic responses in terms of acceleration, velocity and displacement respectively; and $\mathbf{f}(t)$ is the excitation force vector. If the acceleration time histories are available at each floor, they can be successively integrated to obtain the velocity and displacement time histories.

The diagonal mass matrix of a shear-type building can be represented as:

$$\mathbf{M} = \begin{bmatrix} m_1 & 0 & 0 & \cdots & 0 & 0 \\ 0 & m_2 & 0 & \cdots & 0 & 0 \\ \cdots & \cdots & \cdots & \cdots & \cdots & \cdots \\ 0 & 0 & 0 & \cdots & m_{N-1} & 0 \\ 0 & 0 & 0 & \cdots & 0 & m_N \end{bmatrix}. \tag{2}$$

The corresponding damping and stiffness matrices can be shown to be:

$$\mathbf{C} = \begin{bmatrix} c_1 + c_2 & -c_2 & 0 & \cdots & 0 & 0 & 0 \\ -c_2 & c_2 + c_3 & -c_3 & \cdots & 0 & 0 & 0 \\ \cdots & \cdots & \cdots & \cdots & \cdots & \cdots & \cdots \\ 0 & 0 & 0 & \cdots & -c_{N-1} & c_{N-1} + c_N & -c_N \\ 0 & 0 & 0 & \cdots & 0 & -c_N & c_N \end{bmatrix}, \tag{3}$$

and

$$\mathbf{K} = \begin{bmatrix} k_1 + k_2 & -k_2 & 0 & \cdots & 0 & 0 & 0 \\ -k_2 & k_2 + k_3 & -k_3 & \cdots & 0 & 0 & 0 \\ \cdots & \cdots & \cdots & \cdots & \cdots & \cdots & \cdots \\ 0 & 0 & 0 & \cdots & -k_{N-1} & k_{N-1} + k_N & -k_N \\ 0 & 0 & 0 & \cdots & 0 & -k_N & k_N \end{bmatrix}, \tag{4}$$

where m_i, c_i, and k_i $(i = 1, 2, \ldots, N)$ are the mass, damping, and stiffness, respectively, at the ith DDOF of the building.

The general form of the unknown input excitation force $\mathbf{f}(t)$ vector can be represented as:

$$\mathbf{f}(t) = \begin{Bmatrix} f_1(t) \\ f_2(t) \\ \cdots \\ f_{N-1}(t) \\ f_N(t) \end{Bmatrix}. \tag{5}$$

Some of the elements in Eq. (5) can be zero. To clarify the concept, assume that the shear-type structure is subjected to an unknown excitation at the top floor, i.e. at the Nth degree of freedom. Thus, all the elements except $f_N(t)$ in Eq. (5) will be zero. Equation (1) can be used to identify the sub-structure required to evaluate $f_N(t)$ using the ILS-UI method. The requirement can be mathematically expressed as:

$$f_N(t) = m_N\ddot{x}_N(t) + \begin{bmatrix} -c_N & c_N \end{bmatrix}\begin{bmatrix} \dot{x}_{N-1}(t) \\ \dot{x}_N(t) \end{bmatrix} + \begin{bmatrix} -k_N & k_N \end{bmatrix}\begin{bmatrix} x_{N-1}(t) \\ x_N(t) \end{bmatrix}. \quad (6)$$

The masses at the floor levels are generally assumed to be known. Thus, the response information, in terms of displacement, velocity and acceleration, at the top two floors at the minimum, and the unknown parameters c_N and k_N must be available to evaluate the unknown excitation force $f_N(t)$. To develop sufficient constraints, two simultaneous equations for the $(N-1)$th and Nth floors at time t need to be considered [Wang and Haldar[17,21]] as:

$$\begin{bmatrix} \dot{x}_{N-1} - \dot{x}_{N-2} & \dot{x}_{N-1} - \dot{x}_N & x_{N-1} - x_{N-2} & x_{N-1} - x_N \\ 0 & \dot{x}_N - \dot{x}_{N-1} & 0 & x_N - x_{N-1} \end{bmatrix}\begin{bmatrix} c_{N-1} \\ c_N \\ k_{N-1} \\ k_N \end{bmatrix}$$

$$= \begin{bmatrix} -m_{N-1}\ddot{x}_{N-1}(t) \\ f_N(t) - m_N\ddot{x}_N(t) \end{bmatrix}. \quad (7)$$

Equation (7) indicates that at the minimum the response information at the top three floors are needed to evaluate the unknown input excitation force and to identify four unknown parameters of the system. Thus, Eq. (7) defines the sub-structure required for the first stage of ILS-EKF-UI. Assuming that the responses are measured at a constant time increment of Δt for the total number of sample time points h, Eq. (7) can be rewritten in a matrix form as:

$$\mathbf{A}_{(2.h)\times 4}\mathbf{P}_{4\times 1} = \mathbf{F}_{(2.h)\times 1}, \quad (8)$$

where $\mathbf{A}$ is a matrix composed of the system response vector of velocity and displacement; $\mathbf{P}$ is the vector composed of the unknown stiffness and damping parameters; and $\mathbf{F}$ is the vector composed of input excitation and inertia forces at time t. Since $\mathbf{F}$ is partially known, Eq. (8) cannot be solved to identify the four system parameters in $\mathbf{P}$. ILS-UI is an iterative technique. Since the excitation information is not available, Wang and Haldar[17] suggested that the excitation force $\mathbf{f}(t)$ can be assumed to be zero for p time points to start the iterative process. However, this p time points should be kept to a minimum. They observed that p can be only 2 time points if the structure is excited at any floor. Once the iteration process converged satisfying a pre-determined convergence criterion, the unknown system parameters c_{N-1}, c_N, k_{N-1}, and k_N in $\mathbf{P}$ and the unknown excitation force $f_N(t)$ will be available. The information of the system parameters can be used to completely define the initial state vector. The excitation information is also available. Thus, all the information required to implement Stage 2 is now available.

To implement the EKF-based Stage 2, the state vector can be defined as:

$$\mathbf{X}_t(t) = \begin{bmatrix} \mathbf{X}_1(t) \\ \mathbf{X}_2(t) \\ \mathbf{X}_3(t) \\ \mathbf{X}_4(t) \end{bmatrix} = \begin{bmatrix} \mathbf{X}(t) \\ \dot{\mathbf{X}}(t) \\ \tilde{\mathbf{K}} \\ \tilde{\mathbf{C}} \end{bmatrix}, \tag{9}$$

where $\mathbf{X}_t(t)$ is the state vector at time t; $\mathbf{X}(t)$ and $\dot{\mathbf{X}}(t)$ are the displacement and velocity vectors, respectively, at time t for the whole structure, and $\tilde{\mathbf{K}}$ and $\tilde{\mathbf{C}}$ are the vectors of the element stiffnesses and damping, respectively, for the whole structure that need to be identified. These vectors can be expressed as:

$$\mathbf{X}(t) = \begin{bmatrix} x_1(t) \\ x_2(t) \\ \vdots \\ x_N(t) \end{bmatrix}, \quad \dot{\mathbf{X}}(t) = \begin{bmatrix} \dot{x}_1(t) \\ \dot{x}_2(t) \\ \vdots \\ \dot{x}_N(t) \end{bmatrix}, \quad \tilde{\mathbf{K}} = \begin{bmatrix} k_1 \\ k_2 \\ \vdots \\ k_N \end{bmatrix}, \quad \tilde{\mathbf{C}} = \begin{bmatrix} c_1 \\ c_2 \\ \vdots \\ c_N \end{bmatrix}, \tag{10}$$

where N is the total number of DDOFs as mentioned earlier; $x_i(t)$ and $\dot{x}_i(t)$ are the displacement and velocity at the ith DDOF at time t, respectively.

Equation (1) can be expressed as a state equation:

$$\dot{\mathbf{X}}_t(t) = \begin{bmatrix} \dot{\mathbf{X}}_1(t) \\ \dot{\mathbf{X}}_2(t) \\ \dot{\mathbf{X}}_3(t) \\ \dot{\mathbf{X}}_4(t) \end{bmatrix} = \begin{bmatrix} \dot{\mathbf{X}}(t) \\ \ddot{\mathbf{X}}(t) \\ 0 \\ 0 \end{bmatrix} = \begin{bmatrix} \dot{\mathbf{X}}(t) \\ -\mathbf{M}^{-1}\left(\mathbf{K}\mathbf{X}(t) + \mathbf{C}\dot{\mathbf{X}}(t) - \mathbf{f}(t)\right) \\ 0 \\ 0 \end{bmatrix}, \tag{11}$$

where $\mathbf{K}$ and $\mathbf{C}$ are the global stiffness and mass matrices, respectively, and can be determined using Eqs. (3) and (4).

Equation (11) can be expressed as:

$$\frac{d\mathbf{X}_t(t)}{dt} = f(\mathbf{X}_{t_0}, t), \quad \mathbf{X}_{t_0} \sim N(\hat{\mathbf{X}}_{t_0}, \mathbf{P}_{t_0}), \tag{12}$$

$$\mathbf{Y}_{t_k} = \mathbf{H}\mathbf{X}_t(t_k) + \mathbf{V}_{t_k}, \quad \mathbf{V}_{t_k} \sim N(0, \mathbf{R}_{t_k}), \tag{13}$$

where $\mathbf{X}_t(t)$ is a vector of size $4N \times 1$ at time t for a system with N DDOFs; $\mathbf{X}_{t_0}$ is the initial state vector assumed to be Gaussian random variables with mean $\hat{\mathbf{X}}_{t_0}$ and an error covariance $\mathbf{P}_{t_0}$; $\mathbf{Y}_{t_k}$ is an observational vector of size $(B \times 1)$; B is the total number of displacement and velocity observations (the information on acceleration is not required for the EKF-WGI); $\mathbf{H}$ is a matrix of size $(B \times 4N)$ containing information of measured responses; $\mathbf{V}_{t_k}$ is the observational noise vector of size $(B \times 1)$ assumed to be Gaussian white noise with zero mean and a covariance $\mathbf{R}_{t_k}$, t_k is $k\Delta t$; and Δt is the constant time interval between measurements.

Using the information on the initial state vector $\hat{\mathbf{X}}_{t_0}(t_0/t_0)$ and the error covariance matrix $\mathbf{P}(t_0/t_0)$, the recursive process of the EKF algorithm can be carried out in the following steps:

(i) Start with the initial state vector $\hat{\mathbf{X}}_{t_0}(t_0/t_0)$ and its error covariance $\mathbf{P}(t_0/t_0)$. The initial state vector can be obtained from Eqs. (9) and (10). The velocity and displacement time histories in Eq. (10) may not be available at all DDOFs. They are obtained by successively integrating the acceleration time histories available for the identification purpose. $\tilde{\mathbf{K}}$ and $\tilde{\mathbf{C}}$ vectors are the initial values of the stiffness and damping, respectively. Information on c_{N-1}, c_N, k_{N-1}, and k_N obtained from Stage 1 can be used to initially define them. The values of parameters c_1 to c_{N-2} and k_1 to k_{N-2} can be initially assumed to be of the same orders as c_N and k_N, respectively. These vectors can be represented as:

$$\tilde{\mathbf{K}}(t_0/t_0) = \begin{bmatrix} k_1(t_0/t_0) \\ \vdots \\ k_{N-2}(t_0/t_0) \\ k_{N-1}(t_0/t_0) \\ k_N(t_0/t_0) \end{bmatrix} = \begin{bmatrix} k_N \\ \vdots \\ k_N \\ k_{N-1} \\ k_N \end{bmatrix},$$

$$\tilde{\mathbf{C}}(t_0/t_0) = \begin{bmatrix} c_1(t_0/t_0) \\ \vdots \\ c_{N-2}(t_0/t_0) \\ c_{N-1}(t_0/t_0) \\ c_N(t_0/t_0) \end{bmatrix} = \begin{bmatrix} c_N \\ \vdots \\ c_N \\ c_{N-1} \\ c_N \end{bmatrix}. \tag{14}$$

Conceptually, any other initial values can be assumed to define Eq. (14). However, our experience suggests that if the initial values are significantly different from the actual values, it may create convergence problem or it may take more iterations to converge reducing the efficiency of the algorithm.

The initial error covariance matrix, $\mathbf{P}(t_0/t_0)$ is a diagonal matrix and can be represented as:

$$\mathbf{P}(t_0/t_0) = \begin{bmatrix} \mathbf{P}_x(t_0/t_0) & 0 \\ 0 & \mathbf{P}_k(t_0/t_0) \end{bmatrix}, \tag{15}$$

where $\mathbf{P}_x(t_0/t_0)$ is a initial error covariance matrix of size $(2N \times 2N)$, corresponding to velocity and displacement. It is a diagonal identity matrix. $\mathbf{P}_k(t_0/t_0)$ is the initial error covariance matrix of size $(2N \times 2N)$ corresponding to the $\tilde{\mathbf{K}}$ and $\tilde{\mathbf{C}}$. It is a diagonal matrix. A value of 1000 in the diagonal is assumed in this study to accelerate the convergence of the EKF algorithm.

(ii) *Prediction phase*: In the context of EKF, evaluate the predicted state $\hat{\mathbf{X}}(t_{k+1}/t_k)$ and its error covariance $\mathbf{P}(t_{k+1}/t_k)$ by linearizing the nonlinear dynamic equation as:

$$\hat{\mathbf{X}}(t_{k+1}/t_k) = \hat{\mathbf{X}}(t_k/t_k) + \int_{t_k}^{t_{k+1}} f[\hat{\mathbf{X}}(t/t_k), t]dt, \tag{16}$$

$$\mathbf{P}(t_{k+1}/t_k) = \Phi[t_{k+1}, t_k; \hat{\mathbf{X}}(t_k/t_k)] \bullet \mathbf{P}(t_k/t_k) \bullet \Phi^T[t_{k+1}, t_k; \hat{\mathbf{X}}(t_k/t_k)], \tag{17}$$

where $\mathbf{I}$ is a unit matrix; $\Phi[t_{k+1}, t_k; \hat{\mathbf{X}}(t_k/t_k)]$ is the state transfer matrix from time t_k to t_{k+1} and can be represented as:

$$\Phi[t_{k+1}, t_k; \hat{\mathbf{X}}(t_k/t_k)] = \mathbf{I} + \Delta t \bullet \mathbf{F}[t_k; \hat{\mathbf{X}}(t_k/t_k)] \tag{18}$$

In which

$$\mathbf{F}[t_k; \hat{\mathbf{X}}(t_k/t_k)] = \left[\frac{\partial f(\mathbf{X}_{t_k}, t_k)}{\partial X_j}\right]_{X_{tk} = \hat{X}(t_k/t_k)} \tag{19}$$

where X_j is the jth component of vector $\mathbf{X}_{tk}$

(iii) *Updating phase*: Since the observations become available at the $(k + 1)$ iteration, the state vector and the error covariance can be updated as:

$$\hat{\mathbf{X}}(t_{k+1}/t_{k+1}) = \hat{\mathbf{X}}(t_{k+1}/t_k) + \mathbf{K}[t_{k+1}; \hat{\mathbf{X}}(t_{k+1}/t_k)]$$
$$\bullet \left\{\mathbf{Y}(t_{k+1}) - \mathbf{H}[\hat{\mathbf{X}}(t_{k+1}/t_k), t_{k+1}]\right\}, \tag{20}$$

$$\mathbf{P}(t_{k+1}/t_{k+1}) = \left\{\mathbf{I} - \mathbf{K}[t_{k+1}; \hat{\mathbf{X}}(t_{k+1}/t_k)] \bullet \mathbf{M}[t_{k+1}; \hat{\mathbf{X}}(t_{k+1}/t_k)]\right\} \bullet \mathbf{P}(t_{k+1}/t_k)$$
$$\bullet \left\{\mathbf{I} - \mathbf{K}[t_{k+1}; \hat{\mathbf{X}}(t_{k+1}/t_k)] \bullet \mathbf{M}[t_{k+1}; \hat{\mathbf{X}}(t_{k+1}/t_k)]\right\}^T$$
$$+ \mathbf{K}[t_{k+1}; \hat{\mathbf{X}}(t_{k+1}/t_k)] \bullet \mathbf{R}(t_{k+1}) \bullet \mathbf{K}^T[t_{k+1}; \hat{\mathbf{X}}(t_{k+1}/t_k)], \tag{21}$$

where $\mathbf{K}[t_{k+1}; \hat{\mathbf{X}}(t_{k+1}/t_k)]$ is the Kalman gain matrix. It can be shown to be:

$$\mathbf{K}[t_{k+1}; \hat{\mathbf{X}}(t_{k+1}/t_k)] = \mathbf{P}(t_{k+1}/t_k) \bullet \mathbf{M}^T[t_{k+1}; \hat{\mathbf{X}}(t_{k+1}/t_k)]$$
$$\bullet \left\{\mathbf{M}[t_{k+1}; \hat{\mathbf{X}}(t_{k+1}/t_k)] \bullet \mathbf{P}(t_{k+1}/t_k)\right.$$
$$\left.\bullet \mathbf{M}^T[t_{k+1}; \hat{\mathbf{X}}(t_{k+1}/t_k)] + \mathbf{R}(t_{k+1})\right\}^{-1}, \tag{22}$$

where

$$\mathbf{M}[t_k; \hat{\mathbf{X}}(t_k/t_k)] = \left[\frac{\partial \mathbf{H}(\mathbf{X}_{t_i}, t_i)}{\partial X_j}\right]. \tag{23}$$

(iv) Take the next time increment and using Eqs. (16), (17), (20), and (21) predict and update the system parameters. This procedure will continue until the all the time points are used.

These iterative steps covering all the time point is generally defined as the local iteration. The local iteration may not produce stable and convergent estimation of the parameters to be identified. Hoshiya and Saito[8] suggested a weighted global iteration technique for this purpose and used in this study.

At the completion of the local iteration using EKF, the estimations of $\hat{\mathbf{X}}^{(1)}(t_h/t_h)$ and $\mathbf{P}^{(1)}(t_h/t_h)$ are available, where superscript (1) represents the first global iteration. In the second global iteration, a weight factor w is introduced to the error covariance matrix to accelerate the convergence. It is worth to mention here that to get better and stable convergence, the value of w should be large positive number.

The state vector and the error covariance matrix can be expressed as:

$$\hat{\mathbf{X}}^{(2)}(t_0/t_0) = \begin{bmatrix} \hat{\mathbf{X}}_1^{(1)}(t_0/t_0) \\ \hat{\mathbf{X}}_2^{(1)}(t_0/t_0) \\ \hat{\mathbf{X}}_3^{(1)}(t_h/t_h) \\ \hat{\mathbf{X}}_4^{(1)}(t_h/t_h) \end{bmatrix} = \begin{bmatrix} \mathbf{X}^{(1)}(t_0/t_0) \\ \dot{\mathbf{X}}^{(1)}(t_0/t_0) \\ \tilde{\mathbf{K}}^{(1)}(t_h/t_h) \\ \tilde{\mathbf{C}}^{(1)}(t_h/t_h) \end{bmatrix}, \tag{24}$$

$$\mathbf{P}^{(2)}(t_0/t_0) = \begin{bmatrix} \mathbf{I} & 0 \\ 0 & w\mathbf{P}_{tk}^{(1)}(t_h/t_h) \end{bmatrix}, \tag{25}$$

where $\mathbf{P}_{tk}^{(1)}(t_h/t_h)$ is the error covariance matrix corresponding to the parameters $\tilde{\mathbf{K}}$ and $\tilde{\mathbf{C}}$ from the first global iteration. With this information, the prediction and updating phases of the local iteration are carried out for all the time points producing the state vector and the error covariance matrix required to initiate the third global iteration.

The global iterations are repeated until a pre-determined convergence criterion is satisfied. The convergence criterion considered in this study can be represented as $|\mathbf{X}^{(i)}(t_h/t_h) - \mathbf{X}^{(i-1)}(t_h/t_h)| \le \varepsilon$. ε is considered to be 0.1. Wang and Haldar[21] verified this algorithm for the shear-type buildings. As will be discussed next, the algorithm cannot be used to identify frames and other types of structures and the efficiency of the algorithm can be improved significantly. The authors proposed a generalized ILS-EKF-UI method. It will be denoted as GILS-EKF-UI in the subsequent discussions.

3. GILS-EKF-UI Method

In developing the ILS-EKF-UI method, Wang and Haldar[21] used viscous damping in the dynamic governing equation. The information on damping of elements is generally not used for any defect evaluation. Furthermore, when viscous damping is used in the formulation, for N DDOFs shear-type buildings, N stiffness and N damping parameters need to be identified. Thus, the total number of parameters to be identified is $2N$. If Rayleigh-type damping is used, the damping is proportional to the mass and stiffness and can be represented by two damping coefficients α

and β. Since the proposed approach is finite element-based, the mass and stiffness matrixes are readily available, and the consideration of Rayleigh-type damping is not expected to create any additional difficulty in the formulation. However, the total number of parameters to be identified is significantly reduced. The two damping coefficients will be identified in Stage 1 and can be assumed to be known in Stage 2. Thus, only N structural parameters need to be identified. As will be discussed later, for a structure consisting of *ne* structural elements, the total number of parameters to be identified is *ne*, significantly increasing the efficiency of the algorithm, particularly for large structural systems.

To identify any structure, the finite element representation needs to be generalized. Without loosing any generality and for the ease of discussion, two-dimensional frame structures can be considered. Two-dimensional beam elements can be used to represent such frames. For such an element, three DDOFs are present at each node. Two are translational DDOFs; one is along the length of the element (x axis) and the other is perpendicular to the x axis, i.e., along the y axis, and the third DDOF represents is the rotation of the node.

The basic concept used for shear-type buildings can also be used to identify any structural system. As mentioned earlier, the first stage is to select a sub-structure so that the ILS-UI procedure can be used to evaluate the unknown excitation forces and the state vector. Katkhuda[23] discussed the process in detail elsewhere. The first step is to identify key node or nodes where the excitation forces are expected to act. Then, the sub-structure can be developed by considering all the elements connected to all the key nodes. For earthquake loading, all nodes can be considered as key nodes, and the sub-structure can be constituted based on the available response information.

For a sub-structure with Rayleigh-type damping, the governing equation of motion can be written as:

$$\mathbf{K}_{sub}\mathbf{x}_{sub}(t) + (\alpha\mathbf{M}_{sub} + \beta\mathbf{K}_{sub})\dot{\mathbf{x}}_{sub}(t) + \mathbf{M}_{sub}\ddot{\mathbf{x}}_{sub}(t) = \mathbf{f}_{sub}(t), \qquad (26)$$

where $\mathbf{K}_{sub}$ and $\mathbf{M}_{sub}$ are the stiffness and mass matrices of the sub-structure, respectively, $\ddot{\mathbf{x}}_{sub}(t)$, $\dot{\mathbf{x}}_{sub}(t)$, and $\mathbf{x}_{sub}(t)$ are the acceleration, velocity and displacement vectors at time t, respectively, α is the mass-proportional damping coefficient, β is the stiffness-proportional damping coefficient, and $\mathbf{f}_{sub}(t)$ is the unknown excitation force vector for the sub-structure. The damping coefficients α and β can be estimated using the first two undamped frequencies of the structure as suggested by Clough and Penzien.[24]

$\mathbf{M}_{sub}$ can be assembled from the mass matrices of all the elements in the sub-structure as:

$$\mathbf{M}_{sub} = \sum_{i=1}^{nesub} \mathbf{M}^{i}, \qquad (27)$$

where $\mathbf{M}_{sub}$ is a matrix of size ($Nsub \times Nsub$); $Nsub$ is the total number of DDOFs in the sub-structure; *nesub* is the total number of elements in the sub-structure, and $\mathbf{M}^i$ is the consistent mass matrix for the ith beam element of uniform cross section and can be represented as [Cook *et al.*[25]]:

$$
\mathbf{M}^i = \frac{\bar{m}_i L_i}{420}
\begin{bmatrix}
140 & & & & & \\
0 & 156 & & & \text{Sym.} & \\
0 & 22L_i & 4L_i^2 & & & \\
70 & 0 & 0 & 140 & & \\
0 & 54 & 13L_i & 0 & 156 & \\
0 & -13L_i & -3L_i^2 & 0 & -22L_i & 4L_i^2
\end{bmatrix},
\tag{28}
$$

where L_i and $\bar{m}_i$ are the length and the mass per unit length of the ith element.

$\mathbf{K}_{sub}$ can similarly be assembled from the information on the stiffness matrices of all the elements in the sub-structure as:

$$
\mathbf{K}_{sub} = \sum_{i=1}^{nesub} \mathbf{K}^i,
\tag{29}
$$

where $\mathbf{K}^i$ is the stiffness matrix for the ith beam element of uniform cross section (constant flexural stiffness or constant EI) and is given by:

$$
\mathbf{K}^i = \frac{E_i I_i}{L_i}
\begin{bmatrix}
A_i/I_i & 0 & 0 & -A_i/I_i & 0 & 0 \\
0 & 12/L_i^2 & 6/L_i & 0 & -12/L_i^2 & 6/L_i \\
0 & 6/L_i & 4 & 0 & -6/L_i & 2 \\
-A_i/I_i & 0 & 0 & A_i/I_i & 0 & 0 \\
0 & -12/L_i^2 & -6/L_i & 0 & 12/L_i^2 & -6/L_i \\
0 & 6/L_i & 2 & 0 & -6/L_i & 4
\end{bmatrix},
\tag{30}
$$

where E_i, I_i, and A_i are the Young's modulus, moment of inertia, and area of the cross-section of the ith element, respectively.

Equation (30) can be rewritten as:

$$
\mathbf{K}^i = k_i \mathbf{S}^i,
\tag{31}
$$

where $k_i = E_i I_i / L_i$ and $\mathbf{S}^i$ is the 6×6 matrix for the ith element shown in the square bracket in Eq. (30). Equation (29) can be rewritten as:

$$
K_{sub} = \sum_{i=1}^{nesub} k_i \mathbf{S}^i = k_1 \mathbf{S}^1 + k_2 \mathbf{S}^2 + \cdots + k_{nesub} \mathbf{S}^{nesub}.
\tag{32}
$$

Also, Eq. (26) can be rewritten in the matrix form as:

$$
\mathbf{A}_{(Nkey \cdot h) \times Lsub} \mathbf{P}_{Lsub \times 1} = \mathbf{F}_{(Nkey \cdot h) \times 1},
\tag{33}
$$

where matrix $\mathbf{A}$ is a matrix of size $(Nkey \cdot h) \times Lsub$; $Nkey$ is the total number of DDOFs for the key node(s) in the sub-structure, h is the total number of sample points, and $Lsub$ is the total number of unknown parameters in the sub-structure.

The **A** matrix can be expressed as:

$$\mathbf{A}_{(Nkey.h) \times Lsub} = [\mathbf{S}^1 \mathbf{x}(t) \mathbf{S}^2 \mathbf{x}(t) \cdots \mathbf{S}^{nesub} \mathbf{x}(t) \mathbf{S}^1 \dot{x}(t) \mathbf{S}^2 \dot{\mathbf{x}}(t) \cdots$$
$$\mathbf{S}^{nesub} \dot{\mathbf{x}}(t) \mathbf{M} \dot{\mathbf{x}}(t)]. \tag{34}$$

The displacement and velocity vectors, at time t, can be expressed as:

$$\mathbf{x}(t) = [x_1, y_1, \theta_1, x_2, y_2, \theta_2, \ldots, x_{nnsub}, y_{nnsub}, \theta_{nnsub}]^T, \tag{35}$$

and

$$\dot{\mathbf{x}}(t) = \left[\dot{x}_1, \dot{y}_1, \dot{\theta}_1, \dot{x}_2, \dot{y}_2, \dot{\theta}_2, \ldots, \dot{x}_{nnsub}, \dot{y}_{nnsub}, \dot{\theta}_{nnsub} \right]^T, \tag{36}$$

where $nnsub$ is the total number of nodes in the sub-structure.

P vector in Eq. (33) contains the following unknown system parameters:

$$\mathbf{P} = [k_1, k_2, \ldots, k_{nesub}, \beta k_1, \beta k_2, \ldots, \beta k_{nesub}, \alpha]^T. \tag{37}$$

All the parameters have been defined earlier.

F matrix in Eq. (33) can be expressed as:

$$\mathbf{F}_{(Nkey.h) \times 1} = \mathbf{f}_{(Nkey.h) \times 1} - \mathbf{M} \ddot{\mathbf{x}}(t)_{(Nkey.h) \times 1}, \tag{38}$$

where $\ddot{\mathbf{x}}(t)$ vector contains the acceleration responses in the sub-structure at time t and can be expressed as:

$$\ddot{\mathbf{x}}(t) = \left[\ddot{x}_1, \ddot{y}_1, \ddot{\theta}_1, \ddot{x}_2, \ddot{y}_2, \ddot{\theta}_2, \ldots, \ddot{x}_{nnsub}, \ddot{y}_{nnsub}, \ddot{\theta}_{nnsub} \right]^T. \tag{39}$$

Equation (33) can be used to solve for the unknown parameters vector **P** using the ILS-UI method discussed earlier. As in the shear-type building, the iteration process needs to be initiated by assuming $\mathbf{f}(t)$ to be zero at some time points. In developing the GILS-EKF-UI method, the authors [Katkhuda, et al.[20]] observed that the algorithm produces better and more accurate results if the excitation information is assumed to be zero at all time points h. As will be shown later, the algorithm accurately produces information on the unknown excitation force, Rayleigh damping coefficients and the stiffnesses of the elements in the sub-structure.

The state vector for framed structures can be represented as:

$$\mathbf{X}_t(t) = \begin{bmatrix} \mathbf{X}_1(t) \\ \mathbf{X}_2(t) \\ \mathbf{X}_3(t) \end{bmatrix} = \begin{bmatrix} \mathbf{X}(t) \\ \dot{\mathbf{X}}(t) \\ \tilde{\mathbf{K}} \end{bmatrix}, \tag{40}$$

where $\mathbf{X}_t(t)$ is the state vector at time t, $\mathbf{X}(t)$ and $\dot{\mathbf{X}}(t)$ vectors contain the limited response information on displacement and velocity at time t, respectively; and $\tilde{\mathbf{K}}$

is the vector of the element stiffnesses for the whole structure that need to be identified and can be expressed here as:

$$\tilde{\mathbf{K}} = \begin{bmatrix} k_1 \\ k_2 \\ \vdots \\ k_{ne} \end{bmatrix}, \tag{41}$$

where ne is the total number of elements in the whole structure.

Equation (40) can be rewritten as:

$$\dot{\mathbf{X}}_t(t) = \begin{bmatrix} \dot{\mathbf{X}}_1(t) \\ \dot{\mathbf{X}}_2(t) \\ \dot{\mathbf{X}}_3(t) \end{bmatrix} = \begin{bmatrix} \dot{\mathbf{X}}(t) \\ \ddot{\mathbf{X}}(t) \\ 0 \end{bmatrix}$$

$$= \begin{bmatrix} \dot{\mathbf{X}}(t) \\ -\mathbf{M}^{-1}(\mathbf{K}\mathbf{X}(t) + (\alpha\mathbf{M} + \beta\mathbf{K})\dot{\mathbf{X}}(t) - \mathbf{f}(t)) \\ 0 \end{bmatrix}, \tag{42}$$

where $\mathbf{K}$ and $\mathbf{M}$ is the global stiffness and mass matrix for the whole structure, respectively. They can be developed from the element-level information using the standard finite element formulation. $\mathbf{f}(t)$ is identified in Stage 1.

Equations (12) and (13) are applicable for any type of structure; however, the size of the matrixes will change. The size of the state vector $\mathbf{X}_t$ will be $(2N+L) \times 1$ and the size of $\mathbf{H}$ will be $[B \times (2N+L)]$. The four iterative steps in Stage 2 [Eqs. (14) to (23)] will remain the same for framed structures except, Eq. (14) needs to be expressed as:

$$\tilde{\mathbf{K}}(t_0/t_0) = \begin{bmatrix} k_1(t_0/t_0) \\ k_2(t_0/t_0) \\ k_3(t_0/t_0) \\ \vdots \\ k_{ne}(t_0/t_0) \end{bmatrix} = \begin{bmatrix} k_1 \\ k_2 \\ k_2 \\ \vdots \\ k_2 \end{bmatrix}. \tag{43}$$

The weighted global iteration technique remains the same for framed structures except Eq. (24) is changed to:

$$\hat{\mathbf{X}}^{(2)}(t_0/t_0) = \begin{bmatrix} \hat{\mathbf{X}}_1^{(1)}(t_0/t_0) \\ \hat{\mathbf{X}}_2^{(1)}(t_0/t_0) \\ \hat{\mathbf{X}}_3^{(1)}(t_h/t_h) \end{bmatrix} = \begin{bmatrix} \mathbf{X}^{(1)}(t_0/t_0) \\ \dot{\mathbf{X}}^{(1)}(t_0/t_0) \\ \tilde{\mathbf{K}}^{(1)}(t_h/t_h) \end{bmatrix}. \tag{44}$$

The theoretical formulation of the GILS-EKF-UI method is now available. The stiffness properties of all the elements can be evaluated with limited noise-contaminated response information and without using any information on excitation. The procedure needs to be verified using several numerical examples.

4. Numerical Examples

4.1. *Example 1*

A five story plane steel frame, shown in Fig. 1, is considered first.

The frame consists of 15 members; 10 columns and 5 beams. The height of the columns is 3.66 m and the bay width is 9.14 m. W18×71 of grade A36 steel section is used for all the members. Assuming the bases are fixed; the structure is represented by 30 DDOFs. The mass of all the beams and columns is 105.65 kg/m. The beam and column stiffnesses are estimated to be 10650 kN-m and 26625 kN-m, respectively. The mass-proportional damping coefficient α and the stiffness-proportional damping coefficient β are evaluated from the information on the first two undamped frequencies of the structure. The first two frequencies, f_1 and f_2, of the structure were found to be 3.703 and 12.279 Hz, respectively. For an equivalent modal damping of 3% of the critical for the first two modes, α and β are found to be 1.072547 and 0.000597503, respectively.

The frame is excited by the north-south component of the El Centro 1940 earthquake. The time history of the earthquake is shown in Fig. 2. Using a commercially available computer program, the theoretical responses, in terms of displacements, velocities and accelerations of the frame, are calculated. After the theoretical responses are evaluated, the information on the input force is completely ignored.

For the sake of discussion, suppose the response information only at nodes 1, 2, and 3 is available. As discussed earlier, since the frame is excited by an earthquake, any part of the frame can be used as a sub-structure depending upon the available response information. Thus, the minimum size of the substructure that can be used is shown in Fig. 1.

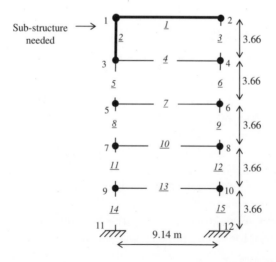

Fig. 1.   Five stories frame excited by El Centro earthquake used in Example 1.

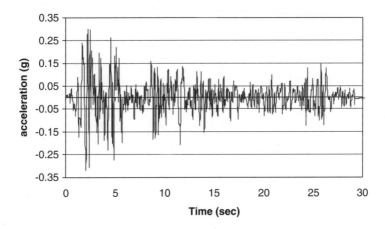

Fig. 2. El Centro earthquake time history.

Using responses from 1.52 to 2.37 sec and assuming that they are recorded at 0.00025 sec time intervals providing 3401 time points, the excitation force, the stiffnesses of the 2 elements, and α and β coefficients are identified using the ILS-UI method. The maximum error in the input excitation force, in term of the peak acceleration, is 0.53%. The maximum errors for α and β are 0.16% and 0.32%, respectively, and for the stiffness of Elements 1 and 2 are 0.0068% and 0.00679%. Obviously, the errors associated with Stage 1 identification are very small. The identified information from Stage 1 is used in Stage 2 to identify the whole frame. The algorithm failed to identify the frame with reasonable error. Two possible reasons are:

(a) Even with the EKF method, minimum amount of response information must be available, particularly for identifying large structural systems as considered in this example. No discussion on the topic can be located in the literature.
(b) The locations of the nodes where responses are available are also expected to be very important. If the responses are available only at the upper floors, the stiffness identification for the lower floors will have high and unacceptable errors.

After an exhaustive study, it was observed that in addition to 9 responses at nodes 1, 2, and 3, six horizontal responses at nodes 4 to 9, giving a total of 15 responses, would identify the stiffnesses of all the elements with a reasonable amount of accuracy. The identified stiffnesses are shown in Table 1. The maximum error in the stiffness identification is found to be 4.54%.

In the above example, the responses are assumed to be have a small amount of noise, i.e. the diagonals $\mathbf{R}$ in the noise covariance matrix in Eq. (13) are of the order of 10^{-4}. However, noise in the response measurements cannot be avoided. To address the issue of noise in the responses, numerically generated white noise with intensity of 5% of the root mean square (RMS) values of the responses observed at

Table 1.   Stiffness (EI/L) identification for five story floors frame excited by El Centro earthquake for Example 1.

Member	Initial theoretical value (kN-m)	Identified small-noise (kN-m)	Error %	Identified large-noise included (kN-m)	Error %
k_1	10650	10699	0.46	10781	1.23
k_2	26625	26484	0.53	26939	1.18
k_3	26625	26705	0.30	26874	0.93
k_4	10650	10590	0.56	10828	1.67
k_5	26625	25866	2.85	27819	4.48
k_6	26625	27011	1.45	27238	2.30
k_7	10650	10733	0.78	10366	2.66
k_8	26625	26059	2.13	27674	3.94
k_9	26625	27066	1.65	27537	3.42
k_{10}	10650	10625	0.23	10462	1.76
k_{11}	26625	26312	1.17	27114	1.83
k_{12}	26625	27814	4.46	28358	6.50
k_{13}	10650	10524	1.18	10850	1.88
k_{14}	26625	27834	4.54	28501	7.04
k_{15}	26625	25816	3.04	28311	6.33

the DDOFs are added to the theoretical responses. This is equivalent to 10^{-2} in the diagonals **R** of the noise covariance matrix. The same procedure was followed by Toki et al.,[10] Koh et al.[13] and Wang and Haldar.[21] With the noise-included response data, the frame is identified and the results are shown in Table 1. As expected, for the noise-contaminated responses, the maximum error increased to 7.04%, but still smaller than other methods [Toki et al.[10] and Koh et al.[13]].

4.2. Example 2

A larger structure represented by a four story two bay plane steel frame excited by one harmonic force, as shown in Fig. 3, is considered in this example. The frame consists of 20 members; 12 columns and 8 beams. The height of the columns is 3.66 m and each bay width is 9.14 m. Steel section of size W18 × 71 of grade A36 is used for all the members. Assuming the bases are fixed; the structure can be represented by 36 DDOFs. The first two frequencies, f_1 and f_2, of the structure were found to be 4.71 and 15.63 Hz, respectively. For an equivalent modal damping of 3% of the critical for the first two modes, α and β are found to be 1.365105 and 0.000469435, respectively.

The frame is excited by a harmonic force, $f(t) = 44.4 \sin(20t)$ kN applied horizontally at the top floor at node 1. As before, the theoretical responses of the frame in terms of displacement, velocity, and acceleration are calculated using a commercially available computer program. After the theoretical responses are evaluated, the information on the input force is completely ignored. The responses from 0.02 to 0.78 sec, recorded at 0.00025 sec time interval, providing a total of 3041 time points, are used to identify the frame.

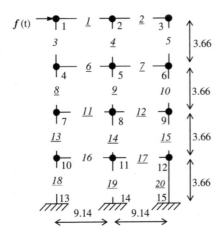

Fig. 3.   Four stories frame excited with harmonic load used in for Example 2.

Node 1 is the key node in this example. For the optimal sub-structure, the responses at 18 DDOFs, i.e. 3 responses at nodes 1, 2, 4 and horizontal responses at nodes 3 and 5 to 12 are necessary to identify the whole frame with reasonable accuracy. The identified stiffness values of all the elements with small and 5% RMS noises in the responses are shown in Table 2. The maximum errors for the two noise levels are found to be 3.12% and 5.89%, respectively.

Table 2.   Stiffness (EI/L) identification for the frame excited by harmonic load and using responses of 18 DDOFs for Example 2.

Member	Initial theoretical value (kN-m)	Identified small-noise (kN-m)	Error %	Identified large-noise included (kN-m)	Error %
k_1	10650	10678	0.26	10727	0.72
k_2	10650	10515	1.26	10976	3.06
k_3	26625	26673	0.18	26820	0.73
k_4	26625	26530	0.36	26880	0.96
k_5	26625	27082	1.72	27562	3.52
k_6	10650	10618	0.30	10552	0.92
k_7	10650	10745	0.90	10860	1.97
k_8	26625	26125	1.88	27459	3.13
k_9	26625	26693	0.25	26426	0.75
K_{10}	26625	26900	1.03	27389	2.87
K_{11}	10650	10820	1.60	11073	3.97
K_{12}	10650	10442	1.95	10991	3.20
K_{13}	26625	26600	0.094	26504	0.46
K_{14}	26625	26887	0.98	27286	2.48
K_{15}	26625	26388	0.89	27200	2.16
K_{16}	10650	10318	3.12	11277	5.89
K_{17}	10650	10957	2.88	11107	4.29
K_{18}	26625	26872	0.92	27118	1.85
K_{19}	26625	26847	0.83	27078	1.70
K_{20}	26625	26162	1.74	27597	3.65

Table 3.   Stiffness (EI/L) identification for the frame excited by
harmonic load and using responses of 21 DDOFs for Example 2.

Member	Initial theoretical value (kN-m)	Identified small-noise (kN-m)	Error %
k_1	10650	10645	0.047
k_2	10650	10654	0.034
k_3	26625	26619	0.022
k_4	26625	26632	0.024
k_5	26625	26641	0.058
k_6	10650	10650	0.003
k_7	10650	10646	0.033
k_8	26625	26647	0.082
k_9	26625	26639	0.052
k_{10}	26625	26614	0.042
k_{11}	10650	10650	0.003
k_{12}	10650	10657	0.064
k_{13}	26625	26597	0.107
k_{14}	26625	26638	0.048
k_{15}	26625	26591	0.13
k_{16}	10650	10648	0.016
k_{17}	10650	10637	0.12
k_{18}	26625	26632	0.026
k_{19}	26625	26630	0.017
k_{20}	26625	26646	0.079

Suppose, the responses are available at 21 DDOFs, i.e. 3 responses at nodes 1 to 6, and horizontal responses at nodes 10 to 12. The frame is again identified and the results are shown in Table 3, assuming the amount of noise in the response information is small. The maximum error is found to be only .0.13%, indicating a significant improvement in the identification.

4.3. *Example 3*

Defect-free relatively large frames excited by earthquake loading applied at the support level or a load applied at the super-structure were identified using limited response information. The GILS-EKF-UI algorithm can also identify defective structures and it is demonstrated in this example.

The same two bays four-story steel frame considered in Example 2 and shown in Fig. 3 is considered again. A beam (Element 11) and a column (Element 15) are assumed to contain defects. To consider the presence of defects, their stiffnesses are reduced by 5% of their original values. The stiffnesses of the defective Elements 11 and 15 are 10 117 and 25 294 kN-m, respectively. The theoretical responses of the defective frame excited by the harmonic force applied at node 1 are calculated. After the theoretical responses are evaluated, the information on the input force is completely ignored. The responses from 0.02 to 0.78 sec, recorded at 0.00025 sec time interval, providing a total of 3041 time points, are used to identify the frame.

Using the substructure containing responses at the same 18 DDOFs considered in Example 2, the whole frame is identified. The results are summarized in Table 4.

Table 4. Stiffness (EI/L) identification for the defected frame excited by harmonic load and using responses of 18 DDOFs for Example 3.

Member	Initial theoretical value (kN-m)	Identified small-noise (kN-m)	Error %	Identified large-noise included (kN-m)	Error %
K_1	10650	10637	-0.12	10594	-0.52
K_2	10650	10661	$+0.10$	10817	$+1.57$
K_3	26625	26634	$+0.034$	26804	$+0.67$
K_4	26625	26653	$+0.10$	27006	$+1.43$
K_5	26625	26678	$+0.19$	27162	$+2.01$
K_6	10650	10673	$+0.22$	10940	$+2.72$
K_7	10650	10614	-0.34	10574	-0.71
K_8	26625	26919	$+1.10$	27540	$+3.43$
K_9	26625	26683	$+0.22$	27090	$+1.74$
K_{10}	26625	26339	-1.07	25903	-2.71
K_{11}	**10650**	**10024**	**-5.87**	**9841**	**-7.59**
K_{12}	10650	10734	$+0.79$	11095	$+4.17$
K_{13}	26625	26657	$+0.12$	26766	$+0.53$
K_{14}	26625	26699	$+0.28$	26957	$+1.24$
K_{15}	**26625**	**25149**	**-5.54**	**24514**	**-7.93**
K_{16}	10650	10745	$+0.89$	11163	$+4.81$
K_{17}	10650	10562	-0.83	10493	-1.48
K_{18}	26625	26481	-0.54	26068	-2.09
K_{19}	26625	26695	$+0.26$	27030	$+1.52$
K_{20}	26625	26658	$+0.12$	27365	$+2.77$

For the low level noise in the responses, k_{11} and k_{15} decreased by 5.87% and 5.54%, respectively, significantly more than the other elements indicating the defects are in Elements 11 and 15. For relatively large noise level in the responses, the stiffnesses of these elements reduced by 7.59%, and 7.93%, respectively, as shown in Table 4, indicating the defects are present in them. Katkhuda[23] considered many other examples to demonstrate the efficiency and accuracy of defect-free and defective frames considering several levels of noises in the responses. The available results indicate that the GILS-EKF-UI is a significant improvement over the ILS-EKF-UI method and can be used to identify any structure that can be represented by finite elements with the help of limited noise-contaminated response information and without using excitation information.

5. Conclusions

A generalized system identification procedure is presented to identify structural stiffness parameters at the element level using only limited noise-contaminated response information and completely ignoring the excitation information. The authors called it a GILS-EKF-UI method. The structures are represented by finite elements. The procedure detects defects by tracking the changes in the stiffness property of each element. The method can identify defect-free and defective structures even in the presence of relatively large amount of noise in the responses. Defects could be minor in nature. The method is very robust and can identify defects caused by different types of loadings including seismic loading. The research team at the University of Arizona is in the process of developing a nondestructive defect assessment procedure for existing structures and the GILS-EKF-UI procedure will be an essential component of that effort.

Acknowledgments

This chapter is based on work partly supported by University of Arizona Foundation under a small grant program. Any opinions, findings, conclusions, or recommendations expressed in this paper are those of the authors and do not necessarily reflect the views of the sponsor.

References

1. S. Doebling, C. Farrar, M. Prime and D. Shevitz, *Damage Identification and Health Monitoring of Structural and Mechanical Systems from Changes in Their Variation Characteristics: A Literature Review*, Report No. LA-13070-MS, Los Alamos National Laboratory (1996).
2. G. Housner, L. Bergman, T. Caughey, A. Chassiakos, R. Claus, S. Masri, R. Skelton, T. Soong, B. Spencer and J. Yao, Structural control: Past, present and future, *Journal of Engineering Mechanics, ASCE* **123**, 9 (1997) 897–971.
3. S. Alampalli and G. Fu, Signal versus noise in damage detection by experimental modal analysis, *Journal of Structural Engineering, ASCE* **123**, 2 (1997) 237–245.
4. S. S. Law, Y. Z. Shi and L. M. Zhang, Structural damage detection from incomplete and noisy modal test data, *Journal of Engineering Mechanics, ASCE* **124**, 11 (1998) 1280–1288.
5. H. F. Lam, S. Katafygiotis and N. C. Mickleborough, Application of a statistical model updating approach on phase I of the IASC-ASCE structural health monitoring benchmark study, *Journal of Engineering Mechanics, ASCE* **130**, 1 (2004) 34–48.
6. L. Barroso and R. Rodriguez, Damage detection utilizing the damage index method to a benchmark structure, *Journal of Engineering Mechanics, ASCE* **130**, 2 (2004) 142–151.
7. C. Yun and M. Shinozuka, Identification of nonlinear structural dynamic systems, *Journal of Structural Mechanics, ASCE* **8**, 2 (1980) 187–203.
8. M. Hoshiya and E. Saito, Structural identification by extended Kalman filter, *Journal of Engineering Mechanics, ASCE* **110**, 12 (1984) 1757–1770.
9. M. Hoshiya and O. Maruyama, Identification of running load and beam system, *Journal of Engineering Mechanics, ASCE* **113** (1987) 813–824.
10. K. Toki, T. Sato and J. Kiyono, Identification of structural parameters and input ground motion from response time histories, *Journal Structural. Engineering/Earthquake Engineering* **6**, 2 (1989) 413–421.
11. A. Oreta and T. Tanabe, Localized identification of structures by Kalman filter, *Journal of Structural/Earthquake Engineering* **9**, 4 (1993) 217–225.
12. A. Oreta and T. Tanabe, Element identification of member properties of framed structures, *Journal of Structural Engineering, ASCE* **120**, 7 (1994) 1961–1976.
13. C. Koh, L. See and T. Balendra, Estimation of structural parameters in time domain: A substructure approach, *Journal of Earthquake Engineering and Structural Dynamics, ASCE* **20**, 8 (1991) 787–801.

14. C. Koh, L. See and T. Balendra, Damage detection of buildings: Numerical and experimental studies, *Journal of Structural Engineering, ASCE* **121**, 8 (1995).
15. M. Hoshiya and A. Sutoh, Kalman filter-finite element method in identification, *Journal of Engineering Mechanics, ASCE* **119**, 2 (1993) 197–210.
16. C. Loh, C. Lin and C. Huang, Time domain identification of frames under earthquake loadings, *Journal of Engineering Mechanics, ASCE* **126**, 7 (2000) 693–703.
17. D. Wang and A. Haldar, An element level SI with unknown input information, *Journal of the Engineering Mechanics Division, ASCE* **120**, 1 (1994) 159–176.
18. X. Ling and A. Haldar, Element level system identification with unknown input with Rayleigh damping, *Journal of Engineering Mechanics, ASCE* **130**, 8 (2004) 877–885.
19. P. Vo and A. Haldar, Health assessment of beams — Theoretical and experimental investigation, *Journal of Structural Engineering* **31**, 1 (2004) 23–30.
20. H. Katkhuda, R. Martinez and A. Haldar, A novel defect identification and structural health assessment technique, *Journal of Structural Engineering* **31**, 1 (2004) 1–8.
21. D. Wang and H. Haldar, System identification with limited observations and without input, *Journal of Engineering Mechanics, ASCE* **123**, 5 (1997) 504–511.
22. P. S. Maybeck, *Stochastic Models, Estimation, and Control*, Vol. 1, Academic Press (1979).
23. H. Katkhuda, *In-service Health Assessment of Real Structures at the Element Level with Unknown Input and Limited Global Responses*, Report-CEEM Department, University of Arizona (2004).
24. R. W. Clough and J. Penzien, *Dynamics of Structures*, 2nd edition, McGraw-Hill, Inc. (1993).
25. R. D. Cook, *Concepts and Applications of Finite Element Analysis*, 4th edition, John Wiley and Sons (2001).

INDEX